Mathematics Education in the Digital Era

Volume 17

Series Editors
Dragana Martinovic, University of Windsor, Windsor, ON, Canada
Viktor Freiman, Faculté des sciences de l'éducation, Université de Moncton, Moncton, NB, Canada

Editorial Board
Marcelo Borba, State University of São Paulo, São Paulo, Brazil
Rosa Maria Bottino, CNR – Istituto Tecnologie Didattiche, Genova, Italy
Paul Drijvers, Utrecht University, Utrecht, The Netherlands
Celia Hoyles, University of London, London, UK
Zekeriya Karadag, Giresun Üniversitesi, Giresun, Turkey
Stephen Lerman, London South Bank University, London, UK
Richard Lesh, Indiana University, Bloomington, USA
Allen Leung, Hong Kong Baptist University, Kowloon Tong, Hong Kong
Tom Lowrie, University of Canberra, Bruce, Australia
John Mason, Open University, Buckinghamshire, UK
Sergey Pozdnyakov, Saint-Petersburg State Electro Technical University, Saint-Petersburg, Russia
Ornella Robutti, Università di Torino, Torino, Italy
Anna Sfard, USA & University of Haifa, Michigan State University, Haifa, Israel
Bharath Sriraman, University of Montana, Missoula, USA
Eleonora Faggiano, Department of Mathematics, University of Bari Aldo Moro, Bari, Bari, Italy

The Mathematics Education in the Digital Era (MEDE) series explores ways in which digital technologies support mathematics teaching and the learning of Net Gen'ers, paying attention also to educational debates. Each volume will address one specific issue in mathematics education (e.g., visual mathematics and cyber-learning; inclusive and community based e-learning; teaching in the digital era), in an attempt to explore fundamental assumptions about teaching and learning mathematics in the presence of digital technologies. This series aims to attract diverse readers including: researchers in mathematics education, mathematicians, cognitive scientists and computer scientists, graduate students in education, policy-makers, educational software developers, administrators and teachers-practitioners. Among other things, the high quality scientific work published in this series will address questions related to the suitability of pedagogies and digital technologies for new generations of mathematics students. The series will also provide readers with deeper insight into how innovative teaching and assessment practices emerge, make their way into the classroom, and shape the learning of young students who have grown up with technology. The series will also look at how to bridge theory and practice to enhance the different learning styles of today's students and turn their motivation and natural interest in technology into an additional support for meaningful mathematics learning. The series provides the opportunity for the dissemination of findings that address the effects of digital technologies on learning outcomes and their integration into effective teaching practices; the potential of mathematics educational software for the transformation of instruction and curricula; and the power of the e-learning of mathematics, as inclusive and community-based, yet personalized and hands-on.

Submit your proposal:

Please contact the publishing editor at Springer: natalie.rieborn@springer.com

Forthcoming titles:

- Mathematics Education in the Age of Artificial Intelligence- Philipe R. Richard, M. Pilar Velez and Steven van Vaerenbergh (eds.)
- Mathematical Work in Educational Context- Alain Kuzniak, Elizabeth Montoya-Delgadillo and Philippe R. Richard (eds.)
- The Mathematics Teacher in the Digital Era: International Research on Professional Learning and Practice (2nd Edition) – Alison Clark-Wilson, Ornella Robutti and Nathalie Sinclair (eds.)
- 15 Years of Mathematics Education and its Connections to the Arts and Sciences- Claus Michelsen, Astrid Berckman, Victor Freiman, Uffe Thomas Jankvist and Annie Savard (eds.)
- Quantitative Reasoning in Mathematics and Science Education - Gülseren Karagöz Akar, İsmail Özgür Zembat, Selahattin Arslan and Patrick W. Thompson (eds.)
- The Evolution of Research on Teaching Mathematics - Agida Manizade, Nils Fredrik Buchholtz and Kim Beswick (eds.)- Open Access!
- Mathematical Competencies in the Digital Era - Uffe Thomas Jankvist and Eirini Geraniou (eds.)

More information about this series at https://link.springer.com/bookseries/10170

Philippe R. Richard · M. Pilar Vélez ·
Steven Van Vaerenbergh
Editors

Mathematics Education in the Age of Artificial Intelligence

How Artificial Intelligence can Serve Mathematical Human Learning

Springer

Editors
Philippe R. Richard
Département de didactique
Université de Montréal
Montréal, QC, Canada

M. Pilar Vélez
Departamento de Ingeniería Industrial
Universidad Antonio de Nebrija
Madrid, Spain

Steven Van Vaerenbergh
Departamento de Matemáticas, Estadística
y Computación
Universidad de Cantabria
Santander, Spain

ISSN 2211-8136   ISSN 2211-8144   (electronic)
Mathematics Education in the Digital Era
ISBN 978-3-030-86908-3   ISBN 978-3-030-86909-0   (eBook)
https://doi.org/10.1007/978-3-030-86909-0

© The Editor(s) (if applicable) and The Author(s), under exclusive license to Springer Nature Switzerland AG 2022
This work is subject to copyright. All rights are solely and exclusively licensed by the Publisher, whether the whole or part of the material is concerned, specifically the rights of reprinting, reuse of illustrations, recitation, broadcasting, reproduction on microfilms or in any other physical way, and transmission or information storage and retrieval, electronic adaptation, computer software, or by similar or dissimilar methodology now known or hereafter developed.
The use of general descriptive names, registered names, trademarks, service marks, etc. in this publication does not imply, even in the absence of a specific statement, that such names are exempt from the relevant protective laws and regulations and therefore free for general use.
The publisher, the authors and the editors are safe to assume that the advice and information in this book are believed to be true and accurate at the date of publication. Neither the publisher nor the authors or the editors give a warranty, expressed or implied, with respect to the material contained herein or for any errors or omissions that may have been made. The publisher remains neutral with regard to jurisdictional claims in published maps and institutional affiliations.

This Springer imprint is published by the registered company Springer Nature Switzerland AG
The registered company address is: Gewerbestrasse 11, 6330 Cham, Switzerland

Foreword

AI *for* the Learning of Mathematics

It would have been good to have a precize and clear definition of Artificial Intelligence (AI) unanimously accepted. Unfortunately, it is not the case today as it was not the case formerly.[1] The common criterion, AI is a property of machines "exhibiting certain behaviours which strikes as intelligent",[2] reminds us that this is a judgement underpinned by a kind of human empathy. Looking closer, it appears that such a judgement assesses both the task which has been achieved by the machine and the way it has been achieved: a behaviour is striking because the task is acknowledged complex and/or the way in which it has been achieved looks *smart*.

Let's be back to the first *Smart mathematical machines*. In the middle of the seventeenth century, the Pascal's calculator performed four arithmetic operations, achieving tasks that only humans could perform until then. With no doubt, it looked striking as intelligent for some, but not to the eyes of Pascal himself who noticed that his invention lacks "willingness" which cannot be separated from "spirit".

The judgement of intelligence requires that the task considered is not merely achievable by a definite technique; a technique described by an algorithm fully specified and determinist, even if it is difficult to implement (and to learn). What I call a complex task is nothing but a *problem* which requires solving strategies and heuristics with the associated risk to fail. Then by AI, I mean that a machine has *knowledge* and *reasoning capabilities* to find *solutions to problems*. As mathematicians we are interested in what AI can do for mathematics, and as mathematics educators we are interested in what it can do for mathematics education. This book touches both, but it insists on the latter for which it opens original perspectives. Before coming to them, I will add a few words with an historical stance.

AI raised hopes in the 1970s with the mainstream research program on Intelligent Tutoring Systems (ITS) inspired by one-to-one tutoring model, that would

---

[1] e.g. Schank (1987); Wang (2008, 2019).
[2] Steven Van Vaerenbergh and Adrián Pérez-Suay, this book, Chap. 12 MS p. 2.

allow personalized learning for all and, as a consequence, possibly make the teacher unnecessary. The scientific program has evolved since then, but it keeps its genetic signature: individualization of learning and autonomy vis-a-vis human interventions. The methodological guideline was mimicking teachers' strategies and behaviours. This research in AI and Education really took off in the 1980s, and since then has been very active and productive. Projects improved with the progress of AI research and the advancement of cognitive sciences. Some ITS went to the mathematics classroom with evidence of success. Nevertheless, the dissemination of AI-based learning environments remained limited and still is, although with differences among the mathematical domains. Because of the prominence of their algorithmic dimension, teaching itself focusing on learning technics, Arithmetic and elementary Algebra received most of the attention. Geometry proved to be more challenging. In the early 1970s, building an artificial geometer seemed accessible but the dream fell short.[3] Modelling human reasoning even in a knowledge domain as well formalized as is mathematics encountered difficulties impossible to overcome without a drastic limitation of the problem-space.

The solutions to the problem of computational knowledge modelling, essentially rule based systems in the early period of ITS research, raised an epistemological problem that was pointed by Allan Newell who criticized the *de facto* AI confusion between knowledge and its representation. The ITS behaves as if learning a piece of knowledge were learning its representation—writing and grammar—and the associated skills—performing correctly procedures on and via representations. One must acknowledge that this reduction is more pragmatic than theoretical, several researchers saw in ITS research a theoretical stake. The classical Etienne Wenger survey of AI and ITS had the subtitle: "computational and cognitive approaches of communication of knowledge"[4]; the author chose not to define knowledge asserting that the objective of ITS research is to understand it.

To understand mathematical knowledge is an old and complex issue often considered as specific when compared with the same issue for other scientific knowledge domains; remember that the philosopher Karl Popper preferred to leave mathematics aside. In actuality, mathematics is the best example which backs Allan Newell's claim: "Knowledge remains forever abstract and can never be actually in hand".[5] Mathematics is forever *abstract*: mathematical knowledge has no referent in the material *world*: research in mathematics explores a world which is already mathematical.[6] Yet, smart mathematical machines bring to reality mathematical objects and offer to human perception physical manifestations of their properties. They open the way to experimental studies of mathematical objects and the exploration of their

---

[3] e.g. Balacheff & Boy de la Tour (2019).

[4] Wenger (1986).

[5] Newell (1982, p. 125).

[6] This does not mean that there are no tight relations between mathematic and reality, but these relations have a heuristic value for mathematics, as it is the case for the scientific disciplines which use tools that mathematics provides them for knowing—with their own concepts and their own methodologies—the world in which we live.

property. Paraphrasing Alan Newell, one could say that knowledge serves as a specification of what the representation systems should be able to display,[7] but these representations are *not* the knowledge it refers to. This is a classical semiotic tension that we need to understand and to overcome. I suggest that a solution is to bring back problems on stage as an epistemological solution.

Problems are the *raison d'être* of knowledge. The relations between problems and knowledge are dual and dialectic: problems are the source of knowledge and of its evolution, and conversely.[8] But, just as knowledge cannot be equated to a representation, problems cannot be identified to a statement, especially in the learning context because both knowledge and problematization are there under construction. It is the *role of situations*[9] to give birth to a problem by setting the scene for interactions between students and a material and social space for actions, and creating the circumstances to stimulate students' engagement. Such situations can be designed to make students experience the relevance and efficiency of a piece of mathematical knowledge by solving problems. How could AI contribute to designing learning situations? ITSs have not proved being appropriate to provide an answer to this question, but there is another line of research which is exactly addressing it.

In the beginning of the 1970s, Marvin Minsky coined the concept of *microworlds* when looking for a strategy to make breakthroughs in robotic. He extended it, in collaboration with Seymour Papert,[10] to nourish a proposal for the learning of mathematics inspired by a firm critic of the evolution of mathematics teaching at the time of the new math movement, and of educational models privileging instruction. They aimed at providing learners—not students—with a rich environment offering the possibility to explore a mathematical domain in a way not foreign to the concrete experience of the world, and evolving with the learners' knowledge. Logo, the seminal microworld they designed, had two faces: a programming language and the image of a turtle moving in *the space of the screen*.[11] A program driving the turtle could be turned into a tool to be used by a new program, that is a "procedure". We understand that a procedure reflects the knowledge constructed by the learner. This knowledge has multiple representations: at the symbolic level (a program) and at the phenomenal level (drawings on the screen). The feedback to learners is a consequence of the internal logical structure of the microworld and not from a tutoring decision. The learner has all the benefit of an open environment with epistemic characteristics favouring the evolution of his or her knowledge in a way coherent with his or her project. As a matter of fact, Logo was not an environment specifically thought for the learning of geometry, but it had interesting potential for this purpose. The gap between Logo's geometry and the geometry of the curricula hindered its full dissemination. But the screen as a field of experience[12] offering genuine mathematical discovery

---

[7] Newell (1982, p. 100).
[8] Vergnaud (1990, 2009).
[9] Brousseau (1986, 1997, 1970).
[10] A tribute to Seymour Papert (Balacheff, 2017).
[11] There was also a version with a concrete turtle moving on the floor.
[12] Boero (1989, p. 65).

was there, the next step was to fill in the gap with school mathematics. It came in the 1980s with *Dynamic Geometry Environment* (DGE); *this* is the contemporaneous legacy of Logo. While Logo is primarily a programming language, a radical change occurred in 1985 when Jean-Marie Laborde designed the first DGE, Cabri-géomètre, introducing the revolution of direct access to manipulating and constructing objects on the screen as if they were real and not mere representations (Laborde, 1995). The behaviours of the objects are a consequence of their construction under the constraints of geometrical primitives. The perceived *visual invariants* on the screen, when messing up constructions, are *theorems*. Denying intelligence, one could claim that *a DGE does not know geometry but that it is its materialization* based on a computational model brought by analytic geometry. This is too quick a judgement.

In the seminal project, a microworld "is very schematic; it talks about a fairyland in which things are so simplified that almost every statement about them would be literally false if asserted about the real world."[13] In effect, Minsky had the idea of microworld as a response to the complexity of the robotic problems he was working on; in short, he decided that if he could not find a comprehensive model for coupling the eye and the hand, then he had to look for solutions of these problems in "simplified worlds". The membership of this approach to AI has been criticized by some, and sometimes it has been denied.[14] However, Logo is not a microworld in the original sense, it is a programming language which opens access to an unlimited universe starting from a few primitive actions which semantic is moving around in a flat land. If Logo is a "fairyland", a land conducive of mind storms, this land has been designed so that the behaviour of the drawings the turtle "leaves" on the screen offers a terrain on which geometry could grow as could other mathematical or algorithmic concepts. The invention of DGEs goes a step further. What is special with them, is that it is the land where geometrical figures thrive: a drawing is more than what you see, it is what you get when manipulating it, that is all drawings which satisfy the constraints imposed by its construction. DGEs materialize a world whose inhabitants are geometrical figures, not only their shadows (which are drawings), whose laws conform Euclidean Geometry.

The DGE *materialization* of Geometry is more than *visualisation*. Borrowing the words of Allan Newell, it is a symbol system which encodes a body of knowledge. It is a semiotic instrument to grasp the objects of Geometry and to discover their properties in *a space ruled by a rational principle*: a visual invariant when manipulating directly a free object on the screen is a graphical representation of a theorem; this is a perfect translation of the *theorems in action*, as Gérard Vergnaud conceptualized them. Moreover, learners can make the DGE evolve in parallel to the evolution of their own knowledge by creating graphical procedures with specified inputs and outputs and giving them a name. A DGE is a smart mathematical machine opening a field of mathematical experiences, but... with no didactical agenda.

---

[13] Minsky & Papert (1970, p. 36) section 5 of the report: "Why we are studying knowledge and learning".

[14] Dreyfus & Haugel (1981).

On the one hand, *AI machines*, the intelligent tutoring systems, have proved promising results for the acquisition of technical skills but they are limited when coming to problems, what limits their impact on developing an understanding of mathematics. On the other hand, *Smart mathematical machines* are efficient tools to design problem situations thanks to the possibility they offer to create fields of mathematical experience, but they have no didactical functions to direct students towards the intended teaching objective. Moreover, one may emphasize that these machines complexify the work of teachers. To get the best of both, the project QED-Tutrix ambitions to design a platform which provides students with a space to explore a problem and an artificial supervisor to feedback on the proof under construction to ultimately validate a solution. There is another route to achieve the same, here illustrated by the case of GeoGebra, which consists in augmenting a smart mathematical machine with reasoning competences and the capacity to provide didactical feedback. This is the project of designing an environment augmenting a DGE with automated reasoning tools (ART).

The combination of smart machines and AI systems is more than an addition, the properties of the whole emerge from the interactions among the parts. Interaction is also a keyword of human-centred computer science. One may remember this claim of Joseph Conrad: "One writes only half the book; the other half is with the reader." This applies nicely to AI *for* the learning of mathematics. The interactions between the humans and the systems, as well as among the humans, have to be considered from the beginning of the design of the learning environment. Hence, its educational relevance and efficiency is not a property of the technology itself but of the orchestration of its use and the dynamic of the multiple interactions and feedback loops.

Eventually, reading a book on Artificial Intelligence and Education at the time of a major pandemic cannot be done without a thought for problems which challenge education, that is teaching and learning. The obligation to shut down classes—or even schools—has put on the fore all expectations decision-makers and the society have on the role that Technology Enhanced Learning (TEL) environments can play. A large panel of technology is deployed since the beginning of 2020. It consists of video communication channels, digital teaching platforms and MOOCs, mail and web-based resources and software specific to the content to be taught. This unprepared intense use of TEL has a mitigated success, but this is no surprise given the sudden and urgent radical change in practice for both students and teachers, and families as well. I don't doubt that this is the start of a new era which will be marked by a refreshed vision of TEL, maybe a less revolutionary than formerly claimed but a more pragmatic and efficient one. A lesson we can already keep for the future is the need students have for social interactions with other students, including sharing the learning experience, its challenges and successes. Another lesson, not the least, is that *teaching is a profession* that parents, technology and internet resources cannot replace. As a profession, it requires specific knowledge, skills and attitudes. It needs instruments and tools as well. Each new knowledge technology since the invention of writing has inspired the design of new teaching tools and learning material. The same is true in the fully digital era in which we live nowadays, where technology not only improves representing, memorizing and communicating knowledge, but allows manipulating and treating it in a way which is not mechanical but *intelligent*. Here is the challenge for research on AI and mathematics education of the future. The book arranged by Philippe R. Richard, M. Pilar Vélez, Steven Van Vaerenbergh—and the team of editors they brought together for this project—will be the source of questions and inspirations to contribute to take up this challenge.

Nicolas Balacheff
Laboratoire d'informatique de Grenoble
Université Grenoble Alpes, CNRS, Grenoble INP
Grenoble, France

References

Balacheff, N. (2017). Seymour Papert (1928–2016) Aux sources d'une pensée innovante et engagée. *Recherches en Didactique des Mathématiques*, *37*(2/3), 383–396. ⟨hal-01716631⟩.

Balacheff, N., & Boy de la Tour, T. (2019). Proof Technology and Learning in Mathematics: Common Issues and Perspectives. In G. Hanna, D. Reid, & M. de Villiers (Eds.), *Proof Technology in Mathematics Research and Teaching*. Springer.

Boero, P. (1989). Mathematical Literacy for All: Experiences and Problems. *Proceedings of PME-XIII*, *1*, 62–76.

Brousseau, G. (1986). Fondements et méthodes de la didactique des mathématiques. *Recherches En Didactique Des Mathématiques*, *7*(2), 33–115.

Brousseau, G. (1970). Processus de mathématisation. *La Mathématique à l'école Élémentaire*, 428–457.

Brousseau, G. (1997). *Theory of didactical situations in mathematics*. Kluwer Academic Publishers.

Dreyfus, H. L., & Haugel, J. (1981). From micro-worlds to knowledge representation: AI at an impasse. In *Mind Design* (pp. 142–182). MIT Press. https://cogsci.ucsd.edu/~coulson/203/dreyfus.pdf.

Laborde, J.-M. (1995). Des connaissances abstraites aux réalités artificielles, le concept de micromonde Cabri. In D. Guin, J.-F. Nicaud, & D. Py (Eds.), *Environnements Interactifs d'Apprentissage avec Ordinateur* (pp. 28–41). Eyrolles, Paris.

Minsky, M., & Papert, S. (1970). *Proposal to ARPA for research on artificial intelligence at M.I.T., 1970-971* (No. 185; Artificial Intelligence, p. 53). MIT AI laboratory.

Newell, A. (1982). The knowledge level. *18*, 87–127.

Schank, R. C. (1987). What Is AI, Anyway? *AI Magazine*, *8*(4), 59–59. https://doi.org/10.1609/aimag.v8i4.623.

Vergnaud, G. (1990). La théorie des champs conceptuels. *Recherches en Didactique des Mathématiques*, *10*(2/3), 133–170.

Vergnaud, G. (2009). The Theory of Conceptual Fields. *Human Development*, *52*, 83–94.

Wang, P. (2008). What Do You Mean by "AI"? *Frontiers in Artificial Intelligence and Applications*, *171*(1), 362–373.

Wang, P. (2019). On Defining Artificial Intelligence. *Journal of Artificial General Intelligence*, *10*, 1–37. https://doi.org/10.2478/jagi-2019-0002.

Wenger, E. (1986). *Artificial intelligence and tutoring systems: Computational approaches to the communication of knowledge*. Kaufman Publishers Inc., Los Altos, CA. https://www.osti.gov/biblio/6742589.

General Introduction

The relationship between humans and machines is very old. As early as the Neolithic period, machines emerged as sophisticated artefacts that could transform one movement into another, for example, to enable a small group of people to move heavy loads and to set them up for a variety of uses. In English, the word machine comes from the Middle French "machine" which is itself a borrowing from the classical Latin "machina", meaning ingenious invention, device, trick. As technology has developed, machines have become increasingly sophisticated. They have become stronger, faster, more precise, more efficient and more tireless tools, capable of performing tasks that man could no longer perform without them. They were equipped with a memory and a program, usually computer-controlled, designed to perform several specific tasks autonomously and in a coordinated manner. Then, their field of application opened up infinitely, being able to perform complex tasks that were once thought to be reserved for human intelligence, even surpassing it with a capacity for calculation, prediction and learning that forces us to completely rethink the nature and effects of the relationship we have with machines. If there has already been man–machine for better or for worse, there is now artificial intelligence which extends the panorama.

In offering a book that straddles the line between artificial intelligence and mathematics education, another linguistic issue arises. In Latin, Germanic or Slavic languages, we have «didactique des mathématiques» and «informatique» as substantives of reference, but in standard English, *didactics* and *informatics* are adjectives. And if we speak, even partially, of «preuves» or «démonstration» in geometry, we know that English prefers the action-oriented gerund *proving* rather than the noun, which primarily evokes the idealized concept. What kind of artificial intelligence will we be able to talk about? We have no choice but to integrate various perspectives and warn the reader when we take some linguistic liberties. Before the term «informatique» came into use, the syntagm «cerveau électronique» was flirted with. In French, the *electronic brain* is still visible in books and media from the 1950s to 1960s: it only reappeared a little later, in the 1990s, in connection with the idea of computer intelligence. While the comparison between human and machine continues, it is clear that it is always posed in terms of the nature of things, their characterization and qualification, from the conceptual aspects to the processes that these engage.

Whether it is natural or artificial, whether it is addressed to a brain or a machine, the notion of intelligence is polysemous. We have identified two trends. The first is based on the principle that if we know what intelligence is in humans, for example, because we have done our homework in philosophy or neuropsychology, then we can characterize the intelligence of the machine by comparison with human intelligence. This is a well-known position that is based on an induction hypothesis, just as mathematicians do. The second is based on the ability of a living being to adapt to a new situation, to understand and solve certain difficulties, to make sense of the things around it, to act with discernment. It is intelligence in an ecosystem, or more precisely within an interaction that constitutes an evolving system in itself. Intelligence is then seen on the level of knowledge and on that of mutual or reciprocal adaptation, if it is a question of a human–machine interaction which aims at the same end, which can be the fact or proceed by imitation, ruptures, accommodation, decision, calculation, etc.

These two trends complement each other in the orientation of our book. Its spirit is radically different from today's 'machinocentric' viewpoint, which is mainly formulated by computer scientists or policy makers, the opposite of the anthropocentrism of yesteryear where humans were compared to machine systems. We often look at artificial intelligence (AI) in terms of the effects of computer science, but we do not question the causes that govern the choices made by decision-makers: humans are always users of AI proposed by computer scientists and they seem to be in thrall to the uses decided by third parties. However, our starting point is the teaching and learning of mathematics and it concerns the interaction between the human and the machine in both directions, as well as the emergence of the human–machine system to characterize instrumented mathematical work. This means that the machine we are interested in is the one that is a partner in the construction of knowledge, and for this to be the case, many questions need to be asked about the modelling of knowledge and the learning of the machine that aims to serve the human. In other words, if we were to undertake an ethical discourse on AI, we would have to start by asking questions about modelling and learning choices, and continue the questioning when the human is interacting with his or her partner.

At the same time, how can we not celebrate the revolution of machine intelligence to make the world a better place? When it comes to using machine learning algorithms in a traditional way or through deep learning, one surely thinks of automatic driving (autonomous vehicles), pattern recognition or computer pattern identification (speech analysis, image search, etc.). One also thinks of decision support in commerce or banking (intelligent assistants for routine work), measuring and predicting of ecological disasters (prevention and reduction of damage risks), content search engines (Google, WolframAlpha, etc.) or the possibility of asking more or less complex questions on one's mobile phone (Siri or equivalent). However, if a neural network simulates human functioning, it is by analogy within a mathematical model executed by a machine, and not a kind of laboratory reproduction of a reality controlled by data. When we take a modern definition of AI, such as that found in the Montreal declaration for a responsible development of artificial intelligence (2018):

General Introduction

Artificial intelligence (AI) refers to the series of techniques which allow a machine to simulate human learning, namely to learn, predict, make decisions and perceive its surroundings. In the case of a computing system, artificial intelligence is applied to digital data.

It is clear that the human is a source of inspiration but does not seem to be a partner at all. Even if this is an entry that deserves to be highlighted, we must first ask ourselves: artificial intelligence by whom, for whom? The 'for whom' of course refers to possible uses, possibly in an instrumented perspective, while the 'by whom' cannot be limited to the world of industry and commerce. Despite their stratospheric means, which, like a space program, can trickle down useful techniques into other fields, their objectives are diametrically opposed to the public good and the development of the mind. Mathematics education is first and foremost a matter of generosity in a collegial dynamic. And it is up to us, authors and multidisciplinary colleagues, wearing different hats and accustomed to reflecting on the relationships between teacher/learner, trainer/student, administrator/colleague, politician/researcher and computer scientist/user, to put forward the principles of *mathematical human learning* and the vision of the mathematics education profession.

It must be said that the historical link between AI and didactics of mathematics is well established. One example is the book *Didactique et intelligence artificielle*, published almost 25 years ago at *Éditions de la pensée sauvage* in the series *Recherches en didactique des mathématiques*, which stated that advances in AI had paved the way for a vigorous stream of research into the development of computer environments for human learning and technology-enhanced learning. But after an initial period of enthusiasm, when we imagined that we would be able to do incredible things very quickly, we went through a somewhat more sombre period of disillusionment when we realized that we had somewhat underestimated the difficulties. Today, in a paradoxical turn of events, it seems that AI is joining didactics with its non-routine problem-solving approaches, which involve learning, modelling and prediction phases that evoke both mathematical work and the design of solutions by specialists. If AI has a role to play in fostering academic success and providing support for learning outcomes, any collaboration must begin with the consideration of didactics in AI models to understand the needs of the student and the teacher, and to be at their service. It is up to the system to adapt to the human and not the other way around.

From this follows a guiding principle: the creation of any digital artefact that respects didactic needs, while remaining epistemologically and cognitively sound, requires the establishment of an ongoing dialogue in human–machine interaction that values mathematical literacy. To be truly intelligent, these systems must first remain flexible enough to adapt to the natural evolution of mathematical work in the classroom, and then leave the parameterization to the teacher or trainer. Furthermore, if we know enough about the resources, goals and orientations of a class, we can come to understand, explain and model actions and decisions that seem unusual or abnormal to the external eye. Intelligence is then revealed in an iterative process of convergence between a priori and observed effects that is progressively refined in use, hopefully releasing the bright side of uncertainty. In mathematics education, it seems then that AI cannot be formulated solely in terms of the machine, but above all in terms of a finalized activity with humans. In a way, we are close to idoneism in the sense of Ferdinand Gonseth, in relation to a double concern of truth and reality.

Our book weaves together "machine thinking" and "human thinking" by using artificial intelligence in the continuation of the scientific work and achievements of the book's authors. The first part looks more at the machine and the third at the human, as if they were the two vectors of the same interaction. As for the second part, in a good intermediate section, it looks more at interaction as such, even if at times a vectorial view is imposed.

More specifically, the first part introduces the reader to "machine thinking" through different novel AI systems, always with a focus on the human use of them. Scientific calculators, Computer Algebra Systems (CAS) and Dynamic Geometry Systems (DGS) are widely extended and used in mathematics teaching and learning, most of the time, as tools to perform arithmetic or algebraic calculations or to visualize mathematical concepts or properties. Nevertheless, their potential extension to become AI systems for "smart learning" has been developed by researchers with different purposes but oriented to Mathematics Education. This part presents some of these systems and their integration in the classroom, mainly in cooperation or symbiosis with CAS or DGS. For instance, DGS provided with reasoning capabilities to decide if a theorem is true or false, derive conclusions or discover properties starting from a geometric configuration in a mathematically rigorous way; a tutorial system to guide students in their solving problems performance; a system where AI allows discovering the hidden mathematics in a monument and the relation with its architectural style; or a system of knowledge organization for the Mathematics curriculum.

The second part deals with new ways of building mathematical knowledge through the interaction between the learner and different AI systems. The didactic implications and the development of digital competences when an AI system intervenes are approached from different points of view: the design of a system based on the experience of teachers and learners, the impact on the mathematical activity and on the learning process or the learner–machine interaction in the development of activities and problem-solving. It also analyses the knowledge and skills required for this transition towards the empowerment of "human thinking" in relation to "machine thinking".

The third part focuses on the human point of view when integrating technology into learning spaces or generating digital *milieus* inspired by classic manipulative resources. The reader can find some several years of studies such as a tertiary level experience in technology integration through the use of CAS, teacher training using expert systems able to simulate classroom situations and mentor–teacher interaction, or classroom use of DGS from paper/pencil constructions to tasks especially designed to take advantage of DGS potentialities. Furthermore, Virtual Reality (VR) and educational approach to learning that uses Science, Technology, Engineering, the Arts and Mathematics (STEAM) are present: an empirical study on the potentialities for visualization and manipulation using a VR system in the classroom, and a STEAM project carried out in cooperation with mathematics and technology teachers who use 3D printing tools or programming environments.

General Introduction

The theme of the book started from an incessant questioning of the authors of the natural links between computer science, mathematics and their didactics at the age of artificial intelligence. But the real trigger for the book came during several informal discussions during the Applications of Computer Algebra conference, held in Montreal in July 2019. We would like to thank Michel Beaudin for providing us with a room and resources at the École de technologie supérieure (ÉTS Montréal) for an off-program meeting where ideas, partnerships and first drafts of the project were initiated. In February 2020 on the other side of the Atlantic Ocean, just before the pandemic restrictions, the International Centre for Mathematical Encounters (CIEM) of the Universidad de Cantabria, in Castro Urdiales, hosted us so that we could crystallize the editorial project with consensual and effective guidelines. We would like to thank Springer Nature and, in particular, the Mathematics Education in the Digital Era series editors, Dragana Martinovic and Viktor Freiman, for their unstinting support from the beginning.

Although the conception and planning of our work was carried out in face-to-face meetings, the editorial process, developed in parallel with the COVID-19 pandemic, has been online. Nevertheless, the digital coldness has been compensated by warm and friendly (online) constructive discussions between the editors, the coordinators of the sections and the emeritus contributors. Our thanks and acknowledgements go to the coordinators Pedro Quaresma, Jana Trgalová and Jean-baptiste Lagrange, who have ensured the quality and internal consistency of each section, as well as, to the emeritus contributors, Nicolas Balacheff and Tomás Recio, for sharing their extensive knowledge, experience and expertise. We are also certainly grateful that all of them kept up with us editors throughout the entire project despite our editorial extravagances, such as insisting to keep the communication between editors in Spanish, to conduct the coordination meetings in French and to organize all other correspondence in English.

Each chapter has undergone a review process by at least two experts, selected from among the authors and external specialists depending on the topic, in addition to a review by the section coordinators. Thanks also to the reviewers for their essential, generous and hidden work. To complement the spirit of collaboration, the introductions to each section were reread by the authors. In short, we would like to emphasize that this book is the result of a collective effort carried out by all these authors and collaborators in a dynamic where the research of some, the experience of others and the reflection of all have led us to the realization of a lasting work. We are very grateful.

Finally, we wish the reader, who has this book in front of her or his eyes, a pleasant and fruitful reading.

Philippe R. Richard
M. Pilar Vélez
Steven Van Vaerenbergh

Contents

Creation of AI Milieus to Work on Mathematics

Evolution of Automated Deduction and Dynamic Constructions in Geometry 3
Pedro Quaresma

Automated Reasoning Tools with GeoGebra: What Are They? What Are They Good For? 23
Zoltán Kovács, Tomás Recio, and M. Pilar Vélez

Intelligence in QED-Tutrix: Balancing the Interactions Between the Natural Intelligence of the User and the Artificial Intelligence of the Tutor Software 45
Ludovic Font, Michel Gagnon, Nicolas Leduc, and Philippe R. Richard

A Decision Making Tool for Mathematics Curricula Formal Verification 77
Eugenio Roanes-Lozano and Angélica Martínez-Zarzuelo

A Classification of Artificial Intelligence Systems for Mathematics Education 89
Steven Van Vaerenbergh and Adrián Pérez-Suay

AI and Mathematics Interaction for a New Learning Paradigm on Monumental Heritage 107
Álvaro Martínez-Sevilla and Sergio Alonso

AI-Supported Learning of Mathematics

Using Didactic Models to Design Adaptive Pathways to Meet Students' Learning Needs in an Online Learning Environment 141
Brigitte Grugeon-Allys, Françoise Chenevotot-Quentin, and Julia Pilet

Combining Pencil/Paper Proofs and Formal Proofs, A Challenge for Artificial Intelligence and Mathematics Education 167
Julien Narboux and Viviane Durand-Guerrier

Interaction Between Subject and DGE by Solving Geometric Problems 193
Jiří Blažek and Pavel Pech

Creative Use of Dynamic Mathematical Environment in Mathematics Teacher Training 213
Roman Hašek

Experimental Study of Isoptics of a Plane Curve Using Dynamical Coloring 231
Thierry Dana-Picard and Zoltán Kovács

Teaching Programming for Mathematical Scientists 251
Jack Betteridge, Eunice Y. S. Chan, Robert M. Corless, James H. Davenport, and James Grant

The present and future of AI in ME: Insight from empirical research

CAS Use in University Mathematics Teaching and Assessment: Applying Oates' Taxonomy for Integrated Technology 283
Daniel Jarvis, Kirstin Dreise, Chantal Buteau, Shannon LaForm-Csordas, Charles Doran, and Andrey Novoseltsev

Modeling Practices to Design Computer Simulators for Trainees' and Mentors' Education 319
Fabien Emprin

Exploring Dynamic Geometry Through Immersive Virtual Reality and Distance Teaching 343
José L. Rodríguez

Historical and Didactical Roots of Visual and Dynamic Mathematical Models: The Case of "Rearrangement Method" for Calculation of the Area of a Circle 365
Viktor Freiman and Alexei Volkov

Implementing STEM Projects Through the EDP to Learn Mathematics: The Importance of Teachers' Specialization 399
Jose-Manuel Diego-Mantecón, Zaira Ortiz-Laso, and Teresa F. Blanco

Digital Technology and Its Various Uses from the Instrumental Perspective: The Case of Dynamic Geometry 417
Jana Trgalová

Conclusions 431

Epilogue 437
**Appendix: Photographs of the Book Project and Some
    of the Authors** 441
Index 447

Contributors

Sergio Alonso Universidad de Granada, Granada, Spain

Jack Betteridge University of Bath, Bath, England

Teresa F. Blanco Universidad de Santiago de Compostela, Santiago de Compostela, Spain

Jiří Blažek University of South Bohemia, České Budějovice, Czech Republic

Chantal Buteau Brock University, St. Catharines, Canada

Eunice Y. S. Chan Western University, London, Canada

Françoise Chenevotot-Quentin Institut national supérieur du professorat et de l'éducation Lille Haut de France, LDAR, Villeneuve-d'Ascq, France

Robert M. Corless Western University, London, Canada

Thierry Dana-Picard Jerusalem College of Technology, Jerusalem, Israel

James H. Davenport University of Bath, Bath, England

Jose-Manuel Diego-Mantecón Universidad de Cantabria, Santander, Spain

Charles Doran University of Alberta, Edmonton, Canada

Kirstin Dreise Brock University, St. Catharines, Canada

Viviane Durand-Guerrier IMAG, Université de Montpellier, CNRS, Montpellier, France

Fabien Emprin Université de Reims Champagne-Ardenne, Reims, France

Ludovic Font École Polytechnique de Montréal, Montréal, Canada

Viktor Freiman Université de Moncton, Moncton, Canada

Michel Gagnon École Polytechnique de Montréal, Montréal, QC, Canada

James Grant University of Bath, Bath, England

Brigitte Grugeon-Allys Université Paris Est-Créteil, LDAR, Créteil, France

Roman Hašek University of South Bohemia, České Budějovice, Czech Republic

Daniel Jarvis Nipissing University, North Bay, Canada

Zoltán Kovács Private University College of Education of the Diocese of Linz, Linz, Austria

Shannon LaForm-Csordas Nipissing University, North Bay, Canada

Nicolas Leduc École Polytechnique de Montréal, Montréal, Canada

Álvaro Martínez-Sevilla Universidad de Granada, Granada, Spain

Angélica Martínez-Zarzuelo Universidad Complutense de Madrid, Madrid, Spain

Julien Narboux Université de Strasbourg,CNRS, Strasbourg, France

Andrey Novoseltsev University of Alberta, Edmonton, Canada

Zaira Ortiz-Laso Universidad de Cantabria, Santander, Spain

Pavel Pech University of South Bohemia, České Budějovice, Czech Republic

Julia Pilet Université Paris Est-Créteil, LDAR, Créteil, France

Adrián Pérez-Suay Universitat de València, València, Spain

Pedro Quaresma Universidade de Coimbra, Coimbra, Portugal

Tomás Recio Universidad Antonio de Nebrija, Madrid, Spain

Philippe R. Richard Université de Montréal, Montréal, Canada

Eugenio Roanes-Lozano Universidad Complutense de Madrid, Madrid, Spain

José L. Rodríguez Universidad de Almería, Almería, Spain

Jana Trgalová Université Claude Bernard, Lyon, France

Steven Van Vaerenbergh Universidad de Cantabria, Santander, Spain

M. Pilar Vélez Universidad Antonio de Nebrija, Madrid, Spain

Alexei Volkov National Tsing-Hua University, Hsinchu, Taiwan

External Collaborators

Josep Maria Fortuny Universitat Autònoma de Barcelona, Barcelona, Spain

Josep Gascón Universitat de Barcelona, Barcelona, Spain

Gloriana González University of Illinois Urbana-Champaign, Champaign and Urbana, USA

Predrag Janičić Université de Belgrade, Belgrade, Serbia

Alexander P. Karp Columbia University, New York, USA

Kotaro Komatsu University of Tsukuba, Tsukuba, Japan

Yves Kreis Université du Luxembourg, Luxembourg, Luxembourg

Zsolt Lavicza Johannes Kepler Universität Linz, Linz, Austria

María Cristina Naya Universidade da Coruña, A Coruña, Spain

Giorgos Psycharis National and Kapodistrian University of Athens, Athène, Greece

Hans-Stefan Siller Julius-Maximilians-Universität Würzburg, Würzburg, Germany

Creation of AI Milieus to Work on Mathematics

Introduction to Section 1 by the Coordinator Pedro Quaresma

In the past decade, artificial intelligence (AI) technologies have accomplished several breakthroughs in solving complex tasks, most notably in computer vision and the development of autonomous agents. These achievements are driven mainly by advances in machine learning and deep learning, and the availability of large computing power and extensive databases.

Currently, modern AI techniques are starting to find their way into several aspects of mathematical work and mathematics education. In interactive learning environments, for instance, AI can be used to extract mathematical knowledge from the real world to generate new methods of content creation. On a more abstract level, AI is a promising technology for automated learner modelling, motivated by results from current research in AI for abstract mathematical reasoning. These technologies are, furthermore, expected to contribute to more intelligent tutoring systems, as employed in online learning environments, which, at present, already use data mining techniques to extract quantifiable insights from the learner's actions.

One of the first approaches to AI, in the 1950s, was the construction of automated theorem provers for geometry, combining the axiomatic theories and forward chaining deduction, with AI techniques. This combination was crucial to manage the complexity of the synthetic proofs, using the geometric constructions as models. In this way it was expected to be able to produce readable proofs, a goal with strong impact in education.

In the first three chapters the question of formal deduction in geometry is addressed. First, Quaresma gives us an account of the the history of automated deduction, from the early development of automated theorem provers for geometry, one of the early applications of AI, and from the emergence of the dynamic geometry systems, to the current status where different application systems combine dynamic geometry and automated deduction to create mathematical milieus where formal deduction tools help in the pursue of mathematical rigour. In the next two chapters, two such systems are described. Kovács et al. give us an account of the, recently included in *GeoGebra*, tools for the mathematically rigorous proof and discovery in geometry, and reflect on the potential educational impact of these new features. Font et al. describe *QED-Tutrix* intelligent tutor, a computational platform that includes an automated proof generator, allowing the students to solve high school geometry problems, tutored, with mathematical rigour, in any possible situation of their resolutions, rooted in a sound educational approach.

The next chapter continues this path, a clever combination of automated deductive and AI techniques at the service of education. Roanes-Lozano and Martínez-Zarzuelo show us a decision-making tool, inspired by a rule-based expert system verification approach, that, if an "official curriculum" exists, then it automatically checks the completeness and soundness of a given "development of the official curriculum". It will allow to produce better (error-free), instances of the curriculum, e.g. textbooks, project-based learning proposals.

The next two chapters deviate from this deductive tools path to show us more generic AI techniques, techniques that can be put at service in the construction of milieus to work on mathematics. First, Van Vaerenbergh and Pérez-Suay provide us with a bird's-eye view of AI systems for teaching and learning mathematics, proposing a classification that can help to choose the different "intelligent" components needed for a milieu construction.

Last, but not in any way, the least, the chapter by Martnez-Sevilla and Alonso Burgos describe a milieu where many AI techniques are in use to help its users to achieve a better understanding of mathematics behind our artistic and monumental heritage.

Evolution of Automated Deduction and Dynamic Constructions in Geometry

Pedro Quaresma

1 Introduction

> Logic appears in a *sacred* and in a *profane* form. The sacred form is dominant in proof theory, the profane form in model theory
>
> D. *van Dalen*, Logic and Structure (van Dalen, 1980)

Sacred Form

It can be said that the sacred form began circa 300 BC with the writing of Euclid's Elements. The Elements can be seen as a seminal work, establishing the basis for proof theory, with a collection of definitions, postulates, propositions (theorems and constructions) and mathematical proofs of the propositions. For centuries, it was included in the curriculum of the majority of the universities. For example, in 1692 Tirso de Molina, Superior General of the Jesuit Order wrote a letter with very specific orders to improve the teaching of Mathematics at the Portuguese province (i.e. Universities of Coimbra and Évora). One of the suggestions was the reproduction of the figures in the Elements in such a way that all the students could see those figures and discuss about the geometric properties behind those figures. The combination of that letter with the Portuguese tradition of tilling ("azulejos") gave rise to a collection of tiles with faithful representation of many of the figures in the *Elementa Geometriæ*

This work is funded by national funds through the FCT—Foundation for Science and Technology, I.P., within the scope of the project CISUC—UID/CEC/00326/2020 and by European Social Fund, through the Regional Operational Program Centro 2020.

P. Quaresma (✉)
CISUC, Mathematics Department, Universidade de Coimbra, Coimbra, Portugal
e-mail: pedro@mat.uc.pt

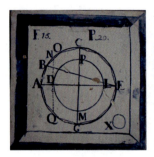  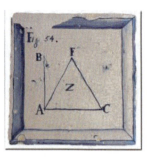

Fig. 1 *Azulejos* with figures from *Elementa Geometriæ* by André Tacquet

by the Jesuit priest André Tacquet (1612–1660) (see Fig. 1) (Gabriel Silva, 2017; Simões, 2007).[1]

David Hilbert in his book *Grundlagen der Geometrie* (1899) established the foundation for a modern axiomatic treatment of Euclidean geometry (Hilbert, 1977). Later, Alfred Tarski built a decision method for elementary algebra and geometry (Tarski, 1951) allowing constructive and even automated approaches for geometry (Beeson, 2015; Quaife, 1989) and, more recently, (1995) Jan von Plato proposed another constructive approach to geometry (von Plato, 1995).

From a computational perspective, the history of geometry automated theorem provers (GATP) began with the early computers and the birth of Artificial Intelligence, in the 1960s. The different sets of axioms of Euclidean Geometry attracted researchers to an attempt to implement synthetic methods, such as the approaches by Gelernter (1995, 1960), Nevis (1975), Elcock, Greeno et al. (1977), Coelho and Pereira (1979, 1986), Chou et al. (1993, 1995). The difficulties found with the synthetic methods, where the need to find a suitable rule to be applied lead to a combinatorial explosion regarding all the possible choices. This resulted in the exploration of other approaches, algebraic, semi-synthetic and logical approaches.

The algebraic style approach is characterized by the translation of geometric problems to algebraic problems, and subsequent development of the proof by the application of algebraic manipulations. The characteristic set method, also known as Wu's method (Chou, 1985; Wu, 1984), the elimination method (Wang, 1995), the Gröbner bases method (Kapur, 1986a, b) and the Clifford algebra approach (Li, 2000) are examples of practical methods of this type. The algebraic approach led to efficient implementations, but, given that all the proofs are developed by algebraic means, the geometric meaning is lost, i.e. apart from a yes/no answer, it is not possible to have a correspondent geometric proof where the axioms of geometry are used. This led to the development of methods capable of, at least partially, combine the geometric readability of synthetics methods with the efficiency of algebraic methods.

---

[1] Unfortunately most of the tiles were lost after the expelling of the Jesuits from Portugal by the Marquis of Pombal in 1759, the subsequent reform of the University of Coimbra and the construction of new buildings on the expense of the old ones.

The semi-synthetic methods use a set of specific *geometric quantities*, e.g. the *ratio of parallel directed segments* and *signed area*, to build an axiom system where the geometric relations and properties can be represented and the proofs developed using a set of geometric lemmas and simple algebraic manipulations. Examples of such methods are the area method (Chou et al., 1996a; Janičić et al., 2012) and the full-angle method (Chou et al., 1996b). These methods combine the readability of synthetic methods and the efficiency of algebraic methods, being able to prove many geometric theorems, efficiently and with geometric, readable, proofs.

More recently (2000–till present), new synthetic approaches are being proposed; the geometric deductive database method combines the full-angle axiom system with the techniques of deductive databases to develop an efficient GATP capable to prove a large set of geometric problems (Chou et al., 2000). Also tutorial systems like the *QED-Tutrix* (Gagnon et al., 2017; Tessier-Baillargeon et al., 2017) are proposed to address the problem in a more contained form, i.e. instead of trying to implement a generic GATP, the goal is to have an efficient and capable of readable geometric proofs GATP, to specific areas of geometry.

Also to be considered are the logical approaches, like the quantifier elimination method of Tarski (Collins, 1975; Tarski, 1951), or the use of axiom systems for geometry (e.g. Tarski, Quaife (1989)) and then using generic automated theorem provers (ATP) to develop the proofs. Many efficient and capable of proving many geometric conjectures ATP are available, but, like in the algebraic approach, the proof has no correspondence with any form of geometric reasoning. From the view point of a geometer, it is difficult to follow (geometrically) the formal proofs produced by the ATP.

Profane Form

The *Profane form* came with programs that allow to build and explore geometric figures. The 1988 Turing prize was awarded to Ivan Sutherland for his pioneering work in the area of computer graphics. The program Sketchpad changed the way people interacted with computers, from non-graphical to graphical (Sutherland, 1963, 2003). While the original aim was to make computers more accessible, introducing graphical manipulations, while retaining the powers of abstraction that are critical to programmers, the direct manipulation interfaces have since succeeded by reducing the levels of abstraction exposed to the user.

The program Sketchpad can be considered as the point of origin for today's computer-aided graphic design programs (CAD). Not detracting from CAD programs, they are of little interest for the geometry practitioner, they are very high-precision tools to draw figures, e.g. for architects, drawing building plans, but they miss the step from drawings (static object) to figures (geometric construction), i.e. a set of objects and geometric relations between then (dynamic object). Meanwhile, dynamic geometry systems (DGS) allow building geometric constructions from free objects and elementary constructions. It became possible to manipulate the free objects (objects universally quantified), preserving the geometric properties of the construction.

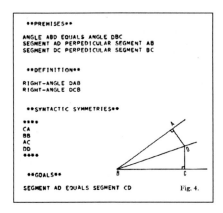 

Fig. 2 Gelernter—Angle bisection

The first software packages that can be classified as dynamic geometry systems were Geometer's Sketchpad (Jackiw, 2001), which appeared first in 1989, and Cabri Géomètre (Laborde & Strässer, 1990), dating back to 1988, and they started another revolution: computers could be used in school for teaching geometry. Since then DGS become mature tools used by millions of users all over the world.

Dynamic geometry systems gave us the profane side of proofs in geometry. For example Fig. 2 (if done with a modern DGS) would have the points A, B and C as free points and point D as a constructed point (intersection of lines), moving the free points we can conjecture that the segments AD and CD are equal in length, i.e. we are exploring "all" possible configurations for a given geometric construction in the Cartesian model. Although those manipulations are not formal proofs because only a finite set of positions are considered and visualization can be misleading, they provide a first clue to the truthfulness of a given geometric conjecture.

The DGS and GATP are in a collision course and that is a good thing. From the development of GATP and DGS as completely separated tools, to the implementation of some GATP method in a DGS (e.g. *Cinderella*) or graphical components into a GATP (e.g. *GCLC* and *JGEX*) to the integration of GATP and DGS (e.g. *GeoGebra*). The fully integration of automated deduction components in other software is becoming a reality and it is expected that in a near future it will be possible to have those components broadly available.

Overview of the chapter In Sect. 2, the evolution of automated deduction in geometry is presented and in Sect. 3 the integration of GATP and DGS is discussed. In Sect. 4 other lines of research are presented and in Sect. 5 conclusions are drawn.

2 Automated Deduction in Geometry

For the last five decades, automated deduction in geometry, the sacred form, has been an important field in the area of automated reasoning. Various methods and techniques have been studied and developed for automatically proving and discovering geometric theorems (Chou, 1987; Chou & Gao, 2001; Chou et al., 1994).

2.1 Synthetic Methods

Adapting general-purpose reasoning approaches developed in the field of artificial intelligence (in the 60s of the twentieth century), synthetic methods, such as the approaches by Gelernter (1995, 1960, Nevis (1975), Elcock (1977), Greeno et al. (1979), Coelho and Pereira (1979, 1986), Zhang et al. (1995), were dedicated to automating traditional proving processes (Chou & Gao, 2001). Making use of axiomatic systems close to the ones used in secondary schools these systems tried to provide readable (by students and teachers) proofs. See Fig. 2 for an example of such proofs, from the GATP by Gelernter.

In many of these first attempts, the diagrams where used as a model (Coelho & Pereira, 1986):

- the diagram as a filter (acting as a counter-example);
- the diagram as a guide (acting as an example, suggesting eventual conclusions).

As a filter the diagram permits to test the non-provability of a candidate sub-goal, pruning the proof tree.

As a guide the diagram can be used as a positive indication. Quoting from Coelho et al. (1986) Coelho and Pereira (1986) (see Fig. 3):

> We want to prove two equal segments $UV = XY$, by congruent triangles. Suppose triangle XYZ exists, and our purpose is to find a triangle UVW on UV to compare to triangle XYZ. We need to search for existing or generated triangles on UV. The first thing is to find a convenient third point W, which must be different from U and V. The possible coordinates of the sought point W are computed from the coordinates of X, Y, Z, U and V, and a check is made in the diagram to see if a point with such coordinates exists. The diagram is used in a positive way for computing the possible coordinates for W.

Fig. 3 Coelho et al. (1986)—Diagram as a guide

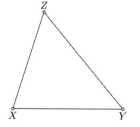

The possibility of having geometric proofs, with natural language and (eventually) visual renderings, is a key aspect of this approach. Unfortunately, the combinatorial explosion while applying postulates implied the use of suitable heuristics that narrow the scope of the GATP and prevent the development of a general-purpose efficient GATP.

New synthetic approaches are being proposed. The geometric deductive database method combines the full-angle axiom system with the techniques of deductive databases, to develop an efficient GATP capable of proving a large set of geometric problems (Chou et al., 2000). A coherent logic[2] based GATP, *ArgoCLP*, is being developed which can be used to generate both readable and formal (machine verifiable) proofs in various theories, primarily geometry. The possibility of, using a top-down approach (from the conjecture to the conclusion), producing natural language proofs is a positive point, but, efficiency considerations are still a major concern (Stojanović et al., 2011).

2.2 Algebraic Methods

A different approach is given by the algebraic style methods, given by the translation of geometric problems to algebraic problems and the subsequent development of the proof by the application of algebraic manipulations. The characteristic set method, also known as Wu's method (Chou, 1985; Wu, 1984), the elimination method (Wang, 1995), the Gröbner bases method (Kapur, 1986a, b) and the Clifford algebra approach (Li, 2000) are examples of practical methods based on the algebraic approach.

Let us consider, for example, the Euler's Line theorem.

Theorem 1 (Euler's Line Theorem) *In any given triangle, the orthocentre, the centroid and the centre of the circumscribed circle are collinear (Fig. 4).*

Transcribing it to algebraic form we get (GATP: *JGEX*, Wu's method):

```
The Algebraic Form:
A: (0,0)    B: (0,x4)    C: (x5,x6)    D: (x7,x8)    E: (0,x10)
F: (x11,x12)  G: (x13,x14)  H: (x15,x16)  I: (0,x18)
J: (x19,x20)  K: (x21,x22)  L: (x23,x24)  M: (x25,x26)

The Equational Hypotheses:
1: D : midpoint(BC)    2x8 - x6 - x4 = 0     2x7 - x5 = 0
2: E : midpoint(BA)    2x10 - x4 = 0
3: F : midpoint(AC)    2x12 - x6 = 0         2x11 - x5 = 0
:
11: LF ⊥ CA    x6x24 + x5x23 - x6x12 - x5x11 = 0
12: M : on line EL    x23x26 + (-x24 + x10)x25 - x10x23 = 0
```

---

[2] Coherent logic is a fragment of (finitary) first-order logic which allows only the connectives and quantifiers ∧ (and), ∨ (or), ⊤ (true), ⊥ (false), ∃ (existential quantifier).

Fig. 4 Euler's line theorem, JGEX construction

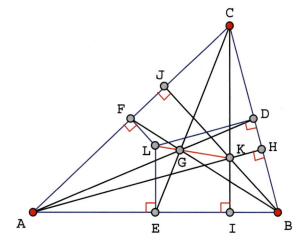

At this stage the geometric/algebraic connection is still possible, each geometric element has a correspondent set of algebraic equations and vice versa. In the proof itself, pure algebraic methods are used. In Wu's method, this implies calculating triangular systems and pseudo-remainders (Chou, 1985; Wu, 1984).

The JGEX's proof (Wu's method) is:

```
The Triangularized Hypotheses (TS):
h0: 2x7 - x5 = 0
h1: 2x8 - x6 - x4 = 0
h2: x9 = 0
:
h18: 2x5x25 + (-2x6 + 2x4)x8 + x4x6 - x5^2 - x4^2 = 0
h19: (2x6 - 2x4)x26 + 2x5x25 + (-2x6 + 2x4)x8 - x5^2 = 0

The Conclusion (CONC):
L, G, K are collinear
(x21 - x13)x24 + (-x22 + x14)x23 + x13x22 - x14x21 = 0

Successive Pseudo Remainder of CONC wrpt TS :
R_18 = [x24, 6]
R_17 = prem(R_18, h_17) = [x23, 6]
:
R_7 = prem(R_8, h_7) = [x13, 6]
R_6 = prem(R_7, h_6) = [0, 0]
Remainder = R_6 = 0
 The conclusion is true
```

GATP based on these methods are efficient and able to proof a large number of geometric conjectures. The price to pay is the absence of geometric proofs and the algebraic proofs, if any, are only (barely) readable by experts (Chou & Gao, 2001).

2.3 Semi-Synthetic Methods

In order to combine the readability of synthetic methods and the efficiency of algebraic methods, some approaches, such as the area method (Chou et al., 1996a; Janičić et al., 2012) and the full-angle method (Chou et al., 1996b), represent geometric knowledge in the form of expressions with respect to geometric invariants.

For stating and proving conjectures, these methods use a set of specific *geometric quantities* that enable treating geometric relations defining an axiom system. Considering the area method, we have:[3]

- *ratio of parallel directed segments*, denoted $\overline{AB}/\overline{CD}$. If the points A, B, C and D are collinear, $\overline{AB}/\overline{CD}$ is the ratio between the lengths of directed segments AB and CD. If the points A, B, C and D are not collinear, and it holds $AB \| CD$, there is a parallelogram $ABPQ$ such that P, Q, C and D are collinear and then $\frac{\overline{AB}}{\overline{CD}} = \frac{\overline{QP}}{\overline{CD}}$.
- *signed area* for a triangle ABC, denoted $\mathcal{S}_{ABC}$ is the area of the triangle ABC, multiplied by -1, if ABC has the negative orientation.
- *Pythagoras difference*,[4] denoted $\mathcal{P}_{ABC}$, for the points A, B, C, defined as $\mathcal{P}_{ABC} = \overline{AB}^2 + \overline{CB}^2 - \overline{AC}^2$.

An axiom system based on these three geometric quantities allows expressing (in form of equalities) geometry properties such as collinearity of three points ($\mathcal{S}_{ABC} = 0$), parallelism of two lines ($\mathcal{S}_{ABC} = \mathcal{S}_{BCD}$), equality of two points ($\mathcal{P}_{ABA} = 0$), perpendicularity of two lines ($\mathcal{P}_{ACD} = \mathcal{P}_{BCD}$), etc. (Chou et al., 1996a; Janičić et al., 2012).

To prove a given conjecture, we have to express the hypotheses of a theorem using a set of starting ("free") points and a set of constructive statements, each of them introducing a new point, and to express the conclusion by an equality between polynomials in the geometric quantities of the method (without considering Cartesian coordinates). The proof is developed by eliminating, in reverse order, the points introduced before, using for that purpose a set of appropriate lemmas (see Fig. 5). After eliminating all the introduced points, the goal equality of the conjecture collapses to an equality between two rational expressions involving only free points. This equation can be further simplified to involve only independent variables. If the expressions on the two sides are equal, the conjecture is a theorem, otherwise it is not.

In the example above (see Theorem 1), the conjecture can be expressed using the signed area of triangles, $\mathcal{S}_{LGK} = 0$, i.e. the signed area of the triangle LGK is zero, so we are considering a degenerate triangle, so the points are collinear. Using the area method, both *GCLC* and *JGEX* failed to prove this conjecture (only using the

---

[3] Negative and positive orientation are only a syntactic convention to disambiguate between "different" geometric constructions built from the same set of points.

[4] The *Pythagoras difference* is a generalization of the Pythagoras equality regarding the three sides of a right triangle, to an expression applicable to any triangle. For a triangle ABC with the right angle at B, it holds that $\mathcal{P}_{ABC} = 0$.

(1) $$\frac{\overrightarrow{HT}}{\overrightarrow{TO}} = 2, \quad \text{by the statement}$$

(2) $$\left(-1 \cdot \frac{\overrightarrow{TH}}{\overrightarrow{TO}}\right) = 2, \quad \text{by geometric simplifications}$$

(3) $$\left(-1 \cdot \frac{S_{TAA_2}}{S_{TAOA_2}}\right) = 2, \quad \text{by Lemma 37, first case — assuming points } T, A, \text{ and } A_2 \text{ are not collinear (point } H \text{ eliminated)}$$

(4) $$\frac{(-1 \cdot S_{TAA_2})}{S_{TAOA_2}} = 2, \quad \text{by algebraic simplifications}$$

(5) $$\frac{\left(-1 \cdot \frac{((P_{ABC} \cdot S_{TAC}) + (P_{ACB} \cdot S_{TAB}))}{P_{BCB}}\right)}{S_{TAOA_2}} = 2, \quad \text{by Lemma 31 (point } A_2 \text{ eliminated)}$$

(6) $$\frac{((-1 \cdot (P_{ABC} \cdot S_{TAC})) + (-1 \cdot (P_{ACB} \cdot S_{TAB})))}{(P_{BCB} \cdot S_{TAOA_2})} = 2, \quad \text{by algebraic simplifications}$$

(7) $$\frac{((-1 \cdot (P_{ABC} \cdot S_{TAC})) + (-1 \cdot (P_{ACB} \cdot S_{TAB})))}{\left(P_{BCB} \cdot \frac{((P_{ABC} \cdot S_{TAOC}) + (P_{ACB} \cdot S_{TAOB}))}{P_{BCB}}\right)} = 2, \quad \text{by Lemma 31 (point } A_2 \text{ eliminated)}$$

(8) $$\frac{((-1 \cdot (P_{ABC} \cdot S_{TAC})) + (-1 \cdot (P_{ACB} \cdot S_{TAB})))}{((P_{ABC} \cdot S_{TAOC}) + (P_{ACB} \cdot S_{TAOB}))} = 2, \quad \text{by algebraic simplifications}$$

(9) $$\frac{((-1 \cdot (P_{ABC} \cdot S_{TAC})) + (-1 \cdot (P_{ACB} \cdot S_{TAB})))}{((P_{ABC} \cdot (S_{TAO} + S_{CTO})) + (P_{ACB} \cdot (S_{TAO} + S_{BTO})))} = 2, \quad \text{by geometric simplifications}$$

$$\vdots$$

(1082) $\quad 0 = 0, \quad$ by algebraic simplifications

NDG conditions are:
$S_{ACC_1} \neq S_{A_1CC_1}$ i.e., lines AA_1 and CC_1 are not parallel (construction based assumption)
$S_{M_a^0 M_c^2 T_c^3} \neq S_{T_a^1 M_c^2 T_c^3}$ i.e., lines $M_a^0 T_a^1$ and $M_c^2 T_c^3$ are not parallel (construction based assumption)
$S_{ACC_2} \neq S_{A_2CC_2}$ i.e., lines AA_2 and CC_2 are not parallel (construction based assumption)
$P_{TOT} \neq 0$ i.e., points T and O are not identical (conjecture based assumption)
$S_{ACC_1} \neq 0$ i.e., points A, C and C_1 are not collinear (cancellation assumption)
$S_{ACB} \neq 0$ i.e., points A, C and B are not collinear (cancellation assumption)
$S_{TAA_2} \neq 0$ i.e., points T, A and A_2 are not collinear (made for Lemma 37)

Number of elimination proof steps: 386; Number of geometric proof steps: 876; Number of algebraic proof steps: 5982; Total number of proof steps: 7244; Time spent by the prover: 4.902 seconds; Number of pages: 169

Fig. 5 Euler's line theorem, *GCLC* proof (area method)

algebraic methods they were capable of proving it), but for the same construction the conjecture $\overline{GK}/\overline{LG} = 2$ is provable. The proof itself is a (readable) sequence of steps preforming algebraic simplifications, geometric simplifications and the application of lemmas allowing to eliminate the free points, but with 169 pages this formal proof cannot be considered readable (see Fig. 5).

The GATP implementing these methods are efficient and capable of producing formal, but readable (in some cases, small) proofs. Given the fact that these methods do not use the usual axioms systems used in secondary schools, the proofs produced are readable, even by secondary schools' students and teachers, but not without a previous study of the axiom system used by the GATP.

2.4 Generic First-Order Provers

First-order ATP must also be considered. Provers like *Vampire*, *Leo* and many others[5] are very efficient[6] and capable of proving theorems in many areas. Using axioms systems for geometry, e.g. Tarski, they can be used in geometry also. For example, the *Thousands of Problems for Theorem Provers* (TPTP) repository (Sutcliffe, 2017) has the *GEO* domain with many problems specified in the first-order logic (FOL), using the Tarski Geometry (Quaife, 1989) or the geometric deductive database method (GDDM) (Chou et al., 2000)[7] axiom systems. The proof of those conjectures can then be attempted by the different ATP.

For example, the Euler's line theorem can be expressed in the following form (using the GDDM):

```
%----Include Geometry Deductive Database Method axioms
include('geometryDeductiveDatabaseMethod.ax').

fof(exemploeulerLineGEO0315a,conjecture,(
  ! [ A,B,C,D,E,F,G,H,I,J,K,L,M ] :
  ((midp(D,B,C) & midp(E,B,A) & midp(F,A,C) & coll(G,B,F)
    & coll(G,C,E) & perp(H,A,B,C) & coll(H,B,C)
    & perp(I,C,A,B) & coll(I,A,B) & perp(J,B,A,C)
    & coll(J,A,C) & coll(K,A,H) & coll(K,C,I) & perp(B,A,M,E)
    & perp(C,B,L,D) & perp(C,A,L,F) & coll(M,D,L))
   =>
   (coll(L,G,K))) )).
```

the specification of the problem is done in *First-Order Form* (FOF). Using the ATP *Vampire*, the conjecture is proved by refutation; below an excerpt of the proof is given:

```
% Refutation found. Thanks to Tanya!
% SZS status Theorem for eulerLineGEO0315a
% SZS output start Proof for eulerLineGEO0315a
1. ! [X0,X1,X2] : (coll(X0,X1,X2) => coll(X0,X2,X1)) [input]
2. ! [X0,X1,X2] : (coll(X0,X1,X2) => coll(X1,X0,X2)) [input]
                           ⋮
2754. $false <- (74) [resolution 1962,1558]
```

---

[5] https://en.wikipedia.org/wiki/Automated_theorem_proving.
[6] http://www.tptp.org/CASC/.
[7] This is a recent contribution, by the author, to TPTP.

```
2765.  ~74 [avatar contradiction clause 2754]
2766.  $false [avatar sat refutation 1963,2749,2765]
% SZS output end Proof for eulerLineGEO0315a
% ------------------------------
% Version: Vampire 4.5.1
% Termination reason: Refutation

% Memory used [KB]: 6652
% Time elapsed: 0.062 s
```

Using this approach, the proof script (if any) would be a logical proof (e.g. using the resolution method). As can be seen above, in the case of the proof done by the ATP *Vampire*, the proof contains information about the lemmas used and, if some post-processing is used, can be readable, but the connection with the geometric construction is difficult, at best.

An advantage of this approach is the availability of many generic ATPs with very efficient implementations.[8] The unavailability of geometric proofs is the drawback.

2.5 Other Approaches

Rule-based approaches explore the possibility of building a sound, not necessarily complete, axiom system. The idea is to have a minimal set of axioms, lemmas and rules of inference that can characterize a given sub-area of geometry (Pambuccian, 2004).

One example of such an approach is given by the tutorial system *QED-tutrix*. Exploring the logic programming language *Prolog*[9] (Clocksin & Mellish, 2003) a set of axioms, lemmas and rules of inference, adapted to the type of problems at hand, are implemented and explored by the *Prolog* rule-based logical query mechanism. The *QED-tutrix* tutorial system builds the *Hypothesis, Properties, Definitions, Intermediate results and Conclusion graph* (HPDIC-graph). The HPDIC-graph contains all possible proofs for a given problem, using a given set of axioms. Having that (possibly, very large) graph the system can help the learner, validating the steps already taken and providing hints for the next steps (Font et al., 2018; Gagnon et al., 2017).

A project, still in its early stages, uses *Maude*[10] (Clavel et al., 2007), an equational (and rewriting) logic system to implement the Tarski axiom system as described by Art Quaife (1989) (see Fig. 6).

---

[8] For the Euler line theorem, *GCLC*'s Wu Method took 0.012s, *Vampire* took 0.062s.
[9] https://en.wikipedia.org/wiki/Prolog.
[10] http://maude.cs.illinois.edu/w/index.php/The_Maude_System.

```
*** System Tarski over G3cp
***
fmod FORMULA is
pr QID . *** Maude's Qualified Identifiers ('a,'b, etc).
sort Prop . *** Atomic propositions
sort Formula . *** Formulas
sort Point . *** Points
sort Segment . *** Segments
subsort Qid < Prop < Formula .
subsort Qid < Point .
(...)
*** Tarski geometry primitive relations
op p : -> Point [ctor] .
op pl : -> Point [ctor] .
op pll : -> Point [ctor] .
op _*_ : Point Point -> Segment [ctor comm] .
op betweenness : Point Point Point -> Prop [ctor] .
op equidistance : Segment Segment -> Prop [ctor comm] .
op extension : Point Point Point Point -> Point . *** extension
op innerPasch : Point Point Point Point Point -> Point .
op euclides1 : Point Point Point Point Point -> Point .
op euclides2 : Point Point Point Point Point -> Point .
op continuity : Point Point Point Point Point Point -> Point .
op _==_ : Point Point -> Formula [ctor comm ] .
endfm
mod Tarski is
***
*** Tarski' Geometry (Art Quaife (1989), JAR 5, 97--118.
(...)
*** A7 Inner Pasch
rl [ip1] : C , betweenness(U,V,W) , betweenness(Y,X,W) |-- betweenness(V,innerPasch(U,V,W,X,Y),Y) , C' =>
      proved .
rl [ip2] : C , betweenness(U,V,W) , betweenness(Y,X,W) |-- betweenness(X,innerPasch(U,V,W,X,Y),U) , C' =>
      proved .
(...)
*** A10 Euclid's axiom
rl [eucl1] : C , betweenness(U,W,Y) , betweenness(V,W,X) |-- U == W , betweenness(U,V,euclides1(U,V,W,X,Y)) ,
      C' => proved .
rl [eucl2] : C , betweenness(U,W,Y) , betweenness(V,W,X) |-- U == W , betweenness(U,X,euclides2(U,V,W,X,Y)) ,
      C' => proved .
rl [eucl3] : C , betweenness(U,W,Y) , betweenness(V,W,X) |-- U == W , betweenness(euclides1(U,V,W,X,Y),Y,
      euclides2(U,V,W,X,Y)) , C' => proved .
(...)
```

Fig. 6 Implementation, as a Maude module, of Tarski axiom system, as described by Art Quaife

Both approaches share the use of logic programming languages where the introduction of specific lemmas, for specific purposes, can be easily made, e.g. SSS, SAS, ASA and AAS lemmas for the congruence of triangles, or the alternate interior and exterior angles of parallel lines.

As stated above, this systems can be used to implement sound, but not necessarily complete, axiom systems. These kind of system can be useful in specific situations, e.g. in secondary schools mathematics classes.

3 Dynamic Geometry

Dynamic Geometry Systems

Dynamic geometry can be characterized by the construction of geometric figures (as opposed of fixed drawings), built from free objects, i.e. universally quantified objects (e.g. points) and elementary properties preserving constructions.

Analysing the example given in Fig. 7, the construction has A, B, D and E as free points, these can be moved freely in the plane, points C and F belongs to AB and DE, respectively, they have only one degree of freedom, being able to move in the line they belong and, finally, points G, H and I are the intersections of two lines, they do not have any degree of freedom.

Dynamic geometry systems give us the profane side of proofs in the Cartesian model of Euclidean geometry. By moving the free points, we can conjecture that the points G, H and I are collinear (red line), i.e. we are exploring "all" possible configurations for that geometric construction. Although these manipulations are not formal proofs because only a finite set of positions are considered and because visualisation can be misleading, they provide a first clue to the truthfulness of a given geometric conjecture.

Dynamic geometry systems are now mature software tools with a very large users base. The dynamic character of these programs give to its users the possibility of building dynamic geometric constructions, exploring conjectures about them, so

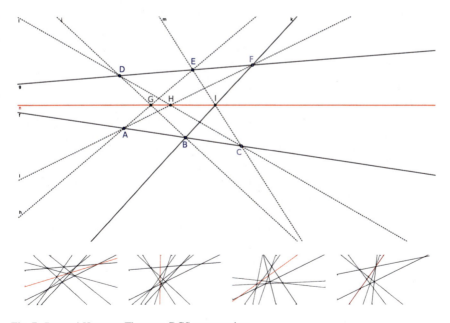

Fig. 7 Pappus' Hexagon Theorem, DGS construction

they give a first, informal, opportunity to explore geometric proofs. But to avoid the visual proofs pitfall, the connection with geometric automated theorem provers (GATP) should be considered.

Dynamic Geometry Systems and Geometry Automated Theorem Provers

There are already some systems combining DGS and GATP. The following list is organized as follows: from GATP and GATP with rendering capabilities to DGS with some automatic proving capabilities built-in, ending with environments with DGS and/or GATP capabilities.

Open Geometry Prover   *The Open Geometry Prover (OGP)* is an open source project,[11] aiming to implement various geometry automated theorem provers. It can be used as a stand-alone tool but can also be integrated into other geometry tools, such as dynamic geometry software. In its current state, OGP implements the Wu's algebraic methods. Some partial work has been made in the semi-synthetic methods: the area method and the full-angle method (Baeta & Quaresma, 2013; Botana et al., 2015; Petrović et al., 2012).

GCLC   *Geometry Constructions → LaTeX Converter (GCLC)*, an open source GATP with a graphics engine, for the Wu, Gröbner bases and area methods. It is possible to add a conjecture to a given geometric construction (with a graphical rendering) and ask for its proof with natural language rendering (Janičić, 2006).

JGEX   *Java Geometry Expert* (JGEX), a GATP with graphics engine, for the Wu, Gröbner bases, area, full-angle and deductive database methods. It is possible to add a conjecture to a given geometric construction and ask for its proof with natural and visual language renderings (Ye et al., 2011).

Cinderella   The *Interactive Geometry Software Cinderella* is a DGS with a randomized theorem checker (Kortenkamp & Richter-Gebert, 2004; Richter-Gebert & Kortenkamp, 1999).

The last one has a different approach from the first three, it is not a prover capable of a formal proof, but more a model checker, capable of generating many random instances for a given geometric configuration. So, for a given conjecture, we obtain a probabilistic answer to its validity (Kortenkamp & Richter-Gebert, 2004).

Apart from the library OGP, these systems[12] are monolith systems, i.e. even if the GATP are modules that can be used by themselves (e.g. the *GCLC*), the DGS and the GATP are tightly integrated. The DGS are not able to use external GATP and the GATP are not prepared to be integrated in other DGS.

A modular approach is beginning to make its way, i.e. an approach where DGS and GATP can be developed by different teams and nevertheless be combined. Examples of such systems are:

---

[11] The OpenGeometryProver github project: https://github.com/opengeometryprover/.
[12] JGEX is currently not being supported/developed.

GeoProof   A DGS that interfaces with the *Coq* Prover Assistant (Bertot & Castéran, 2004), allowing to check proofs built interactively (Narboux, 2007).

GeoGebra   The DGS embedded prover system chooses one of the available methods and translates the problem specified by the end user as the input for the selected method, similarly to portfolio solvers.[13] The available GATP implement the Wu's method, the Buchberger–Kapur method, the area method and Recio's exact check method. The separation between the GATP and the DGS opens the possibility of using third-party GATP. For example, the *Open Geometry Prover* Wu's method GATP. From the DGS interface the user can ask for the validation of a given geometric conjecture, related to the construction, e.g. `Prove AreCollinear(G, H, I)` is the command that a user may use to prove, formally, that the points G, H and I in the construction of Fig. 7 are collinear. For the moment, this is only a formal validation, no proof script is available (Botana et al., 2015; Kovács, 2015; Kovács & Recio, 2020; Nikolić et al., 2019).

This project can be visited at https://www.geogebra.org/m/McEqwQNb.

When applied to areas such as education GATP and DGS are being combined in environments that use both tools for the purpose of learning.

QED-Tutrix   *QED-Tutrix* is an intelligent tutor for geometry (see Sect. 2.5) offering an interface to help high school students to freely explore the geometric problems and their proofs.

For each geometric conjecture, the system builds the tree of all possible proofs, allowing the students to try to prove the conjecture, with the system helping in each step of the proof (Font et al., 2018; Gagnon et al., 2017).

Geometriagon   *Geometriagon*[14] is a project to explore geometric constructions made with ruler and compass. The dynamic geometry software supporting the project is C.a.R.[15] It provides an extensive list of problems to be solved by registered users, validating (or not) the solution found.

A similar approach, not in the area of geometry, is given by *Edukera*,[16] a Web-environment to teach Logic and Mathematics (calculus and sets), with the assistance of the *Coq* proof assistant.

---

[13] Portfolio problem solving is an approach in which for an individual instance of a specific problem, one particular, hopefully most appropriate, solving technique is automatically selected among several available ones and used. The selection usually employs machine learning methods.

[14] http://polarprof-001-site1.htempurl.com/geometriagon/.

[15] http://car.rene-grothmann.de/doc_en/index.html.

[16] https://www.edukera.com/.

4 Other Lines of Research

The usefulness of automated deduction methods and tools in geometry does not circumscribe itself to its use in dynamic geometry systems:

Repositories of Geometric Knowledge   In the repository *Thousand of Geometric problems for geometric Theorem Provers* (TGTP),[17] a list of more than two hundred geometric conjectures and their proofs by several GATP are kept (Quaresma, 2011). It implements a protocol for the exchange of geometric information between applications, allowing a direct access to its repository from external tools.

In the repository *Thousands of Problems for Theorem Provers* (TPTP) (Sutcliffe, 2017),[18] a section is dedicated to geometric problems (in this repository the problems are specified as first order logic problems).

Recent efforts are being made to start a *Geometry Automated Provers Competition* (Baeta et al., 2020), in order to help improve the efficiency and usefulness of GATP.

Knowledge Management   Besides the efforts to implement readable formal proofs produced by GATP, there are other avenues of research that are being explored. For example: to evaluate the complexity and also the interestingness of geometric proofs (Gao et al., 2019; Quaresma et al., 2020);[19] to explore the automatic discovery of theorems in elementary geometry (Recio & Vélez, 1999); to implement a semantic geometric search mechanism, i.e. the possibility of, having a given geometric construction, to search for other congruent geometric constructions or even other geometric constructions with some common geometric properties. A prototype of this search mechanism is already implemented in the *TGTP* repository (Haralambous & Quaresma, 2014, 2018).

All these efforts are being made to help answer the questions related to the use of GATP as one more tool in the geometer's toolbox.

5 Conclusions

It is clear that automated deduction methods and tools are beginning to make their way into many different uses, in many different contexts. However, there are many issues to be solved before a more complete integration of automated deductive tools can take place. Examples of problems related with the use of automated deduction in geometry tools are: application programming interface; common formats for information interchange; natural and visual languages renderings of the proofs; repositories of

---

[17] http://hilbert.mat.uc.pt/TGTP/index.php.

[18] http://www.tptp.org/.

[19] Pedro Quaresma and Pierluigi Graziani, Measuring the Readability of a Proof, submitted to publication.

geometric knowledge with powerful search mechanisms; proof discovery; problems and proof classification for complexity and efficiency.

The complexity and the sheer size of the task led to recent efforts to establish a network of researchers working in the area of formal reasoning, knowledge-based intelligent software and geometric knowledge management. The network will need to focus in the creation of an intelligent computational environment in which advanced software tools and deduction mechanisms are embedded for symbolic-numeric geometric computation, interactive or automated geometric reasoning, knowledge validation, knowledge discovery and knowledge management. Such a "superset of a book" of geometric knowledge with embedded tools, freely available in all computational platforms, adaptable, collaborative and adaptive to each and every user's profiles, would bring together a whole new generation of mathematical tools with impact at all levels: exploratory research, applications in mathematics and education.

All these efforts are leading to an integration of formal deduction and dynamic geometry in vivid environments, to be used in the exploration of mathematics in its fullness. It is a complex task, but it is also an exciting task, as new advances in many areas are expected in the near future.

References

Baeta, N., & Quaresma, P. (2013). The full angle method on the OpenGeoProver. In C. Lange, D. Aspinall, J. Carette, J. Davenport, A. Kohlhase, M. Kohlhase, P. Libbrecht, P. Quaresma, F. Rabe, P. Sojka, I. Whiteside, & W. Windsteiger (Eds.), *MathUI, OpenMath, PLMMS and ThEdu Workshops and Work in Progress at the Conference on Intelligent Computer Mathematics, no. 1010 in CEUR Workshop Proceedings*. Aachen. http://ceur-ws.org/Vol-1010/paper-08.pdf.

Baeta, N., Quaresma, P., & Kovács, Z. (2020). Towards a geometry automated provers competition. In *Proceedings 8th International Workshop on Theorem Proving Components for Educational Software, (ThEdu'19)*. Electronic Proceedings in Theoretical Computer Science (Vol. 313, pp. 93–100). Natal, Brazil, 25th August 2019. https://doi.org/10.4204/EPTCS.313.6

Beeson, M. (2015). A constructive version of Tarski's geometry. *Annals of Pure and Applied Logic, 166*(11), 1199–1273. https://doi.org/10.1016/j.apal.2015.07.006.

Bertot, Y., & Castéran, P. (2004). *Interactive theorem proving and program development: Coq'Art: The calculus of inductive constructions*. EATCS: Springer. https://doi.org/10.1007/978-3-662-07964-5.

Botana, F., Hohenwarter, M., Janičić, P., Kovács, Z., Petrović, I., Recio, T., & Weitzhofer, S. (2015). Automated theorem proving in GeoGebra: Current achievements. *Journal of Automated Reasoning, 55*(1), 39–59. https://doi.org/10.1007/s10817-015-9326-4.

Chou, S. (1985). *Proving and discovering geometry theorems using Wu's method*. Ph.D. thesis, The University of Texas, Austin.

Chou, S. C. (1987). *Mechanical geometry theorem proving*. Dordrecht: D. Reidel Publishing Company.

Chou, S. C., & Gao, X. S. (2001). Automated reasoning in geometry. In J. A. Robinson & A. Voronkov (Eds.), *Handbook of automated reasoning* (pp. 707–749). Elsevier Science Publishers B.V. https://doi.org/10.1016/B978-044450813-3/50013-8.

Chou, S. C., Gao, X. S., & Zhang, J. Z. (1993). Automated production of traditional proofs for constructive geometry theorems. In M. Vardi (Ed.), *Proceedings of the Eighth Annual IEEE Symposium on Logic in Computer Science LICS* (pp. 48–56). IEEE Computer Society Press.

Chou, S. C., Gao, X. S., & Zhang, J. Z. (1994). *Machine proofs in geometry*. World Scientific. https://doi.org/10.1142/2196.

Chou, S. C., Gao, X. S., & Zhang, J. Z. (1995). Automated production of traditional proofs in solid geometry. *Journal of Automated Reasoning, 14,* 257–291. https://doi.org/10.1007/bf00881858.

Chou, S. C., Gao, X. S., & Zhang, J. Z. (1996a). Automated generation of readable proofs with geometric invariants, I. Multiple and shortest proof generation. *Journal of Automated Reasoning, 17*(13), 325–347. https://doi.org/10.1007/BF00283133.

Chou, S. C., Gao, X. S., & Zhang, J. Z. (1996b). Automated generation of readable proofs with geometric invariants, II. Theorem proving with full-angles. *Journal of Automated Reasoning, 17*(13), 349–370. https://doi.org/10.1007/BF00283134.

Chou, S. C., Gao, X. S., & Zhang, J. Z. (2000). A deductive database approach to automated geometry theorem proving and discovering. *Journal of Automated Reasoning, 25,* 219–246.

Clavel, M., Durán, F., Eker, S., Lincoln, P., Martí-Oliet, N., Meseguer, J., & Talcott, C. (2007). *All about Maude: A high-performance logical framework.* Lecture Notes in Computer Science (Vol. 4350). Springer. https://doi.org/10.1007/978-3-540-71999-1.

Clocksin, W. F., & Mellish, C. S. (2003). *Programming in Prolog.* Springer, Berlin, Heidelberg. https://doi.org/10.1007/978-3-642-55481-0.

Coelho, H., & Pereira, L. M. (1979). Geom: A Prolog geometry theorem prover. Memórias 525, Laboratório Nacional de Engenharia Civil, Ministério de Habitação e Obras Públicas, Portugal.

Coelho, H., & Pereira, L. M. (1986). Automated reasoning in geometry theorem proving with Prolog. *Journal of Automated Reasoning, 2*(4), 329–390. https://doi.org/10.1007/BF00248249.

Collins, G. (1975). Quantifier elimination for real closed fields by cylindrical algebraic decomposition. In H. Brakhage (Ed.), *Automata Theory and Formal Languages 2nd GI Conference Kaiserslautern.* Lecture Notes in Computer Science (Vol. 33), 20–23 May 1975. Springer, Berlin, Heidelberg. https://doi.org/10.1007/3-540-07407-4_17.

van Dalen, D. (1980). *Logic and structure.* Universitext. Springer-Verlag

Elcock, E. W. (1977). Representation of knowledge in geometry machine. *Machine Intelligence, 8,* 11–29.

Font, L., Richard, P. R., & Gagnon, M. (2018). Improving QED-Tutrix by automating the generation of proofs. In P. Quaresma, & W. Neuper (Eds.), *Proceedings 6th International Workshop on Theorem Proving Components for Educational Software,* Gothenburg, Sweden, 6 Aug 2017, Electronic Proceedings in Theoretical Computer Science (Vol. 267, pp. 38–58). Open Publishing Association (2018). https://doi.org/10.4204/EPTCS.267.3.

Gabriel Silva, J. (2017). Coimbra: uma universidade global, desde o século xvi. *Rua Larga, 50.* Imprensa da Universidade de Coimbra

Gagnon, M., Leduc, N., Richard, P., & Tessier-Baillargeon, M. (2017). Qed-tutrix: Creating and expanding a problem database towards personalized problem itineraries for proof learning in geometry. In *Proceedings of the Tenth Congress of the European Society for Research in Mathematics Education (CERME10).*

Gao, H., Li, J., & Cheng, J. (2019). Measuring interestingness of theorems in automated theorem finding by forward reasoning based on strong relevant logic. In *2019 IEEE International Conference on Energy Internet (ICEI)* (pp. 356–361). IEEE. https://doi.org/10.1109/ICEI.2019.00069.

Gelernter, H. (1995). Realization of a geometry-theorem proving machine. In *Computers & thought,* 2nd edn. (pp. 134–152). MIT Press, Cambridge, MA, USA.

Gelernter, H., Hansen, J. R., & Loveland, D. W. (1960). Empirical explorations of the geometry theorem machine. In Papers presented at the 3–5 May 1960, western joint IRE-AIEE-ACM computer conference, IRE-AIEE-ACM'60 (Western) (pp. 143–149). ACM, New York, NY, USA. https://doi.org/10.1145/1460361.1460381.

Greeno, J., Magone, M. E., & Chaiklin, S. (1979). Theory of constructions and set in problem solving. *Memory and Cognition, 7*(6), 445–461. https://doi.org/10.3758/BF03198261.

Haralambous, Y., & Quaresma, P. (2014). Querying geometric figures using a controlled language, ontological graphs and dependency lattices. In S. W. et al. (Eds.), *CICM 2014*. LNAI (Vol. 8543, pp. 298–311). Springer.

Haralambous, Y., & Quaresma, P. (2018). Geometric search in TGTP. In H. Li (Ed.), *Proceedings of the 12th International Conference on Automated Deduction in Geometry*. SMS International. http://adg2018.cc4cm.org/ADG2018Proceedings.

Hilbert, D. (1977). *Foundations of geometry*, 10th Revised edition. Paul Barnays: Open Court Publishing.

Jackiw, N. (2001). *The Geometer's Sketchpad v4.0*. Key Curriculum Press.

Janičić, P. (2006). GCLC—A tool for constructive Euclidean geometry and more than that. In A. Iglesias & N. Takayama (Eds.), *Mathematical Software, ICMS 2006*. Lecture Notes in Computer Science (Vol. 4151, pp. 58–73). Springer. https://doi.org/10.1007/11832225_6.

Janičić, P., Narboux, J., & Quaresma, P. (2012). The area method: A recapitulation. *Journal of Automated Reasoning, 48*(4), 489–532. https://doi.org/10.1007/s10817-010-9209-7.

Kapur, D. (1986a). Geometry theorem proving using Hilbert's Nullstellensatz. In *SYMSAC'86: Proceedings of the Fifth ACM Symposium on Symbolic and Algebraic Computation* (pp. 202–208). New York, NY, USA: ACM Press. https://doi.org/10.1145/32439.32479.

Kapur, D. (1986b). Using Gröbner bases to reason about geometry problems. *Journal of Symbolic Computation, 2*(4), 399–408. https://doi.org/10.1016/S0747-7171(86)80007-4.

Kortenkamp, U., & Richter-Gebert, J. (2004). Using automatic theorem proving to improve the usability of geometry software. In P. Libbrecht (Ed.), *Proceedings of MathUI 2004*. http://kortenkamps.net/papers/2004/ATP-UI-article.pdf.

Kovács, Z. (2015). The relation tool in GeoGebra 5. In F. Botana & P. Quaresma (Eds.), *Automated deduction in geometry* (pp. 53–71). Springer International Publishing. https://doi.org/10.1007/978-3-319-21362-0_4.

Kovács, Z., & Recio, T. (2020). Geogebra reasoning tools for humans and for automatons. In *Proceedings of the 25th Asian Technology Conference in Mathematics* (pp. 16–30). Mathematics and Technology, LLC. https://doi.org/10.13140/RG.2.2.26851.58407.

Laborde, J. M., & Strässer, R. (1990). Cabri-géomètre: A microworld of geometry guided discovery learning. *International Reviews on Mathematical Education- Zentralblatt fuer Didaktik der Mathematik, 90*(5), 171–177.

Li, H. (2000). Clifford algebra approaches to mechanical geometry theorem proving. In X. S. Gao & D. Wang (Eds.), *Mathematics mechanization and applications* (pp. 205–299). San Diego, CA: Academic Press. https://doi.org/10.1016/B978-012734760-8/50009-0.

Nikolić, M., Marinković, V., Kovács, Z., & Janičić, P. (2019). Portfolio theorem proving and prover runtime prediction for geometry. *Annals of Mathematics and Artificial Intelligence, 85*, 119–146. https://doi.org/10.1007/s10472-018-9598-6.

Narboux, J. (2007). A graphical user interface for formal proofs in geometry. *Journal of Automated Reasoning, 39*, 161–180. https://doi.org/10.1007/s10817-007-9071-4.

Nevis, A. (1975). Plane geometry theorem proving using forward chaining. *Artificial Intelligence, 6*(1), 1–23. http://hdl.handle.net/1721.1/6218.

Pambuccian, V. (2004). The simplest axiom system for plane hyperbolic geometry. *Studia Logica, 77*(3), 385–411. https://doi.org/10.1023/B:STUD.0000039031.11852.66.

Petrović, I., Kovács, Z., Weitzhofer, S., Hohenwarter, M., & Janičić, P. (2012). Extending GeoGebra with automated theorem proving by using OpenGeoProver. In *Proceedings CADGME 2012*, Novi Sad, Serbia.

von Plato, J. (1995). The axioms of constructive geometry. In *Annals of pure and applied logic* (Vol. 76, pp. 169–200). https://doi.org/10.1016/0168-0072(95)00005-2.

Quaife, A. (1989). Automated development of Tarski's geometry. *Journal of Automated Reasoning, 5*, 97–118. https://doi.org/10.1007/BF00245024.

Quaresma, P. (2011). Thousands of geometric problems for geometric theorem provers (TGTP). In P. Schreck, J. Narboux & J. Richter-Gebert (Eds.), *Automated deduction in geometry*. Lecture

Notes in Computer Science (Vol. 6877, pp. 169–181). Springer. https://doi.org/10.1007/978-3-642-25070-5_10.

Quaresma, P., Santos, V., Graziani, P., & Baeta, N. (2020). Taxonomy of geometric problems. *Journal of Symbolic Computation, 97,* 31–55. https://doi.org/10.1016/j.jsc.2018.12.004.

Recio, T., & Vélez, M. P. (1999). Automatic discovery of theorems in elementary geometry. *Journal of Automated Reasoning, 23,* 63–82. https://doi.org/10.1023/A:1006135322108.

Richter-Gebert, J., & Kortenkamp, U. (1999). *The interactive geometry software Cinderella.* Springer.

Simões, C. (2007). Azulejos que ensinam: Entrevista a António Leal Duarte. *Gazeta de Matemática, 153,* 4.

Stojanović, S., Pavlović, V., & Janičić, P. (2011). A coherent logic based geometry theorem prover capable of producing formal and readable proofs. In P. Schreck, J. Narboux & J. Richter-Gebert (Eds.), *Automated deduction in geometry.* Lecture Notes in Computer Science (Vol. 6877, pp. 201–220). Berlin, Heidelberg: Springer. https://doi.org/10.1007/978-3-642-25070-5_12.

Sutcliffe, G. (2017). The TPTP problem library and associated infrastructure. From CNF to TH0, TPTP v6.4.0. *Journal of Automated Reasoning, 59*(4), 483–502. https://doi.org/10.1007/s10817-017-9407-7.

Sutherland, I. E. (1963). *Sketchpad, a man-machine graphical communication system.* Ph.D. thesis, Massachusetts Institute of Technology, Lincoln Laboratory.

Sutherland, I. E. (2003). *Sketchpad: A man-machine graphical communication system.* Technical Report. UCAM-CL-TR-574, University of Cambridge, Computer Laboratory. http://www.cl.cam.ac.uk/techreports/UCAM-CL-TR-574.pdf.

Tarski, A. (1951). *A decision method for elementary algebra and geometry.* Technical Report. RAND Corporation.

Tessier-Baillargeon, M., Leduc, N., Richard, P., & Gagnon, M. (2017). Étude comparative de systèmes tutoriels pour l'exercice de la démonstration en géométrie. *Annales de Didactique et de Sciences Cognitives, 22,* 91–117.

Wang, D. (1995). Reasoning about geometric problems using an elimination method. In J. Pfalzgraf & D. Wang (Eds.), *Automated pratical reasoning* (pp. 147–185). New York: Springer.https://doi.org/10.1007/978-3-7091-6604-8_8.

Wu, W. T. (1984). On the decision problem and the mechanization of theorem proving in elementary geometry. In *Automated theorem proving: After 25 years.* Contemporary Mathematics (Vol. 29, pp. 213–234). American Mathematical Society.

Ye, Z., Chou, S. C., & Gao, X. S. (2011). An introduction to java geometry expert. In T. Sturm & C. Zengler (Eds.), *Automated deduction in geometry.* Lecture Notes in Computer Science (Vol. 6301, pp. 189–195). Berlin, Heidelberg: Springer. https://doi.org/10.1007/978-3-642-21046-4_10.

Zhang, J. Z., Chou, S. C., & Gao, X. S. (1995). Automated production of traditional proofs for theorems in Euclidean geometry I. The Hilbert intersection point theorems. *Annals of Mathematics and Artificial Intelligence, 13,* 109–137.https://doi.org/10.1007/BF01531326.

Automated Reasoning Tools with GeoGebra: What Are They? What Are They Good For?

Zoltán Kovács, Tomás Recio, and M. Pilar Vélez

1 Introduction

We are immersed in the society of digitalization, automation, data science, and artificial intelligence. The interaction of this scenery with mathematics is twofold. On the one hand, mathematics is the hidden support for this whole technological panorama; on the other, personal computers provide digital tools that perform incredible calculations (including most of the tasks required in the current mathematics curriculum) or facilitate drawing dynamic graphs that help visualize mathematical objects. Yet, it seems that the current mathemathics education landscape does not respond to, or remains far from, what the new reality itself demands. "No mainstream school maths curriculum has yet been based on the (obvious) assumption that computers exist." (Wolfram 2020, p. 4). This incoherent absence is perhaps one of the reasons for the current widespread debate within the educational community about what, how, and why to teach and learn mathematics, at all educational levels.

Digitalization brings to math education new tools that require a new curriculum, a new design of tasks, and a greater interaction with other disciplines (as in the STEAM approach). But, what is more important in this new context is the leading role of the student—with the help of digital tools—regarding his/her own learning process. For example, working through open-ended tasks, that can be defined as

---

Z. Kovács
The Private University College of Education of the Diocese of Linz, Linz, Austria
e-mail: zoltan@geogebra.org

T. Recio (✉) · M. P. Vélez
Universidad Antonio de Nebrija, Madrid, Spain
e-mail: trecio@nebrija.es

M. P. Vélez
e-mail: pvelez@nebrija.es

© The Author(s), under exclusive license to Springer Nature Switzerland AG 2022
P. R. Richard et al. (eds.), *Mathematics Education in the Age of Artificial Intelligence*,
Mathematics Education in the Digital Era 17,
https://doi.org/10.1007/978-3-030-86909-0_2

"…tasks where students are asked to explore objects and to discover and investigate their mathematical properties…" (Ulm 2011, p. 23)

In this new educational context, Dynamic Geometry Systems (DGS) can be considered as specifically appropriate, useful tools. Indeed, these software environments were, from their conception, meant mainly as a tool for fostering students' ability regarding geometric visualization and experimentation. Nowadays, modern DGS have started to include features for automated reasoning that allows for the automatic and mathematically rigorous verification and discovery of geometric theorems Kovács et al. (2018), say a kind of "geometry calculator" for AI ("augmented intelligence"[1]), that can improve the teaching and learning of reasoning and proof Sinclair et al. (2016) using an open-ended task methodology.

Thus, the popular dynamic mathematics program GeoGebra has always offered, as has any other Dynamic Geometry System, some remarkable possibilities to improve the construction and visual exploration of geometric objects. This is done by dragging the elements in a figure, thereby allowing the student to perceive the resulting changes and permanent relations by dragging the elements in a figure and, then, facilitating the student to perceive the resulting changes and the permanent relations. GeoGebra's potential for problem solving, reasoning, and the influence of task design is analyzed in Olsson (2019), where the reader can find a revision of earlier research into the potential of dynamic software to support problem solving and reasoning.

Yet, more recently, since the computer algebra system Giac was embedded in GeoGebra Kovács and Parisse (2015), GeoGebra has been able to have implemented automated proving algorithms based on the algebraic approach described in Recio and Vélez (1999) and Kovács et al. (2019). The result is a collection of GeoGebra features and commands (the so-called Automated Reasoning Tools (ART)) that allow the rigorous mathematical verification (i.e., the Automatic Theorem Proving (ATP)) and the automatic discovery of general propositions about Euclidean geometry figures built by the user.

Although mathematics education experts have long been aware of the existence of dynamic geometry programs that offer, on an experimental basis and with a limited number of users, certain ATP features, the accessibility (as it is free of charge) and portability of GeoGebra; its availability on tablets, smartphones, and computers; its online and off-line accessibility; its worldwide diffusion-especially in the educational field; and the inclusion of automatic deduction and discovery tools, makes the use of these revolutionary ART techniques in GeoGebra a qualitatively different phenomenon with an unusually high potential for academic impact.

The purpose of this paper is, firstly, to make a summary presentation of the ART functions in GeoGebra through some illustrative examples of tasks showing how ART could be used within an educational context, helping students to develop "augmented intelligence" skills by reasoning in collaboration with the computer. Then we will

---

[1] "Augmented intelligence is a design pattern for a human-centered partnership model of people and artificial intelligence (AI) working together to enhance cognitive performance, including learning, decision-making, and new experiences, cf. https://www.gartner.com/en/information-technology/glossary/augmented-intelligence".

reflect on the possible educational use of these new features, in particular through the analysis of some results of recent experiences we have developed with our students regarding the use of ART techniques.

Lastly, we will think about the advantages and disadvantages that this novelty could bring to the learning and teaching of geometry. As a final conclusion, we argue how that in order to fully benefit from this impressive tool, it will require that its use will become embedded in a larger ecosystem that should be developed by the scientific and teaching community, globally covering different aspects of computer-supported geometric reasoning Kovács et al. (2020).

2 GeoGebra Automated Reasoning Tools

This section introduces, describes, and exemplifies the technical features of some recently implemented Automated Reasoning Tools (ART) in the dynamic mathematics software GeoGebra. As mentioned above, these tools (given by a button in the Menu) and commands (to be introduced in the Command Line) allow the user to automatically conjecture, discover, and prove statements concerning different elements of a given geometric construction. Basic automated reasoning features are available since GeoGebra, version 5; yet, certain ART improvements and advanced characteristics can be found in *GeoGebra Discovery*, an experimental version of GeoGebra, available at https://github.com/kovzol/geogebra-discovery, operating on top of GeoGebra Classic 5, for computers and laptops, on Windows, Mac, or Linux operating systems; and the GeoGebra Classic, version 6, for browsers, accessible at http://autgeo.online/geogebra-discovery/, therefore valid also on tablets and smartphones.

Examples have been chosen that, on the one hand, illustrate the functionalities of ART tools in GeoGebra, and on the other hand, show the necessary interaction between human and machine reasoning, synthesizing into what we refer to here as "augmented intelligence" Semenov and Kondatriev (2020).

First of all, let us enumerate the list of GeoGebra's Automated Reasoning commands:

- The `Relation` command and tool, for the automatic finding of properties that relate certain objects in a construction, that can be used, for instance, to check geometric conjectures and for the verification or denial of these conjectures. A complete list of the properties between geometric objects that `Relation` is able to obtain can be found at https://wiki.geogebra.org/en/Relation_Command.
- The `LocusEquation` command, which calculates the implicit equation of a certain semi-free point such that a given property holds.
- The `Prove` and `ProveDetails` commands, which decide if a statement is true in general and, eventually, give some additional conditions for its truth, avoiding degenerate cases.

- The `Discover` tool and command which finds a collection of statements holding true and involving a certain element selected by the user in the figure. This is a new feature, only available in GeoGebra Discovery.

  Let us also mention that a detailed tutorial can be found in Kovács et al. (2018).

2.1 The `Relation` Tool and Command

This basic automated reasoning tool in GeoGebra is the symbolic extension of the previously existing `Relation` command. Initially, this command was purely numerical (see Kovács (2015a, b)): after the user has selected two geometric objects in a construction and invoked the `Relation` command (between the two objects), GeoGebra answered by asserting the possibility, or not, that certain relationships would occur between them, such as perpendicularity, collinearity, parallelism, equality, or incidence, as long as the numerical verification of such properties exceeded a certain threshold, with the user being warned in a message that the reached conclusion was only numerically valid.

In the current version of GeoGebra, the `Relation` command allows the user to click an additional button, labeled "More...", in the output message. By pressing this button, a symbolic calculation process is launched within the ART system of GeoGebra, translating the given geometric figure into a collection of polynomial equations and considering, systematically, as a thesis, the algebraic translation of the possible relations we have referred to, between the chosen geometric elements.

For instance, points can be considered as collinear for `Relation`, if by taking the line between two of them, the third one will turn out to be "approximately in the same line", where "approximately" depends on the number of digits that the user has chosen in the application preferences (namely, in *Tools ▷ Rounding* in the user menu system of GeoGebra Classic 5, or in *Settings ▷ Global ▷ Rounding* in version 6) to perform calculations in the session with GeoGebra.

Example 1 Figure 1 shows a diagram with three collinear points A, B, and C, a free point O, and the midpoints D, E, and F of the segments OA, OB, and OC, respectively. In the Input Bar, the command `Relation({D, E, F})` is introduced to study the existence, if any, of some property between the points D, E, and F.

The left part of Fig. 2 includes the response message to this command, which indicates that the points D, E, and F are, at least for this figure and approximately, collinear. Finally, Fig. 2, to the right, shows the result of pressing the icon "More...": it is the rigorous check of the general validity of the theorem that says that the midpoints of the different segments from a point to different points in a fixed line are collinear. It also warns the user that it is true except for a degenerate construction. Such verification is based on the execution of certain algorithms that involve, without the user perceiving it, several aspects of advanced computational algebraic geometry. See, for more technical details, Botana et al. (2015) or Recio and Vélez (1999).

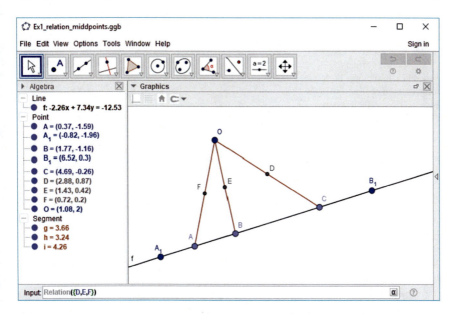

Fig. 1 GeoGebra construction of the midpoints of the segments between a point O and three collinear points A, B, and C, and Relation($\{D, E, F\}$) inserted in the command line

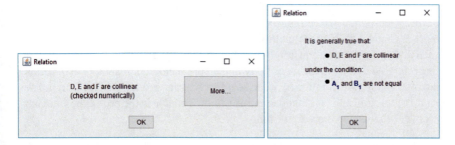

Fig. 2 The numerical answer to the command Relation($\{D, E, F\}$) on the left and, to the right, the rigorous result after clicking on the icon "More…"

The computational power of GeoGebra's ART is not limited to basic geometric constructions, but is able to find and test much more sophisticated geometric relationships. For instance, we have considered a problem from one of the exams of the Spanish recruitment method to become a civil servant math teacher for the secondary school system. It is a method that requires passing and receiving the best grades on a series of public exams (also known as "oposiciones"). In one of these recent tests, the candidates were requested to solve the following elementary geometry question, which asked the test-writer, to conjecture, formulate and, then, to prove, the ratio holding between two particular segments in a given figure (see Fig. 3 from Gamboa et al. (2019)).

Fig. 3 Let ABC be a triangle with a right angle at B and with angles 60° and 30° at the other vertices. It has been rotated twice, both times centered at A and with equal rotation angle. Find the ratio $B'C/AN$

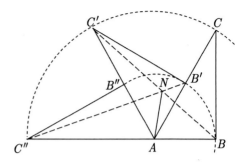

We used GeoGebra ART to perform this task in order to demonstrate how much it simplifies the solving of the problem, as well as how it involves a different way of reasoning through interaction with the computer.

Example 2 See the statement in Fig. 3, as given to the candidates. Then, the first challenge is to reproduce this picture as a GeoGebra construction.

To build the initial right triangle we have taken advantage of the fact that a right triangle, whose non-right angles are of 60° and 30°, is half of an equilateral triangle (thus allowing us to start building a regular 3-sided polygon with the corresponding GeoGebra command Polygon). Then we have rotated vertices B and C 60° counterclockwise twice, till C'' becomes aligned with A, B, as shown in Fig. 4.

Now we use the Relation tool to ask about any possible relation between segments $B'C$ and AN. Notice that this tool is usually utilized to find relations between exactly two objects in the construction; asking for any relation involving more objects, it is preferable to type *Relation* directly in the command line. It should also be noted that the computation of ratios between segment lengths through Relation is a new feature of this tool and command, currently only available in the prototype GeoGebra Discovery that we are here describing.

In Fig. 4 we show the position of the *Relation tool* in the toolbar. After clicking on the corresponding icon $a \overset{?}{=} b$ we select segments $B'C$ and AN, labeled as r and q respectively, and we obtain the numerical answer in a pop-up window, see the left of Fig. 5. After clicking on the "More..." icon, a symbolic answer is displayed, as shown on the right of Fig. 5.

We leave to the reader the investigation of other surprising results concerning ratios of segments in this construction, for instance, the ratio between BC' and AN.

2.2 The Prove and ProveDetails Commands

The Prove, ProveDetails commands work in a similar way. Unlike Relation, the user must enter the conjectured thesis (for instance, that the ratio between r and q is $\sqrt{7}/2$), obtaining as an answer the truth or falsity of their conjecture and, in

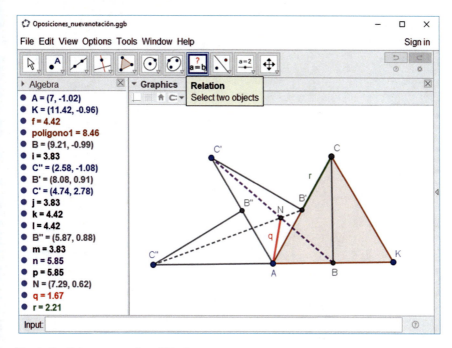

Fig. 4 GeoGebra construction of Fig. 3

Fig. 5 The numerical and symbolic answers to Relation(*r*,*q*)

the affirmative case, providing some additional geometric conditions that must be verified so that the given statement is generally true. These are the so-called non-degeneracy conditions, which usually prescribe that certain input objects (e.g., the free vertices of a triangle) must not coincide or be aligned, etc., for the conjectured statement to be true.

Figure 6 confirms the above-obtained relationship (Fig. 5) for Example 2. It means that the ratio $r/q = \sqrt{7}/2$ for the given construction holds true, except if points A and K (that were taken as free) coincide.

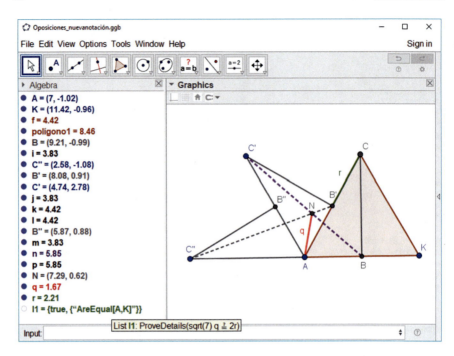

Fig. 6 After introducing `ProveDetails(sqrt(7)q==2r)` in the command line of GeoGebra, the obtained answer (in green) appears in the last line of the Algebra window

2.3 The `LocusEquation` Command

Another function that is made possible through the ART tools in GeoGebra is the discovery of new theorems, looking for complementary hypotheses for a certain thesis to hold. For instance, trying to generalize Example 2, we will now consider as initial hypotheses a similar construction, but starting with more general triangles, and we will try to find where to place vertex C in order to have same segment lengths ratio between AN and $B'C$.

Example 3 Let us start with a general (not necessarily equilateral) triangle ABC, and then let us roughly follow the previous construction, but with some modifications in order to end up with three, regularly rotated, triangles, ending up with A, B, and C'' aligned. Thus, instead of rotating vertex B and C, we are going to reflect vertex K with respect to the midpoint of AC and with respect to vertex A, yielding vertices C' and C'', respectively (see Fig. 7). Notice that this construction ends up obtaining the same figure as the one in Example 2 that was built using rotations, and starting with an equilateral triangle.

Now, the command `LocusEquation` allows us to discover for which triangles ABC the ratio $r/q = \sqrt{7}/2$ holds; more specifically, we ask GeoGebra, through the

Automated Reasoning Tools with GeoGebra: ... 31

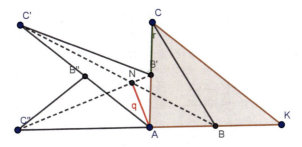

Fig. 7 GeoGebra construction starting from a general triangle and taking midpoints and point reflections

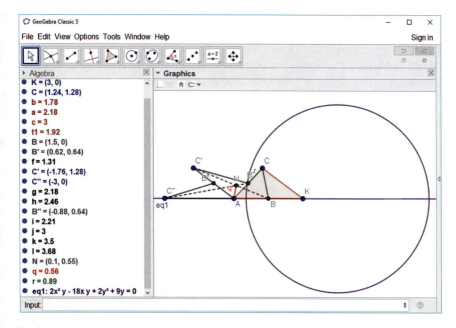

Fig. 8 Typing `LocusEquation(sqrt(7)q==2r,C)` in the command line we obtain the equation of a curve in the Algebra window and its display (in purple) in the Graphics window

`LocusEquation(sqrt(7)q==2r,C)` command, where to put vertex C such that the required ratio holds.

In Fig. 8 the reader can see GeoGebra's answer: to have the segment ratio $r/q = \sqrt{7}/2$, the vertex C must be placed in a cubic curve which is the union of a circle and a line (a degenerate case). Now it is the machine's turn to challenge us to discover the geometric characteristics of this circle; GeoGebra cannot provide any further information about it. For instance, it seems that the center of the circle is in the reflection of point B with center K. But, what can we say about the radius?

2.4 The Discover Tool and Command

The Discover tool and command is a new feature currently available only in GeoGebra Discovery. This command searches, in an automatic and combinatorial way, for a whole series of possible geometrical relationships between the elements of the construction in which a point (pointed out by the user) is included, and then verifies its truth or falsity.

The use of this tool is illustrated with an old and challenging example that may be the object of an open-ended activity in which it will be necessary to explore, to discover, to conjecture, to prove, ...while applying some knowledge of elementary geometry, to reveal an enigma: the *Treasure Island Problem*.

Example 4 *(Treasure Island Problem)*
The Treasure Island Problem is described in Wilson (1997) where it is pointed out that *"In 1948, George Gamow wrote a book called "One, Two, Three, ...Infinity". In it, he presents a problem suggested by a treasure map found in a grandfather's attic"*. The problem is stated as follows:

> A young man was going through the attic of his grandfather's house and found a paper describing the location of a buried treasure on a particular island. The note said that on the island one would find a gallows, an oak tree, and a pine tree. To locate the treasure one would begin at the gallows, walk to the pine tree, turn right 90 degrees and walk the same number of paces away from the pine tree. A spike was to be driven at that point. Then return to the gallows, walk to the oak tree and turn left 90 degrees and walk the same number of paces away from the oak tree. Drive a second spike in the ground. The midpoint of a string drawn between the two spikes would locate the treasure.
>
> The young man and his friends mounted an expedition to the island, found the oak tree and the pine tree but no gallows. It had been eliminated years ago without a trace. They returned home with the map above and no treasure.
>
> Show them where to look for the treasure.

Why should we not help the young man and his friends to locate the treasure? Yes, we can try, assisted by GeoGebra ART!

We start by reproducing in GeoGebra the steps narrated in the paper to arrive to the treasure: we take three free points, representing the pine P, the oak O, and the gallows G, then we draw the segment from G to P and rotate it 90° counterclockwise with center P to determine the point S_1. We do the same with the oak point O, but rotating clockwise to obtain the point S_2. Finally, we know that the treasure T is located at the midpoint of points S_1 and S_2 (see Fig. 9). As the story tells us that the gallows has disappeared, we can use GeoGebra's dynamic capabilities to drag point G at random, trying to see the influence of this fading on the location of the treasure (compare the two maps in Fig. 9). Finally, one can become easily convinced, visually, that the position of the treasure does not depend on where we place point G.

Therefore, we conjecture that there must be some theorem linking the position of points P and O with that of T, but not involving a particular situation for G. Let us investigate the possible geometry theorems involving point T using the Discover tool: in the construction of Fig. 9 select the Discover icon in the toolbar and

Automated Reasoning Tools with GeoGebra: ...   33

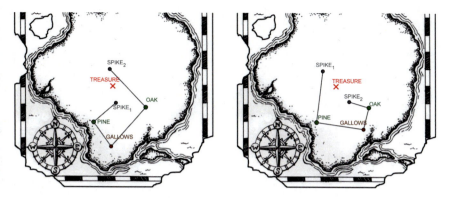

Fig. 9 Using GeoGebra over the map of an island to get possible locations of the treasure for two different positions of the gallows

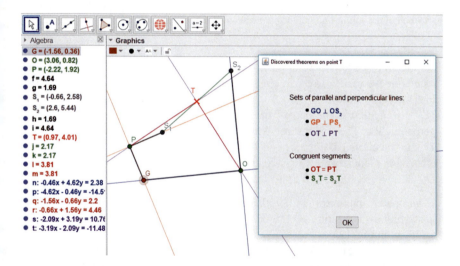

Fig. 10 Two GeoGebra windows after selecting the tool Discover and clicking at point T (equivalently, introducing Discover(T) in the command line). The pop-up window lists different geometric theorems involving T in different colors, and the Graphics Window displays, with the same colors, the geometric objects involved in these relationships

click the point T. GeoGebra computes a series of possible geometrical relationships between the elements of the construction in which point T is included. A pop-up window appears with the obtained theorems involving T (see Fig. 10).

After discarding trivial relationships as well as those involving point G (that we assume is missing on the island), we have found the following conclusions:

- $OT = PT$, that is, the treasure (point T) is at the same distance from both trees (points P and O). The reader surely knows that this means that T lies within the perpendicular bisector of P and O; otherwise, GeoGebra ART can

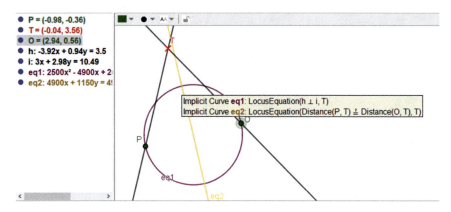

Fig. 11 Start drawing three points P, O and T and ask GeoGebra consecutively where to put T so that `Distance(P,T)==Distance(O,T)` and $OT \perp PT$. It yields T must be at the intersection of the yellow line and the purple circle

help us to discover where to put T such that $OT = PT$ by entering in the command line `LocusEquation(Distance(P,T)==Distance(O,T))` (see Fig. 11, yellow line).

- $OT \perp PT$, that is the paths from the trees (P and O) to the treasure (T) are perpendicular. The reader surely knows also Thales' Theorem and can deduce that T is in the circle passing through P and O having the segment PO as diameter; otherwise, GeoGebra ART can again help us to discover where to put T such that $OT \perp PT$ (see Fig. 11, purple circle).

A geometer "uncomfortable" with the result, could still require the student to prove that, starting from the initial configuration, point T verifies these two properties; this is achieved in Fig. 12.

Now, do not you consider that we could easily tell the young man in the story how to arrive at the treasure point, even not knowing where G was originally placed?

3 Toward an Automated Geometer

In a certain sense, the ART tools we have described in the previous section can be considered as a kind of omniscient teacher, ready to answer whatever questions are posed by a human user—a sort of "geometry calculator." Going a little bit further, a true "automated geometer" should be a kind of machine capable to investigating, without requiring any suggestion from humans, the geometric properties of a figure. Actually, this is already partially accomplished by GeoGebra Discovery, through the `Discover` command we have previously described, although it needs a human user to choose a concrete point in the given figure to focus the beginning of the discovery

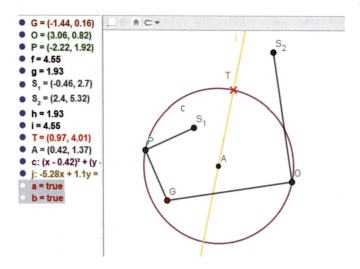

Fig. 12 Start with the construction in Fig. 10 and add the perpendicular bisector j of P and O, and the circle C centered at the midpoint of PO and passing by O. Ask GeoGebra to Prove($T \in j$) and Prove($T \in c$). The answer "true" for both commands appears in the Algebra window

task. Thus, as a further step toward a completely independent performing "automated geometer," we have implemented in GeoGebra a new tool which we have called the *Automated Geometer*, that is a web-based service available at http://autgeo.online/ ag, able to obtain, just by itself, sound relationships in a geometric construction.

Let us illustrate the behavior of the Automated Geometer by considering the ICMI Study series "School Mathematics in the 90s" Howson and Wilson (1986), dated back to 1986, which includes an elementary geometry question, as described in Fig. 13, item a), asking the user to decide if three given segments of the diagonal of a square are equal or not.

Let us roughly sketch how the question appearing in Fig. 13 can be solved by our Automated Geometer.

First, a web version of GeoGebra is loaded in the web browser and the user is asked to construct a geometric figure (or select one of the built-in examples). Five different options are then available to check certain properties holding in the figure: collinearity of three points, equality of distances between two points, perpendicularity of segments defined by two points, parallelism of segments defined by two points, and concyclicity of four points (see Fig. 14).

Once the user has decided among these options, and after having launched the discovery process, the Automated Geometer creates a list of possible statements in a combinatorial way (all possible triples of points are considered as potentially collinear, etc.) and calls the previously described automatic reasoning tools to verify or deny the truth of the considered propositions. For example, in the case of the Fig. 13, out of 1260 possible statements only 119 are true—all of these checks are performed within 9 s on a normal PC with $8 \times i7$ cores and 16 GB RAM, tested in

Consider, for example, the following question (to other aspects of which we shall wish to refer later):

Two lines are drawn from one vertex of a square to the midpoints of the two non-adjacent sides. They divide the diagonal into three segments (see Figure 5.2).

(a) Are those three segments equal?

(b) Suggest several ways in which the problem can be generalised.

(c) Does your answer to (a) generalise?

(d) Can the argument you used in (a) be used in the more general cases?

(e) If your answer to (d) is 'No', can you find an argument which does generalise?

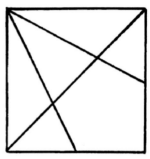

Fig. 5.2

Fig. 13 A question from the so-called "Kuwait" ICMI Study, back in 1986

Google Chrome 86. The obtained statements are visually presented in the program (see Fig. 15) and also a short list of them is printed (see Fig. 16). In particular, we mention that the proposed question is affirmatively answered considering items 21 and 22 in this Fig. 16.

We highlight here that some unexpected results may also be included in the output. For example, concyclicity of points E, F, G, and H may be not completely trivial at the first look—one needs to find the axial symmetry through the diagonal BD to confirm this result. For other input figures, however, many more non-trivial facts can be eventually obtained. Actually, this is what we expect from discovery: we want to be surprised!

4 Discussion and Conclusions

It is clear that the above-considered Kuwait Study problem can be solved in various ways, traditionally by pure geometric means, but—as a more modern approach—the possibility of considering an algebraic solution is already mentioned in the 1986 ICMI Study Howson and Wilson (1986). The following sentences from this book are particularly relevant in our context:

> ...even more challenging, computer-based opportunities for transforming geometry teaching in the 1990's will be provided by computer assisted design software which at the moment has had little impact on schools...As a result of recent curricular changes, the 'few', the mathematically gifted students, have at their disposal powerful algebraic tools, coordinates

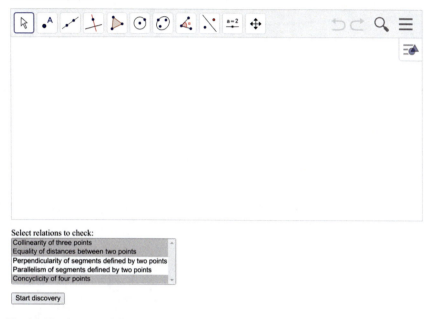

Fig. 14 The Automated Geometer on its startup

and vectors, which they can apply to geometric problems. They, then, might approach such problems in a different manner to school pupils of old—the methods may be less elegant, and provide less scope for creative and penetrating thought—but they offer a more systematic approach. ...whereas the student of the 1950s had only purely geometrical ways of tackling the problem [*the authors refer to question (a) in* Fig. 13], today's student may well be able to apply algebraic methods ...The solution derived by applying a mechanical procedure may be less aesthetically satisfying than a geometrical one, but are there other objections to algebraic methods than that of aesthetics? (Howson and Wilson 1986, p. 58–59)

The quoted text reflects on the pros and cons of the different ways of solving the proposed question and states that an algebraic solution is maybe not as elegant as a geometric one, but "are there other objections to algebraic methods than that of aesthetics?"

As our methods rely heavily on the algebraic side, we sustain a very similar opinion, but in an even more radical form. In fact, the automated discovery process we have implemented in GeoGebra Discovery or in the Automated Geometer does not require any geometrical background in the proving process, since it starts translating mechanically every geometric relation into algebraic equations. Also, the obtained proof has nothing to do with geometry—it usually contains a large amount of vari-

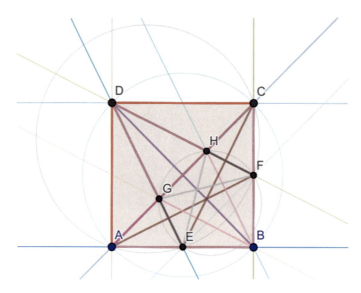

Fig. 15 Visual results of automated discovery. The same colors usually refer to equality or parallelism

The following theorems can be proven:

1. E∈AB	13. AD=BC	26. BE=CF	39. EH=FG	52. AF⊥DE	65. BF⊥CD	77. AC∥CH	88. AG∥EF	99. BG∥DH	110. DE∥EG
2. G∈AC	14. AD=CD	27. BF=CF	40. AB⊥AD	53. AF⊥DG	66. BG⊥CE	78. AC∥EF	89. AG∥GH	100. BG∥FH	111. DF∥DH
3. H∈AC	15. AE=BE	28. BG=BH	41. AB⊥BC	54. AF⊥EG	67. CD⊥CF	79. AC∥GH	90. AH∥CG	101. BH∥DE	112. DF∥FH
4. H∈AG	16. AE=BF	29. BG=DG	42. AB⊥BF	55. AG⊥BD	68. CE⊥DF	80. AD∥BC	91. AH∥CH	102. BH∥DG	113. DG∥EG
5. F∈BC	17. AE=CF	30. BG=DH	43. AB⊥CF	56. AH⊥BD	69. CE⊥DH	81. AD∥BF	92. AH∥EF	103. BH∥EG	114. DH∥FH
6. H∈CG	18. AF=CE	31. BH=DG	44. AC⊥BD	57. BC⊥BE	70. CE⊥FH	82. AD∥CF	93. AH∥GH	104. CG∥CH	115. EF∥GH
7. G∈DE	19. AF=DE	32. BH=DH	45. AD⊥AE	58. BC⊥CD	71. AB∥AE	83. AE∥BE	94. BC∥BF	105. CG∥EF	116. DE∘ABC
8. H∈DF	20. AF=DF	33. CE=DE	46. AD⊥BE	59. BD⊥CG	72. AB∥BE	84. AE∥CD	95. BC∥CF	106. CG∥GH	117. FE∘ACE
9. AB=AD	21. AG=CH	34. CE=DF	47. AD⊥CD	60. BD⊥CH	73. AB∥CD	85. AG∥AH	96. BE∥CD	107. CH∥EF	118. HE∘ACG
10. AB=BC	22. AG=GH	35. CH=GH	48. AE⊥BC	61. BD⊥EF	74. AC∥AG	86. AG∥CG	97. BF∥CF	108. CH∥GH	119. HE∘EFG
11. AB=CD	23. AH=CG	36. DE=DF	49. AE⊥BF	62. BD⊥GH	75. AC∥AH	87. AG∥CH	98. BG∥DF	109. DE∥DG	
12. AC=BD	24. BC=CD	37. DG=DH	50. AE⊥CF	63. BE⊥BF	76. AC∥CG				
	25. BE=BF	38. EG=FH	51. AF⊥BH	64. BE⊥CF					

Finished, found 119 theorems among 1260 possible statements.
Elapsed time: 0h 0m 9s

Restart with a new or the same experiment

Fig. 16 List of obtained true statements of automated discovery

ables and a number of polynomial equations of higher degrees, and then millions of elementary steps are required to obtain the required proof. Only the final translation of the algebraic results requires geometry again—a mechanical translation of the non-degeneracy results (some polynomials that should not be zero) back into geometric terms (e.g., some points should not be aligned).

In this approach, therefore, the whole proving, internal process is a kind of *non-aesthetic* operation and, what is more *unaesthetic*, in most cases, is the fact that it is *completely unreadable* for a human. On the other hand, this approach is *extremely powerful* for solving problems such as the one proposed in the ICMI Study, as it was already noticed by the ICMI Study authors concerning the use of "powerful algebraic tools" vs. the traditional, geometric technique. For example, anyone carrying a laptop, tablet, or smartphone, and using our GeoGebra automated reasoning and discovering tools, can quite easily solve, and even extend, some non-trivial problems, such as the

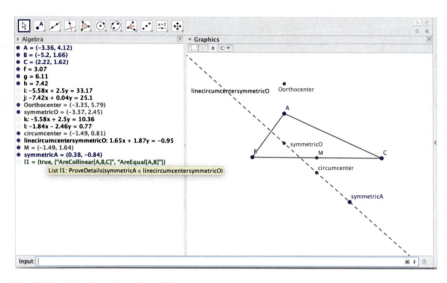

Fig. 17 Proving Chou's problem 230

one listed as number 230 in Chou's collection of 512 mechanically proved theorems, included in his foundational monograph Chou (1998).

Indeed, nowadays it is quite straightforward to prove (with a commonplace laptop and without any appreciable time lapse) Chou's original statement: "Show that the symmetric ($symmetric A$) of vertex A of a triangle (ABC) with respect to the midpoint (M) of the opposite side is collinear with the symmetric ($symmetric O$) of the orthocenter ($Oorthocenter$) with respect to A, and the circumcenter ($circumcenter$) of the triangle." Thus Fig. 17 shows GeoGebra's immediate proof of the statement $symmetric A \in linecircumcenter symmetric O$, where $linecircumcenter symmetric O$ is the line defined by the circumcenter and the symmetric of the orthocenter, declaring it is true except when $A = B$ or A, B, C are collinear, that is, for degenerate cases.

Moreover, GeoGebra allows, as well, to conjecture of a generalized version of this theorem, analyzing if the same alignment thesis would hold for other choices of a center of symmetry D involved in building the point $symmetric A$. Thus, Fig. 18 shows, on top, in red, the locus of the possible positions of D for the collinearity of the three points $symmetric A, symmetric O, circumcenter$, namely, a parallel line to the one defined by the circumcenter and $symmetric O$, going through the midpoint M. In the same figure, below, it is verified that this collinearity holds true except for degenerate cases (the triangle ABC collapses to a line), when choosing as symmetry center any point D in the red line.

It is evident for us, after all these examples that we have described so far, that teaching students having "…at their disposal powerful…tools which they can apply to geometric problems" Howson and Wilson (1986) cannot be a mere repetition of the traditional curriculum (in a broad sense: aims, goals, contents, methods, assessment,

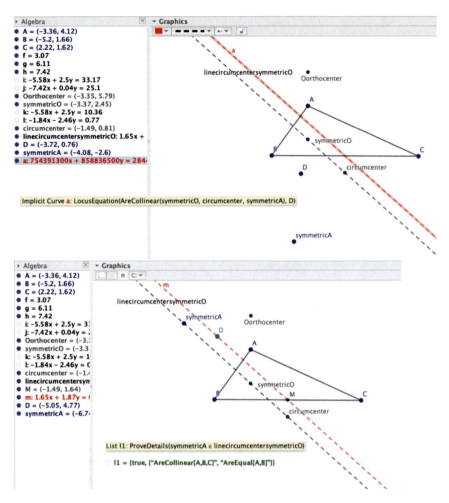

Fig. 18 Generalizing Chou's example 230

evaluation). Obviously, the training of the mind through the traditional approach to proving geometry theorems, has always been a crucial requirement for the development of mathematical skills.

Thus, we could consider addressing ICME-Study question a) in Fig. 13 without any auxiliary instrument, as a mere training task. But, while working on this question, can a student's mind ignore that fact that in the school bag sitting next to her/him, there is a small machine that could answer the posed question in a moment? To put the question in a different context: are we equally motivated in climbing up mountains that have a cable car going to the top, when there are many others that cannot be so easily reached? Is it fun walking up stairs instead of taking the lift?

Thus, to overcome such drawbacks, we could ask ourselves: are there not other worlds of mathematical activities where we can exercise our reasoning techniques, and that are not already (or that can never be) automatized? Can we get profit of the "(obvious) assumption that computers exist" (Wolfram 2020, p. 4)?

We think that the answer to these questions is affirmative and that geometry and dynamic geometry systems with automated reasoning tools remain a very rich context for developing human reasoning skills. However, a context bringing a new, strong focus on open-ended tasks, as remarked in the Introduction: "…tasks where students are asked to explore objects and to discover and investigate their mathematical properties…" (Ulm 2011, p. 23). The idea is not new, but needs to be revisited: indeed, it was already back in 1995 that the father of automated reasoning in geometry, Prof. Wen-tsun Wu, stated that "…geometry problem solving instead of geometry theorem proving should be emphasized…" and that "algebra and geometry should be kept in pace in the teaching," (Wu 1995, p. 72), without, apparently, much success in the "mainstream school maths curriculum." (Wolfram 2020, p. 4)

Some examples regarding how to implement this new approach—i.e., merging open-ended tasks and problem solving in geometry and algebra with automated reasoning programs as fundamental tools—have already been implemented in some of our classrooms (reported in Recio et al. (2019)). A more recent contribution concerning the possible use of DGS reasoning tools in the classroom (i.e., proposing a workflow to incorporate these tools), aiming toward the development of an ecosystem for computer-supported geometric reasoning, appears in Kovács et al. (2020). Again, the repeatedly mentioned Kuwait Study provides another example of this new scenario that we would like to sketch out here as a final contribution. In fact, in the previous sections we have addressed just the first question in Fig. 13, noting that it can be easily solved using GeoGebra. Yet, question b) is of a different kind: it is an open-ended question in which the user is greatly benefited by having a DGS at hand for its exploration, but it is also a question that computers cannot automatically answer.

Obviously, if question a) deals with the division in two parts of the side AB of a square, its natural generalization should address the case of dividing the same side in n equal parts. Yet, GeoGebra reasoning tools cannot deal with a question that depends on n as a parameter…so all we can do is to experiment with different cases $n = 3, 4, 5, \ldots$ and try to find out if there is some common property holding in all these instances among the segments in the diagonal resulting from the intersection with the lines from the upper left vertex of the square to the points dividing the opposite side in $n = 3, 4, 5, \ldots$ parts. See Fig. 19 for the case $n = 3$, where we have displayed only half of the lines, as the whole construction is symmetrical with respect to the DB diagonal and, thus, whatever properties that could be found on the segments AL, LM, MO in the figure could easily be stated for the full set of five segments.

One immediate way of generalizing question a) could be attempting to prove that all of the obtained segments are equal, but is easy to verify (e.g., using numerical approximation with GeoGebra's `Relation` tool) that they are not. Another possibility is to consider if there is some algebraic combination holding among these

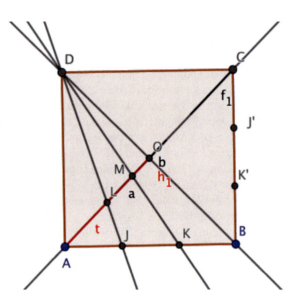

Fig. 19 Kuwait ICMI Study question b) for $n = 3$. By symmetry, only half of the construction is shown

segments, say, as it happens in the case $n = 3$, that the sum of the first two and the last two is 4 times the middle segment (equivalently, that $AL + LM = 4 \cdot MO$, see Fig. 19), something that we can conjecture and quickly verify with the help of automated reasoning tools. But there are quite diverse possible algebraic expressions holding in the same construction (e.g., $AL = LM + MO$) and it is not easy to guess a general formula that takes place for all values of n.

Moreover, the geometric argument that we could have used (in a traditional approach to this question) for solving item a), i.e., that the barycenter of a triangle is twice more distant from the vertex than from the feet of the median, is obviously not applicable to the general case: we know nothing about properties of "tertians" (an invention of ours: lines from a vertex of a triangle to a point in the opposite side after dividing the side into three parts), etc. It could be a good opportunity to address this open-ended task, with the help of our digital tools, but question e) in the Kuwait problem requires us to now find a general formula, much beyond the case $n = 3, \ldots$ We do not want to spoil the solution to this intriguing question for the interested reader, but we can say that it has been very helpful for us to combine,

- DGS constructions, for visual conjecturing,
- Automated Reasoning Tools, for deciding the truth or falsity of our conjectures,
- Computer Algebra Systems (CAS) to handle formulas depending on parameters such as n, equal to the number of parts we divide side AB, or r, related to the point corresponding to the r/n-th part of the side.

A final reflection: we would argue that a wider, wiser, and deeper study of the potential curricular inclusion of these methodological changes is badly needed. Not only regarding GeoGebra Automated Reasoning capabilities, but also considering their connection to GeoGebra's augmented reality tools (https://www.geogebra.org/m/RKYFdQJy) for exploring 3D-objects in the real world, as sketched in Botana et al. (2019).

In the meantime, the increasing, and already large number of GeoGebra users, over 100 million worldwide, is a decisive step toward making true this premonitory sentence from Hohenwarter et al. (2019): "as with pocket calculators, people will probably start using ART for checking geometric facts without the consensus of the pedagogical community on its role."

Our education system cannot stay blind, for another 30 years—as those that have elapsed since the publication of the ICMI Study "School Mathematics in the 90s" Howson and Wilson (1986)—to the existence, advance, and availability of digital tools contributing to the development of mathematical activities.

Acknowledgements The authors are partially supported by FEDER/Ministerio de Ciencia, Innovación y Universidades–Agencia Estatal de Investigación/MTM2017-88796-P.

References

Botana, F., Hohenwarter, M., Janičić, P., Kovács, Z., Petrović, I., Recio, T., & Weitzhofer, S. (2015). Automated theorem proving in GeoGebra: Current achievements. *Journal of Automated Reasoning, 5*(1), 39–59.
Botana, F., Kovács, Z., Martínez-Sevilla, A., Recio, T. (2019). Automatically Augmented Reality with GeoGebra. In: Prodromou, T. (Ed.), *Augmented reality in educational settings*, 347–368, Brill/Sense, Leiden, The Netherlands. https://doi.org/10.1163/9789004408845.
Chou, S. C. (1998). *Mechanical geometry theorem proving*. Dordrecht, Netherlands: D. Reidel Publishing Company.
Gamboa, J. M. (2019). Problemas resueltos de oposiciones. Tomo 9, et al. (2017). *Y 2018* (2nd ed.). Madrid, Spain: Editorial Deimos.
Hohenwarter, M., Kovács, Z., & Recio, T. (2019). Using GeoGebra automated reasoning tools to explore geometric statements and conjectures. In G. Hanna, M. de Villiers, & D. Reid (Eds.), *Proof technology in mathematics research and teaching, series: Mathematics education in the digital era* (Vol. 14, pp. 215–236). Cham: Springer.
Howson, G., & Wilson, B. (1986). *ICMI Study series: School mathematics in the 1990's*. Kuwait: Cambridge University Press.
Kovács, Z. (2015a) *Computer based conjectures and proofs*. Ph.D. Dissertation. Johannes Kepler University, Linz.
Kovács, Z. (2015b) The Relation Tool in GeoGebra 5. In: Botana, F., Quaresma, P. (Eds.), *Proceedings of the 10th international workshop on automated deduction in geometry (ADG 2014)*, July 9–11 2014. Lecture notes in artificial intelligence 9201, pp 53–71. Springer.
Kovács, Z., & Parisse, B. (2015). Giac and GeoGebra: Improved Gröbner Basis Computations. Lecture Notes in Computer ScienceIn J. Gutierrez, J. Schicho, & M. Weimann (Eds.), *Computer algebra and polynomials* (Vol. 8942, pp. 126–138). Cham: Springer.
Kovács, Z., Recio, T., Richard, P. R., Van Vaerenberg, S., Vélez, M. P. (2020) Towards an ecosystem for computer-supported geometric reasoning. *International Journal of Mathematical Education in Science and Technology*. https://www.tandfonline.com, https://doi.org/10.1080/0020739X.2020.1837400.

Kovács, Z., Recio, T., & Vélez, M. P. (2018). Using automated reasoning tools in GeoGebra in the teaching and learning of proving in geometry. *The International Journal for Technology in Mathematics Education, 25*(2), 33–50.

Kovács, Z., Recio, T., & Vélez, M. P. (2019). Detecting truth, just on parts. *Revista Matemática Complutense, 32*(2), 451–474.

Olsson, J. (2019). Relations between task design and student' utilization of GeoGebra. *Digital Experiences in Mathematics Education, 5,* 223–251.

Recio, T., Richard, P. R., & Vélez, M. P. (2019). Designing tasks supported by GeoGebra automated reasoning tools for the development of mathematical skills. *The International Journal for Technology in Mathematics Education, 26*(2), 81–89.

Recio, T., & Vélez, M. P. (1999). Automatic discovery of theorems in elementary geometry. *Journal of Automated reasoning, 23*(1), 63–82.

Semenov, A., Kondatriev, V. (2020). Learners as Extended Minds in the Digital Age. In: M. Noskov, A. Semenov, S. Grigoriev (Eds), *Proceedings of the 4th international conference on informatization of education and E-learning methodology: Digital technologies in education (IEELM-DTE 2020)*, Krasnoyarsk, Russia, October 6–9 2020. http://ceur-ws.org/Vol-2770/paper5.pdf. Cited 13 Jan. 2021.

Sinclair, N., Bartolini Bussi, M. G., de Villiers, M., Jones, K., Kortenkamp, U., Leung, A., & Owens, K. (2016). Recent research on geometry education: An ICME-13 survey team report. *ZDM Mathematics Education, 48,* 691–719.

Ulm, V. (2011). Teaching mathematics: Opening up individual paths to learning. In: *Towards new teaching in mathematics, 3, SINUS International.* http://sinus.uni-bayreuth.de/math/tnt_math_03.pdf. Cited 23 Sept. 2020.

Wilson, J. (1997). Island treasure, mathematics education, EMAT 4600/6600, The University of Georgia. http://jwilson.coe.uga.edu/EMT725/Treasure/Treasure.html. Cited 22 Oct. 2020.

Wolfram, C. (2020). *The Math(s) fix: An educational blueprint for the AI age.* Manitoba, Canada: Wolfram Media Inc.

Wu, W. (1995). Geometry problem-solving and its contemporary significance. In: *Electronic proceedings of the first Asian technology conference in mathematics (ATCM). Innovative use of technology for teaching and research in mathematics.* 18–21 Dec. 1995. The Association of Mathematics Educators, 67–73. Singapore. http://atcm.mathandtech.org/EP1995/ATCM1995.pdf. Cited 3 Nov. 2020.

Intelligence in QED-Tutrix: Balancing the Interactions Between the Natural Intelligence of the User and the Artificial Intelligence of the Tutor Software

Ludovic Font, Michel Gagnon, Nicolas Leduc, and Philippe R. Richard

1 Context

When considering the mathematics competences referential (Niss & Højgaard, 2019), whether from Québec Education Program,[1] the Standards of the National Council of Teachers of Mathematics[2] or some large international studies such as the Programme for International Student Assessment,[3] reasoning and mathematical proofs are always at the forefront. Although the notion of proof is limited, in elementary school, to mathematical reasoning and conviction, and in post-secondary education, to demonstration and written communication, the evolution of the treatment reserved for deductive reasoning in secondary school has the appearance, depending on the region, of an eternally shilly-shallying between valorization and marginalization. This status quo creates a major source of epistemic injustice (Tanswell & Rittberg, 2020), if only for mathematical conceptualization or the development of high-level competences. Therefore, it is a major educational issue at the

---

[1] http://www.education.gouv.qc.ca/enseignants/pfeq/.
[2] https://www.nctm.org/Standards-and-Positions/Principles-and-Standards/.
[3] https://www.oecd-ilibrary.org/docserver/b25efab8-en.pdf.

L. Font (✉) · M. Gagnon · N. Leduc
École Polytechnique de Montréal, Montréal, QC, Canada
e-mail: ludovic.font@polymtl.ca

M. Gagnon
e-mail: michel.gagnon@polymtl.ca

N. Leduc
e-mail: nicolas.leduc@polymtl.ca

P. R. Richard (✉)
Faculté des Sciences de L'éducation, Département de Didactique, Université de Montréal, Montréal, QC, Canada
e-mail: philippe.r.richard@umontreal.ca

© The Author(s), under exclusive license to Springer Nature Switzerland AG 2022
P. R. Richard et al. (eds.), *Mathematics Education in the Age of Artificial Intelligence*, Mathematics Education in the Digital Era 17,
https://doi.org/10.1007/978-3-030-86909-0_3

heart of everything from teaching to learning, including the training of mathematics teachers.

It is already recognized that the field of reasoning and proof is an authentic vector of training during compulsory education (Coutat et al., 2016). In our digital age, the understanding of the field intertwines as never before didactic, mathematical, and computer issues. However, this domain is often the most resilient to technological change, as if the mathematical sciences could only prove with writing. Without denying the crucial role of discourse in mathematical work, we retain, first, an unavoidable strain between traditional and technological mathematics, the latter still fuzzily defined but flourishing under our eyes, and, second, the lack of an epistemological reference, precise and verifiable in the long run, apt to inspire education. If it is possible to link technological tools to discourse with the notion of instrumental proof—notion developed in mathematics education on epistemological bases (Richard et al., 2019)—, our argument is based on the idea of interaction between a human and a machine.

Since Rabardel's work (Rabardel, 1995) on cognitive ergonomics and the cross-look offered in mathematics didactics (Trouche, 2003), the notion of instrument has become detached from its usual meaning of a man-made object used to perform certain operations. The modes of use constructed by a user were then attached to the technical object, thus underlining the idea of a mixed entity consisting of the interaction between an artefact and a subject. What changes is that beyond the direct interaction between a subject and an object, i.e., anything perceived by the senses, new interactions are integrated, such as the interactions between the subject and the instrument, the interactions between the instrument and the object on which it allows acting, as well as the subject-object interactions mediated by the instrument. The instrument then appears as an emergent system which is viewed from different angles, associating with the instrumental genesis two interdependent processes which remain the fact of the subject, i.e., instrumentation and instrumentalisation (Rabardel, 1995). We come to distinguish between those aspects of the process of instrumental genesis which are directed toward the subject itself (instrumentation) and those which are directed toward the artefact (instrumentalization). The intelligence of the system is therefore a shared intelligence where one can see above all the machine intelligence, the human intelligence, or the intelligence of an evolving system, always looking at the same interactions of mathematical activity under three complementary glances.

In order to explore in depth these complex issues, we created the QED-Tutrix project. At its core, it is a software that, first, provides a platform to experiment on the benefits of using technology to teach geometry, and, second, allows the precise analysis of the interactions between student and subject, since those interactions are all done in a digital space where every action can be logged, stored, and automatically included in broad statistics. The aim of this chapter is to provide a standalone report on the goals, stakes (both in computer science and mathematics education aspect), and internal workings of the QED-Tutrix software. This first section explains the context surrounding our project. The second section provides detailed information about the software itself, in particular the various original structures we developed to represent complex situations in a computable form. Finally, the third section contains a report on the ancillary system we created to address the crucial issue of automated proof generation. Most of the content of the latter is similar to one of our previous paper Font et al. (2020), with the notable addition of the validation process.

1.1 Symbiosis Between the Mathematical Work in Schools and Computer Science

One of the prominent theories of mathematics education is the one of mathematical working space (Kuzniak et al., 2021). The theory of mathematical working spaces facilitates the specific study of mathematical work in schooling, both in terms of learning and implementation. Mathematical work is, in a way, the visible part of mathematical thought and it is constructed progressively as a process of bringing together epistemological and cognitive aspects, in accordance with three genetic developments that are jointly engaged, identified in the theory as the semiotic, instrumental and discursive geneses. With respect to our system, this representation separates the epistemological plane, containing the "absolute" theoretical mathematical knowledge, and the cognitive plane, containing the current knowledge of a student at a given time. The interaction of knowledge between these planes is done on three geneses: the semiotic genesis, representing the mathematical concepts and symbols and their meaning; the discursive genesis, representing mathematical organization in a structured form, whether written or oral; and the instrumental genesis, representing the use of material or symbolic artefacts to manipulate or transform concepts, for example, with the help of a ruler, compass and protocol, or, more relevant to our case, tutorial software, and all the tools it engages. The knowledge constructed by a subject (student, teacher, trainer) emerges from this interaction and can, under certain conditions, constitute autonomous instrumented knowledge. This model is summarized in Fig. 1.

In the light of this representation, we can specify the role of the QED-Tutrix software in supporting mathematical work. Thus, for the learner, it favors the creation of knowledge and the development of mathematical competencies; for the teacher, it allows him/her to organize the reference knowledge according to the usual mathematical work in his/her lessons; for the teacher, it makes possible the simulation of mathematical work in a concrete space of necessity, exploring the valence of mathematical work for absent subjects (the learners). This objective has been one of the guiding principles during the whole development process. More generally, QED-Tutrix is the result of a close collaboration between experts in computer science and mathematics education, following the principle of design in use. As a result, all its core functionalities have been validated, implicitly or explicitly, as being relevant for improving the experience of its users, both student, teacher, and trainer. We present these functionalities in Sect. 2.

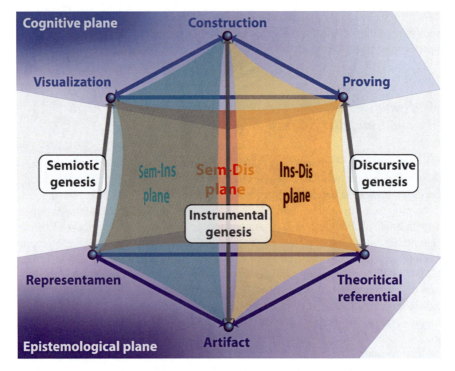

Fig. 1 The main components of the mathematical working space. The vertical planes highlight the coordination of the dominant geneses in a specific mathematical work, for example, to signify certain mathematical competencies at stake or types of instrumental proofs (Richard et al., 2019)

1.2 Existing Tutor Softwares

In Leduc (2016), Leduc analyzed the existing solutions for learning mathematics. He first identified non-tutor systems, divided into four groups: tools for autonomous learning; tools modeling the learning path and curriculum; micro-worlds; and tools for automated proving.

Tools for Autonomous learning, typically websites providing answers to specific questions, such as Mathway (2021) or private tutoring companies. These are fantastic knowledge bases but are either passive tools or non-automated.

Learner Modeling, the student is guided on a learning path and his knowledge is taken into account when giving him new content. Examples include one of the first such systems, ELM-ART (Weber & Brusilovsky, 2001) for learning LISP, and ALEKS (Falmagne et al., 2006), ActiveMath (Melis et al., 2009), and Wayang Outpost (Arroyo et al., 2004) for mathematics specifically. These require a modelization of the user's knowledge, that can be done using various techniques, such as fuzzy logic (Jean-Daubias, 2000) or Knowledge Components (Aleven & Koedinger, 2013).

These systems are based solely on the problems solved, and not on the way the problem was solved, which is one of the foci of QED-Tutrix.

Micro-worlds, the student, unlike in the paper-pencil environment, can manipulate a dynamic geometrical figure, following certain rules, such as Euclid's axioms. A pioneer in this domain is the Logo framework, developed in 1980 (Papert, 2020), but a typical modern example is GeoGebra (Hohenwarter, 2013), a popular dynamic geometry software with an open source code and an active community. These tools are usually dependent on the presence of the teacher, since they offer little to no control over the student's actions, and no help toward the resolution of a problem.

Automated Proof, these systems allow verifying statements, or discover new facts. These systems are discussed in detail in Sect. 3.1.

In a second part, he analyzed in detail 10 tutoring systems for geometry. Each of these systems offers interesting characteristics, but none combines them all. The goal of QED-Tutrix is to offer:

- an interface to explore the problem by allowing the student to freely manipulate the figure;
- freedom in the construction of the proof, allowing the student to construct his proof in any order;
- handling of all possible synthetic proofs, acceptable at a high-school level, i.e., excluding coordinates or complex numbers-based ones;
- a tutor system to help the student, based on the identification of the step of the proof on which he is working;
- autonomy from the teacher, allowing an unsupervised use by the students.

In the remaining of this section, we provide a short summary of the systems analyzed and explain their shortcomings.

One of the first tutor systems developed for geometry is GeometryTutor. It is based on a solid theoretical model, the ACT-R cognitive theory (Anderson, 1996; Anderson & Schunn, 2000). However, it does not allow the student to explore the problem outside of the rigorous path identified by the software.

A few years later, the PACT Geometry Tutor has been developed (Aleven et al., 1998; Aleven & Koedinger, 2000) that evolved into Geometry Cognitive Tutor (Aleven & Koedinger, 2002; Aleven et al., 2006; Roll et al., 2014). This system is limited to problems based on Cartesian coordinates, since it handles elements numerically. In QED-Tutrix, we want to be able to handle any geometry problem.

To include proofs that require an additional construction, the Advanced Geometry tutor was developed by Matsuda (2005), based on the GRAMY theorem prover (Matsuda & Vanlehn, 2004). This system has interesting characteristics, since it is one of the few to handle proofs with intermediate constructions. However, it is quite rigid on the accepted proof and does not allow the student to explore the problem or use a proof that is not the optimal proof calculated by the prover.

The software ANGLE (Koedinger, 1991; Koedinger & Anderson, 1993), unlike the previous ones, aims at helping the student to construct his proof. It is based on the Diagram Configurations (Koedinger & Anderson, 1990) theory, based on interesting configurations of the geometrical figure, used by the experts to produce the proof. It gives the student freedom to explore the solutions. However, the Diagram Configurations are modeled on the work of experts, which can be quite far from the formal geometry taught in high school. Besides, even though the student can explore the problem, there is no interface to manipulate the figure.

An interesting approach is the Baghera system (Balacheff et al., 2003; Webber et al., 2001), providing a web platform. This system offers two sides: one for the teacher, where he can create new problems and follow in real time or replay the progress of the student; and one for the students, who can choose and solve problems. The weakness of this system when compared to our goals is that it offers no automated tutor.

To provide interactive figure manipulation, it is a logical step to use an interactive geometry software. Two systems are based on the Cabri software (Baulac, 1990; Kordaki & Mastrogiannis, 2006), Cabri-DEFI (Luengo & Balacheff, 1998) and Cabri-Euclide (Luengo, 1997; Luengo, 2005). The first one helps the student to plan his proof by asking him questions about the figure that ultimately direct him toward a proof. It is an interesting approach, but it offers no freedom of exploration, since it is the system that asks questions. Besides, there is no tutor system to help the student when he is stuck in his resolution. The second one goes in the opposite direction, by allowing the student to explore freely and enter conjectures, that are later organized in a graph. However, there is no help provided to the student and no mechanism to ensure that the problem is ultimately solved, which can be an issue for unsupervised use.

The system Mentoniezh (Py, 1994, 1996, 2001) offers a novel approach by dividing the proof into four steps: understanding the problem; exploration of the figure; planing of the proof; and redaction. The software helps and directs the student during each step. The division of the proof in steps is one of the foundations of QED-Tutrix. However, the software provides no help to find the next proof element, and a student can therefore encounter an impasse in his resolution that will force him to ask the teacher for help.

Another system dividing the proof into steps is Geometrix (2021), where the student can first construct a figure, and then allows him to create a problem based on that figure and to solve it. It therefore allows the creation of exercises by the teachers, including customized error messages to help the student. However, it remains mainly a demonstration assistant and offers little in the tutoring aspect.

Finally, the Turing system (El-Khoury et al., 2005; Richard et al., 2007), that largely inspired QED-Tutrix, provides an interface for the student that allows him to manipulate a dynamic figure, and to provide statements to construct his proof, in any order. The integrated tutor system analyzes his input and gives feedback depending on the validity of the statement. After a period of inactivity, the tutor gives him a hint to restart his resolution process. However, the hint is based only on the last element provided by the student, which may not be the step on which he is currently working.

Overall, all of these systems, except ANGLE, are based on fully formal geometry, which limits the number of acceptable proofs, even though less formal proofs are typically accepted by the teachers. ANGLE is based on a model of the reasoning of experts, which is quite far from the proofs used in class. Besides, only Mentoniezh keeps the previous work of the student in memory but does not use it to provide hints toward the next step. The systems that provide hints are based on forward or backtracking, limiting their usefulness. Finally, no system allows the student to explore different proof paths at the same time. This illustrates the need that gave birth to the QED-Tutrix project.

2 Genesis of the QED-Tutrix Project

This section focuses on providing an overview of the QED-Tutrix software itself, both in its goals in terms of mathematics education, its user interface, and its internal working.

2.1 Task Description

As we mentioned previously, QED-Tutrix has several well-established goals:
- providing an interface to solve geometry problems by constructing the proof in a way similar to what the student would do in class;
- accounting for the three geneses of mathematical work (semiotic, discursive and instrumental);
- offering a tutoring system to help the student in the problem resolution.

These high-level, conceptual goals directly create low-level constraints in the very conception of the software itself. Indeed, the first goal means that the software must be able to assess, on the fly, the validity of the mathematical proposition the student is entering, both in its form ("The line AB is parallel to point B" is formally invalid) and in its relevance for solving the problem at hand. For instance, if the student begins the proof by providing a theorem that is needed to obtain the conclusion, and is, therefore, at the very end of the proof, the software must be able to assess that this theorem is a relevant element, despite being the first element provided. In turn, this means that the software must know, in advance, the whole set of proofs that a student could provide. Finally, this creates the need for, first, a structure to represent the set of all possible proofs to a problem, and, second, a way to generate such a structure. The definition of such a formal structure formalizes the representation of a proof, and, therefore, ensures that there is not an infinite number of possible proofs. We provide more detail on this structure in Sect. 2.3.

The second goal means that the interface of QED-Tutrix must allow the student to explore the problem, provide mathematical elements (such as results or properties) that are relevant to the proof, and finally organize these elements in a valid, formal proof. These three aspects follow the processes defined by Coutat and Richard (2011) of discovery, validation, and modelization.

The third goal means that the software must be able to determine the proof the student is working on in real time. Indeed, even simple problems can have up to several millions of possible proofs because of the combinatorial explosion when there are several possible variations at every step. This required the creation of several structures to store and manipulate the progression of the student on his or her proof. Furthermore, the messages sent by the tutor to help the student had to be carefully crafted, as well as a way to choose which one to send in every possible situation.

2.2 Software Overview

We now present the interface of the current version of QED-Tutrix. The main window is composed of four core elements: the GeoGebra interface (top-left), the sentence selection menu (center), the redaction section (top-right), and the tutor chatbox (bottom). A view of the interface is provided in Fig. 2.

2.2.1 GeoGebra Interface

The figure section is present on the interface at all times. It displays GeoGebra's (Hohenwarter, 2013) the main window which was incorporated into QED-Tutrix to allow the student to explore the dynamic figure linked to the problem. An important point is that there is a limited communication between the GeoGebra interface and the software internal structures. If the student creates a new element (such as a new line or circle), then this element will not be useable to solve the problem, since it

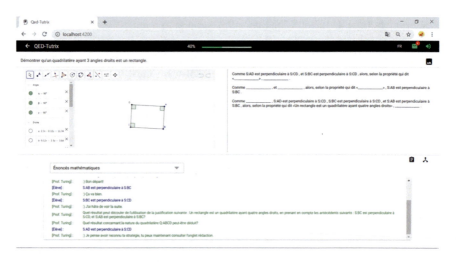

Fig. 2 QED-Tutrix's main interface

Fig. 3 The Parallelogram problem statement

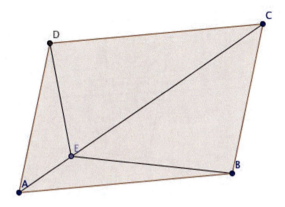

has not been integrated in the set of possible proofs. As a consequence, we must ensure that the GeoGebra figure contains each and every geometrical element that can appear in any proof of the problem that we want the software to be able to handle. We call such a figure the super-figure.

To provide a more detailed example, let us consider the parallelogram problem, as stated in Fig. 3. A possible proof uses the heights of triangles ACD and ABC through D and B, respectively, another uses the lines parallel to AB through E and parallel to AD through E, and another uses the diagonal of $ABCD$ through B and D. Therefore, all these new elements, both lines and points, must have a name, so that the student is able to refer to them in his proof, and the names must correspond between the displayed GeoGebra figure and the internal representation of the software. As a result, these elements are created beforehand and provided to the student in the GeoGebra plugin, forming the super-figure of the problem, as shown in Fig. 4. If, at some point, a student thinks of another way of solving the problem, such as adding a point D' on AB such as $AD' = AD$ and using the coordinates in the DAD' system, then the software would not be able to handle this, since it does not know about any D' point.

2.2.2 Sentence Selection Menu

This menu is accessed by clicking on the "Énoncés mathématiques" button. This opens an interface, displayed in Fig. 5, allowing the student to search a mathematical result, property, or definition following one or more keywords. In this example, the student selected the keywords "Area" and "Base." This combination leads to the displaying of four choices in the right menu.

Selecting one of those leads to the next step, shown by the sentence in the middle of Fig. 6. Indeed, in the case of a mathematical result, the student must indicate the object(s) in which he or she is providing a result. In this example, the sentence is

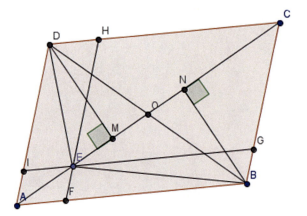

Fig. 4 The Parallelogram problem super-figure

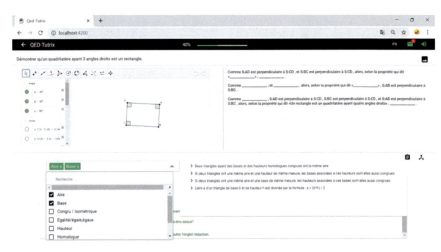

Fig. 5 QED-Tutrix's sentence selection menu

"The angles __ and __ are supplementary angles". To enter this result, the two angles must be specified by entering their names.

Lastly, if the entered names are lexically valid (three letters for an angle, one letter for a point, etc.), the student can click the green validation button to send the proposition to the software. He or she immediately receives a feedback from the tutor.

Intelligence in QED-Tutrix: Balancing the Interactions Between the Natural ...

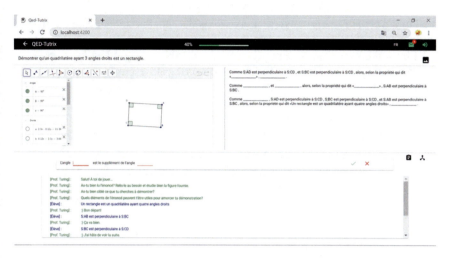

Fig. 6 QED-Tutrix's sentence confirmation prompt

2.2.3 Tutor Chatbox

Present at the bottom of the interface, this section takes the form of a chat between the student and the virtual tutor, Prof. Turing. It is not, however, a true chatbox, since the only way for the student to "speak" is to enter sentences using the interface presented in the previous section. Regarding the tutor, there are two possible interactions.

The first one is an immediate feedback on the entered sentence: a positive smiley face and message ("Good start," "This is going well" or "I can't wait to see what's next") in case of a valid sentence, or a sad face and a feedback message in the other case. The feedback message depends on the error. The most common case is simply entering a sentence that is not useful for the problem, but the student could also have entered the same sentence twice. Lastly, the tutor also handles the case where the student entered the reciprocal of an expected property ("If a quadrilateral is a rectangle, then it has four right angles" instead of "If a quadrilateral has four right angles, then it is a rectangle").

The second type of interaction is a series of help messages in case the student is stuck. The software considers that there is an impasse if the student has not entered any correct sentence recently. Then, until it happens, a help message is sent every minute. To summarize the process, the software finds out which proof the student is likely to work on. Then, it identifies the element of this proof closest to the last valid sentence he or she entered. Then it attempts to help the student to find this element. The sequence of messages is determined by a finite state machine whose details are provided in the work of Leduc (2016). A summary is available in Sect. 2.3.4.

For example, let us go back to the parallelogram problem in Fig. 3. The student has entered the results "(DM) is the height of the triangle AED through D," "(BN) is the height of the triangle AEB through B" and "The area of triangle AED is equal to the area of triangle AEB." Let us consider that the student entered that last sentence most recently. To complete this inference, "The line segments $[DM]$ and $[BN]$ have the same length" and "Two triangles whose height and base are equal have the same area" are missing. The second one is the closest to the most recent valid sentence entered. The tutor will therefore attempt to help the student toward "Two triangles whose height and base are equal have the same area." To do so, it provides messages that depend on the structure of the inference ("What is the property that allows inferring Y from X_1, X_2 and X_3?"), then on the property itself ("How can you obtain the result that two triangles have the same area?"). After the set of possible messages has been exhausted, in general after a dozen or so messages, the tutor recommends the student to consult his or her professor.

This whole series of help messages is reset, and a new series is generated, every time a correct sentence is entered.

2.2.4 Redaction Section

This last section, on the top-right of the interface, allows the student to visualize his or her progression by displaying the written proof the student is working on. When any valid proof of the problem is 40% completed, it is identified as the one he or she is working on and is displayed for the student to see. Prof Turing also indicates in the discussion zone that a strategy has been identified and is shown in the writing tab. Thereafter, if the software determines that the most advanced proof has changed, the writing tab displays this new one. The statements are organized in a forward chaining fashion and contain only the elements that have been entered previously. The remaining elements are replaced by a blank. The student can't interact directly with this section and must go through the statement menu to submit the missing pieces of the proof.

It is not possible to interact with this section. Its only purpose is to help the student toward the completion of the proof, by highlighting the missing elements and handling most of the redaction process.

2.3 The Core Layers of QED-Tutrix

The features presented previously require a precise representation of several aspects of problem resolution. First, we need a structure to represent the set of all possible proofs for a given problem. Second, we need to keep in memory the various resolution steps taken by the student, and to be able to explore these steps to identify which proof he or she is working on. Third, we need to be able to generate, automatically, tutor messages in case the student is stuck in the resolution. In this section, we present the core layers of the software that we developed to tackle these needs.

2.3.1 The HPDIC Graph

The first structure we had to create is a way to store the set of possible proofs and the inferences composing them. Indeed, it is necessary to be able to navigate this set to find out, first, if the element entered by the student is relevant to this problem, and, second, the proof he or she is likely working on. This representation of the set of possible proofs is possible thanks to the structure of HPDIC graph. This graph, that we present in detail in this section, is the first and most central of the four layers that, by interacting together, form the core of the QED-Tutrix software.

This graph includes **H**ypotheses, **P**roperties, **D**efinitions, **I**ntermediary results and a **C**onclusion, hence the HPDIC appellation. This graph is unique for each problem and is built from the inferences deemed acceptable to solve the problem. Therefore, generating this graph is a prerequisite to add a new problem to QED-Tutrix. It is a static structure which is loaded once when a problem is chosen by the user and is referred to during the running of the software.

To build the HPDIC graph, we must consider each inference as a (directed) tree structure, at the center of which is the justification that is linked to parent premises, and to an only child, the consequent. This structure allows inferences to be chained, since the consequent of an inference can be used as a premise to another. Therefore, in this representation, a mathematical proof is a chain of inferences that begin with the problem's hypotheses and ends with the problem's conclusion.

Furthermore, since the problem's hypotheses and conclusion do not change from one proof to another, and since several proofs can use common intermediary steps, it is natural to fuse the inference chains of the various proofs together in a single structure. This process is illustrated in Fig. 7. Subfigure a. represents an inference, subfigure b. represents the creation of a proof by adding two inferences centered on "Justification 2" and "Justification 4," and subfigure c., with the addition of the two inferences centered on "Justification 3" and "Justification 5," represents a set of two proofs for the problem. The gray part in subfigures b. and c. represent the addition of the step compared to the previous one.

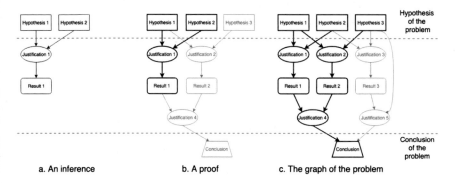

Fig. 7 The informal construction process of an HPDIC graph

Although both an inference and a proof have tree structures, the complete HPDIC graph does not, since it can include cycles in the case of symmetrical inferences. Symmetrical inferences occur when for a same problem to different admissible proofs use two inferences that differ only by the fact that their premises and results are reversed. In other words, this happens when two reciprocal justifications appear in two different proofs ("If a quadrilateral is a rectangle, then it has four right angles" and "If a quadrilateral has four right angles, then it is a rectangle"). This is the simplest case where a cycle can occur, but lengthier cycles also exist.

Overall, the HPDIC graph offers a straightforward means of storing the complete set of possible proofs for a problem. Furthermore, the fact that inferences are shared between proofs allows for an efficient assessment of the progress of the student. Lastly, because of its structure, it is theoretically quite simple to generate a HPDIC graph by forward chaining, i.e., starting from the hypotheses and generating all possible inferences from there, as we will explain in Sect. 3. Some steps must, however, be taken in practice to avoid combinatorial explosion, such as ordering the premises of each inference.

2.3.2 The MIA

The second layer, the MIA or Modèle Interactif de l'Apprenant (Interactive Model of the Learner), is used to modelize the student's progress while she or he drafts a proof. Classic geometry tutorial systems usually use forward or backward chaining motion to predict the next suitable action for the student and limit help to that particular step. In order to achieve this, only the list of activated nodes, without regards to the activation order, is necessary. Sometimes, they use the last activated node to offer more personalized help. With QED-Tutrix, our objective was to provide, when a student is stuck, different solution paths like a teacher would do, as we explained previously. We therefore introduced the MIA which accounts for the student's progress, but also for the chronological order of his or her actions, in order to provide a tailored help that respects the student's cognitive state.

The HPDIC graph being static, we had to define a structure to record the student's progression in the drafting of his or her proof. Therefore, the MIA adds information about the student's actions to every node of the HPDIC graph. For every statement form, the HPDIC (valid action) submitted by the student, we take note, in the corresponding node, a value according to the activation time. This information is constantly updated while the student drafts a proof and is used in the superior layers to generate tailored messages and to identify the proof he or she is most likely working on, according to his or her past actions.

2.3.3 The EDOI

The third layer is the EDOI, Évaluation des Démonstrations et Ordonnancement des Inférences (demonstration evaluation and inference ordering). In order to help the student, QED-Tutrix generates instant messages as well as a series of hints. These aim at reviving a staled solving process without handing the answers to the problem directly. To do this, we must know which of the admissible proofs is most advanced and arrange the inferences according to the student's cognitive state for them to be used by the next layer, the GMD (see Sect. 2.3.4). The EDOI layer is used, in the first place, to estimate the progress in each of the potential strategies explored by the student, but only the most advanced solution is considered to evaluate the general progress and to generate the instant messages and the proof which will be displayed in the redaction section. Then, the ordering algorithm will favor the inferences belonging to this same proof, which is considered as the student's current strategy. IT will combine this information with the data from the MIA to generate a list of inferences which will be updated after each of the student's valid actions and injected into the GMD.

In QED-Tutrix, we define the most advanced proof as the proof for which the submitted statements to statements left to submit ratio is the highest. In the HPDIC graph, we therefore calculate the percentage of activated nodes for each solution. However, this procedure means the ratio for each proof must be recalculated each time the student submits a valid statement, which can be time consuming for problems for which there are millions of possible proofs. To avoid these costing calculations, we designed an algorithm that shares information about the percentage of nodes activated in the HPDIC graph, providing that the graph is a tree. If it is not the case, we create beforehand a set of trees that combined include all the different solutions from the original HPDIC graph. We then can, by backwards tracing, almost instantaneously, find the most advanced proof. Next, an instant message is generated to indicate to the student if the submitted statement is accepted as part of a solution, part of the HPDIC graph, or of no use. In the case of an admissible statement, the displayed message depends on the percentage of completion of the proof.

Then, the inference arranging algorithm is solicited to provide the GMD with hints to submit to the student if he or she gets stalled. With the MIA's data, it is possible to know which inference the student has recently worked on in order to help him or her on this inference first and foremost. In fact, we arrange all the HPDIC's inferences according to a reverse chronological order, from the most recent to the first inference the student worked on. In the case of inferences that have not yet been evoked by the student, they are organized in a random order at the end of the list, since they constitute new paths not yet explored by the student. When the most advanced proof reaches a certain percentage of completion, the inferences that belong to it will be treated first and foremost, even if other inferences were worked on more recently, since we want to spur the student to complete this proof. To do so, first, we organize the inferences of the most advanced proof. We then add to the list, in reverse chronological order, the rest of the graph's inferences. Once the organization is complete, the organized list is provided to the GMD to generate the necessary messages to restart a blocked solving process.

2.3.4 The GMD

The final layer is the GMD, Générateur de Messages Discursif (discursive messages generator). One of QEDX's goals is to help the student the way a teacher would, so, in order to reach this goal, we added the GMD layer which generates Prof Turing's messages. The type and order of the messages to display were determined by analyzing the student-teacher interactions. We uncovered a certain structure in the order in which teachers provide hints as well as in the content of these messages, and we used it as an inspiration in programming the messages for QEDX. The GMD layer, which uses to list of organized inferences, therefore completes Prof Turing's intelligence and constitutes one of the main contributions of our project since it is able to display help messages in a similar fashion to a teacher, while following a predetermined schedule. It is therefore impossible for the student to directly address the tutoring agent with a question or to force it to solve the problem for him or her.

The finite state machine therefore includes the necessary states to take into account most of the behaviors witnessed in class and in our analysis. However, since all states are not used in every situation, it is possible to bypass these states by providing a list of empty messages. Since QED-Tutrix' architecture is guided by data, the messages associated with the different states are defined by didacticians or teachers in a text file. Besides the static messages, it is possible to program dynamic messages which depend on the statements submitted by the student. Also, different levels of messages can be implemented, from the more generic to the most precise. The content designers therefore have much control over Prof Turing's reactions.

To generate these messages, the machine, after the student submits a valid statement, retrieves the new list of organized inferences in the EDOI and positions itself in an initial state. Then, for each state, it displays the messages in the planned order by letting a prescribed delay between each message. At first, it submits generic messages that can be applied to the whole problem. Then, the different states are specific to given forms of inference, the EDOI list. Our finite state machine sequentially treats each help message for a given inference before dealing with the next one. The machine treats the inferences from the EDOI list until it is empty or until a maximum time without valid actions is expired. From there, the machine adopts a terminal help state to tell the student whose statement to submit or by advising him or her to consult a teacher. This last layer, by using the data from the lower layers and messages created by the didacticians, replicates the actions of a teacher whose helping a student solve a geometry problem.

In this section, we summarized the four core layers of QED-Tutrix. The base layer includes the HPDIC graph which contains all the different proofs for a given problem and is used by QED-Tutrix's different components. On top of that is the MIA layer that reports the student's progress while storing the action chronology. From this chronology, the EDOI layer determines the most advanced proof, generates an instant message stating the student's progress, and arranges the inferences in order to identify the help strategy QED-Tutrix will adopt. Finally, with the arranged list of inferences, the GMD layer generates, in the same fashion a teacher would, Prof Turing's messages. Although these layers are the main components of the software, the QED-Tutrix project encompasses more aspects. See also: Richard et al. (2018) about messages; Corbeil et al. (2020) on blockages; Farid (2020) on extraction of knowledge; Cyr (2021) on the theoretical referential system; and Tessier-Baillargeon et al. (2017) about tutorial systems.

3 The Need for Automated Proof Generation

As we explained in the previous section, the HPDIC graph is at the core of QED-Tutrix. Therefore, expanding the available set of geometry problems require providing an HPDIC graph per new problem, meaning that generating those automatically provides a huge boon to the expansion of QED-Tutrix's reach. In this section, we provide a report on the automated proof generator we developed to address this issue.

3.1 Existing Theorem Provers

The main objective of this part of the project can be summarized in a couple of words: to automate the process of generating the HPDIC graph of a given problem. This broad objective can be separated into two tasks: first, to automatically generate the set of possible proofs for a problem, and, second, to translate this set into a HPDIC graph. The second task is quite straightforward and consists entirely of writing down translation rules. Therefore, this chapter focuses on the first one.

Finding a proof to a problem is identical to the task of proving a theorem, since a theorem is, in essence, simply a way to provide a shortcut between some hypotheses and a conclusion. Indeed, let us go back to the example of the parallelogram problem. Finding a solution to this problem is similar to finding a proof to the theorem: "In a parallelogram $ABCD$, with a point E on the diagonal AC, the areas of triangles AEB and AED are equal." Therefore, a natural first step is to explore the avenue of (Geometrical) Automated Theorem Proving, (G)ATP. However, unlike "classical" theorem proving, we have several unique constraints:

- the proofs must be **readable**;
- they must use only properties available at a **high-school level**;
- there must be a way to handle the **inferential shortcuts**, i.e., the inference chains that can be deemed too formal by some teachers and are therefore skipped in a demonstration.

These constraints direct our search for a way to automatically find proofs. Indeed, there currently exist two general research avenues for geometry automated theorem provers: algebraic methods and synthetic, or axiomatic, methods. The first one is based on a translation of the problem into some form of algebraic resolution, and the second one uses an approach closer to the natural, human way of solving problems, by chaining inferences.

One of the main goals of the research community in automated theorem proving is the performance. Since synthetic approaches are typically slower, most solvers are based on an algebraic resolution. Algebraic methods include the application of Gröbner bases (Buchberger, 1988; Kapur, 1986), Wu's method (Chou, 1988; Wu, 1979) and the exact check method (Zhang et al., 1990). Practical applications include the recent integration of a deduction engine in GeoGebra (Botana et al., 2015), which is based on the internal representation of points in the plane as coordinate tuples inside GeoGebra. Other examples include the systems based on the area method (Chou et al., 1996; Janičić et al., 2012), the full-angle method (Chou et al., 1994), and many others. These systems seldom provide readable proofs, and when they do, they are far from what a high-school student would write. Given our readability and accessibility goals, all these systems are not relevant to our interests.

For this reason, we focus on synthetic methods. A popular approach is to use Tarski's axioms, which have interesting computational properties (Braun & Narboux, 2017; Narboux, 2006). However, the geometry taught in high school is not based on Tarski's axioms. Therefore, proofs based on them are quite inaccessible for high-school students, violating our second constraint.

A prover that has very similar goals is GRAMY (Matsuda & Vanlehn, 2004). It is based solely on Euclidean geometry, with an emphasis on the readability of proofs. Besides, it has been developed as a tool for the Advanced Geometry Tutor. It is therefore able to generate all proofs for the given problem. Finally, one of its major strengths is the ability to construct geometrical elements. However, to the best of our knowledge, the source is not accessible, and no work has been done on it since 2004. Furthermore, it does not provide the complete set of proofs.

Another very interesting work is the one of Wang and Su (2015), as it aims at providing proof for the iGeoTutor, and therefore has the same readability and accessibility at a high-school-level objectives. In particular, its template-matching algorithm for finding auxiliary constructions is quite promising. However, it requires the use of an external arithmetic engine, and, more importantly, also focuses on finding one proof.

Overall, the very specific needs dictated by the focus on educational interest considerably limit our options. The two only systems with similar goals are GRAMY and iGeoTutor, and they are not suited to our needs. Therefore, we chose to implement our own system.

3.2 The Choice of Logic Programming to Generate Inferences

In theory, a mathematical inference is quite easy to model in a computer, as it is essentially a combination of premises ("$ABCD$ is a parallelogram" and "$\widehat{ABC}$ is a right angle"), a property ("a parallelogram with a right angle is a rectangle"), and a result ("$ABCD$ is a rectangle"). It can be quite different in practice, however, as we explain in Sect. 3.7, but, for the moment, let us consider only this ideal case. This structure is extremely similar to the inference mechanism in logic programming, where a program is a set of facts and rules, and where we infer new facts (the result) based on a rule (the property) and existing facts (the premises). Then, because the result of an inference can be used as the premise of another, a proof is simply a chaining of inferences, starting at the hypotheses of the problem, and reaching the conclusion. Finally, since the mathematical results can be used in several proofs, we can merge the proofs to create a unique structure containing all the possible proofs for a problem, as we illustrated in the previous section, in Fig. 7. Thus, the core of our problem solver is, in theory, very simple: we create a Prolog fact for each hypothesis of the problem, a Prolog rule for each property, and we let the Prolog inference engine infer every possible new fact (mathematical results) from those. Every time a new fact is inferred, we store it in the form of a mathematical inference (premises + justification = consequent).

3.3 Available Data

Available Properties

Although not detailed in this chapter, an important part of this work includes collecting, compiling, and organizing the mathematical elements that are used by the software. We have analyzed 19 Quebec high-school textbooks ranging from 7th to 11th grade and extracted 2855 mathematical statements, representing 707 properties and definitions that can be translated into an inference (Cyr, 2021). This large databank of definitions and properties is necessary to properly adapt the software. We call *referential* (Kuzniak, 2011; Kuzniak & Richard, 2014) the set of properties and definitions that are allowed or expected in the resolution of a problem. These are known to greatly vary among grades, textbooks, and teachers. Different phrasing might be used from one referential to another even when describing foundational mathematical properties such as *the sum of interior angles in a triangle is 180°*. This variability in phrasing must be accounted for to ensure consistency among the properties and definitions used by QED-Tutrix and to allow teachers to choose their innate preferences to be used by their classes. The final cumulative referential extracted from the various school textbooks includes Euclidean geometry, area and volume formulas, metric relations, transformational geometry, analytic geometry, and basic vector geometry. We do not attempt to isolate a minimal set of axioms that are used in high-school geometry, but instead aim at implementing enough properties and definitions to cover all the material present in high-school geometry textbooks. Our ultimate goal is to encode all of this comprehensive set of properties in Prolog.

Available Problems

To learn mathematics, one must solve problems (Brousseau & Balacheff, 1998). The tasks of finding, adapting, and creating new problems play an important role in a tutoring system such as QED-Tutrix. Currently, our work base is composed of sixty problems covering a vast array of mathematical topics from high-school courses. We divided those into a training set of 19 problems and a validation set of 41 problems. The training set was available during the development of the automated proof generator, and the validation was used after the development to obtain statistics on the coverage of the available proofs.

3.4 Encoding of a Problem

After identifying relevant problems, the next step is to translate them in a Prolog file. An example of such a translation is provided in Fig. 8. The resulting file has several parts.

Implicit Hypotheses

The first eight lines provide names to the geometrical objects present in the problem figure. We refer to these as **implicit hypotheses**, since they are needed for the prover, but are not specified in the problem statement. Furthermore, even though the names

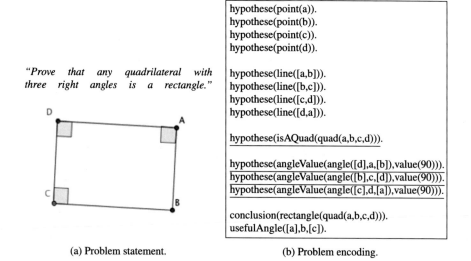

Fig. 8 The translation of the rectangle problem from its statement

of the points are explicit in this case, the Prolog engine needs names for each point, line, etc., even if they are not at all given, neither in the statement nor in the figure. In that case, arbitrary names must be provided (such as $P1$, $P2$, etc.). Lines 1 to 4 provide the set of points present in the figure. Then, lines 5 to 8 indicate which point are linked by a line. Here, we have the four lines (AB), (BC), (CD), and (AD) (we do not differentiate segments and lines). If we wanted to add the line through A and C, we would have to add $hypothese(line([a, c]))$ to the file. The example here is simple enough, but in more complex problems, the encoding of lines can become a delicate issue. Indeed, we must provide, **as soon as the problem is encoded**, all the points that are on the line. This set of points is, as far as Prolog is concerned, the unique identifier for the line. A direct consequence of this implementation is that it becomes impossible to add, during a proof, new points to a line.

Explicit Hypotheses and Conclusion

The next four lines provide the hypotheses in a more general sense, meaning the ones that are usually given explicitly in the problem statement, hence the name **explicit hypotheses**. Here, the statement of the problem, "Prove that any quadrilateral with three right angles is a rectangle," with the addition of the figure, provides four hypotheses: there is a quadrilateral named ABCD, and three of its angles, $\widehat{DAB}$, $\widehat{BCD}$ and $\widehat{CDA}$, are right angles. The problem statement also provides the expected conclusion: $ABCD$ is a rectangle, encoded in the second-to-last line.

Auxiliary Hypotheses

In many problem resolutions, there is, at some step, the need for the construction of additional elements, such as a new line or point. However, the creation of such elements is a difficult issue. Given that our aim in this project is not actually to solve problems in and of itself, but to ease the process of adding new problems to QED-Tutrix, we chose to require the addition of **auxiliary hypotheses**, i.e., elements that are not present on the problem statement directly but are useful in one or several of its resolutions. For instance, in the rectangle problem, one could write a proof using the diagonals of the rectangle. The fact that (AC) is a line would therefore enter in this category. Structurally, auxiliary hypotheses are identical to implicit and explicit hypotheses, it is their origin that differs. When adding these auxilliary elements to the figure of the problem, we create the **super-figure**.

Additional Elements

The last line in our example is neither a hypothesis nor a conclusion, but additional information provided to the prover. It can be of two types: useful angles and

Fig. 9 A geometrical situation with many angles

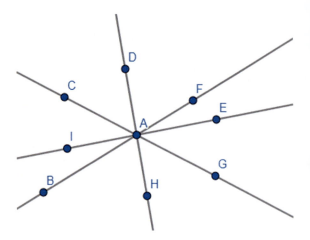

dictionary items. The useful angles are required because in some geometrical situations, such as the one in Fig. 9, the number of possible angles can become quite huge. Here, there are 8 points around the center A, meaning that there are 8 × 7 angles. When considering inference such as "two angles that share a side are adjacent" and "two angles whose sum measures 180 degrees are supplementary angles", the number of possible inferences becomes very impractical. For this reason, to limit the combinatorial explosion that plagues many synthetic automated theorem provers, we chose to impose the following requirement: the problem file must contain the list of all angles on which Prolog is allowed to infer results. In other words, if an angle is not explicitly written in the problem file, no result on this angle will be present in the HPDIC graph. Finally, the dictionary allows the user to provide alternative names for geometrical objects. This is useful in problems where lines, angles, circles, or, more rarely, triangles and quadrilaterals, have alternative names, such as "line l" or "rectangle R." This is not needed for the solver but is important for the tutor software, since the students must be able to enter their sentences with any valid name.

One should note that, as required by the Prolog language, each constant must have its first letter in lowercase, otherwise Prolog would consider it as a variable. This could lead to some collisions if there is a point named A and a point named a, but, in our experience, the conventions used in the statement of high-school geometry problems prevent that situation.

3.5 Generation of the Complete Set of Proofs

Once the problem is encoded in Prolog, we proceed to the proof generation process. To summarize, we proceed to a construction of the graph by forward chaining. We begin with the hypotheses provided in the problem file, such as the one in Fig. 8, and put them in the set of known facts. Then, until this set is empty, we take one fact out of the set and ask Prolog: "what new facts can you infer using this one?" and add all the new facts to the set. Every new fact also leads to the storage of the inference that led to the creation of this fact in the proto HPDIC graph.

When the set of known facts is empty, it means that Prolog is not able to find any new fact. Therefore, it has found all the results that could possibly be inferred from the set of hypotheses, using the allowed properties. If the problem's conclusion is in this set of results, then we have found at least one proof of the problem. If not, there is an error. Otherwise, we explore the generated graph in backward chaining, starting from the conclusion, and marking an element on the graph only if it can be used to infer the conclusion. This last step is necessary, because the inference engine can infer results that are valid, but useless for the resolution of the problem, i.e., not used in any proof that reaches the conclusion. These results are marked, but not removed from the graph. This will be useful when the student works on the problem and enters such a result. In this situation, the software should not say "This result is false," but instead "This result is valid, but not useful for the problem. Try something else." For this behavior to be possible, the "valid but useless" elements must remain in the graph and be identified as such.

Since the generated graph is based on mathematical properties, which typically have a reciprocal, there are usually many cycles in the resulting graph. For example, if ABC is a right triangle in A, then we can infer that $\widehat{BAC}$ is a right angle. And since $\widehat{ABC}$ is a right angle, we can infer that ABC is a right triangle in A. This may be a problem for the calculation of the students' progress in the proof resolution, but it is solved by a clever exploration of the graph. This algorithm, however, is not part of the proof generation process and is not detailed in this chapter.

This algorithm is written in Python and sends a Prolog query every time the program needs to obtain new results from a fact. Because Prolog is a logic programming language, it is not well-suited for handling complex data structures and printing them in a text file, hence the choice of using a more classical language for all the "standard" tasks.

3.6 Validation

Ideally, the generator should be able to generate all the possible proofs for each problem. However, it is impossible to assess if all the proofs have been found without already knowing all the possible proofs beforehand, which is such a time-consuming process that we dedicated this whole project to avoid it. Therefore, we used a two-step validation process.

First, we used the four problems for which we had a manually created HPDIC graph. These four problems are not extremely difficult and remain accessible to high-school students, but have been carefully crafted by experts in mathematics education to offer a large set of possible proofs that encompass many geometry concepts, and therefore have large HPDIC graphs. Two of these four problems have been used as

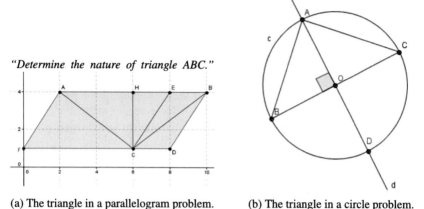

(a) The triangle in a parallelogram problem. (b) The triangle in a circle problem.

Fig. 10 The statements of two of the four intial problems

examples in this chapter, in Figs. 3 and 8. The other two are given in Fig. 10. We then encoded these problems in Prolog and attempted to reproduce the handcrafted HPDIC graph for each of them. Since these graphs have been created, and therefore validated, by mathematics experts, the simple fact that the proof generator is able to re-generate them confirms that the process in itself is sound and that the generator is indeed able to infer all possible results using a given set of properties.

Second, we also had to validate that the proof generator is also able to solve simple, more realistic problems. Indeed, these four problems are doable in high school but are not representative of the simpler problems used in class. Therefore, we extracted 60 problems from various datasets: 27 from the work of Py (2001), 14 from high-school manuals, and 19 were created by an expert in mathematics education to complete this set. Among these 60, we randomly selected 41 to be used exclusively for the validation process, and the remaining 19 were available during the development of the proof generator.

Then, we encoded those problems in Prolog, including the creation of the superfigure, and entered them in the proof generator. If at least one proof was found, i.e., the conclusion was successfully inferred by Prolog, we considered this problem as solved. Since those problems are much simpler than the original four, they should not have more than one or two possible solutions, and therefore considering that generating at least one proof is enough for a success was deemed acceptable.

Out of these 41 problems, the proof generator managed to solve 33 of them, or 80%. Then, we analyzed the reason why the remaining 8 were not solved. Five of them need some algebraic manipulation, for instance being able to "remember" that a segment is 2/3 of the length of another, and to use that information later in the proof, a process that we chose not to implement in the proof generator at the moment. Four of them failed because they needed concepts in a domain of geometry that we had not yet implemented, such as the notion of a perimeter or of a polygon with more than four sides. Lastly, two of them failed because of a missing or incorrect property. It should be noted that these reasons are not mutually exclusive.

Overall, among these 8 misses, only two are due to real issues in the set of implemented properties. The remaining six are simply out of the initial scope of this project. Therefore, we can conclude the following:

- the automated proof generator is able to solve a respectable portion, 80%, of the given problems;
- the initial scope of this project, which does not encompass several geometry concepts such as polygons or perimeters, and excludes the possibility of an algebraic manipulation, is sufficient for 80% of the problems, and by solving the issues of the two missing or incorrect properties, this would reach 85%;
- the usage of logic programming is a sound solution to generate proofs, and by both expanding the scope of encoded properties and solving implementation issues on existing properties, we would potentially be able to solve all the given problems.

It is important to note that performance has not been a consideration at all in our validation process. Therefore, this validation only confirms that logic programming is a sound approach to generate proofs relevant for high-school education, but we make no claim whatsoever about the performance of such a method in the ATP domain as a whole. Nevertheless, we also measured the time spent on proving each of the 33 successful problems, and these results are quite interesting. The average is 259 s, but the median is 1 s. This is explained by the fact that one problem in particular is unreasonably long to process, at 8440 s, even though it is not more complex than the other: "ABC is an equilateral triangle, and E, F and G are the middle points of AB, BC, and AC, respectively. Prove that EFG is also equilateral." The source of this complexity is a typical example of combinatorial explosion. Indeed, even though this problem is quite simple to solve, there is an incredible number of possible variations of the proof, because this problem has no less than 12 angles measuring 60 degrees, meaning 132 (12 · 11) results indicating that two angles have the same measure, each obtainable in around a hundred possible inferences. Furthermore, there are nine segments of identical lengths. By combining the two, we obtain around 600 inferences about equal segment lengths, 900 inferences about equal angle measures, and a thousand inferences related to equal triangles. By comparison, the second most time-consuming problem, at 43.7 s, has a total of around 180 inferences in its HPDIC graph. Therefore, it is not surprising that solving this problem took such a disproportionate amount of time.

If we remove this problem from the calculations, the average goes down to 3.4 s and the median to 0.95. By removing the next two problems, at 14.8 and 43.7 s, the average becomes 1.68 s and the median 0.85. In other words, this proof generator is overall reasonably fast and could be used in real time for most problems in a future application, but it is also very easy to design a problem that is absolutely in the scope of high-school education, but extremely time-consuming for the proof generator.

3.7 Limitations

Despite the encountered educational challenges, logic programming is indeed quite adapted to generate the proofs of high-school geometry problems. However, it is a lengthy task, and our solver still has several limitations whose correction would require work of varying magnitudes: possible in the current implementation; possible but requiring fundamental changes in the implementation; and intrinsically difficult or impossible.

3.7.1 Possible in the Current Implementation

As we mentioned previously, the solver currently does not handle algebra. For instance, it is impossible to represent the result "the length of segment AB is three times the length of the segment CD" without knowing either value. This would require the implementation in Prolog of basic arithmetic operations, but it would be possible, since the use of such results remains in the domain of inferences. Furthermore, we do not handle proofs by contradiction, but, similarly, these kinds of proofs still follow an inferential format. The difference is that the hypotheses represent an impossible situation, and the conclusion is "we reach a contradiction." This would require to implement all the numerous possible inferences that result in a contradiction, such as having a triangle with two parallel sides, but it would fit in our solving process. Those two limitations cause another problem concerning an axiomatic referential. Indeed, Euclid's axioms generate most of the high school's properties, but most of his work is done by proofs by contradiction which is not currently feasible by our system. On the other hand, Clairaut's adaptation of Euclid's work also generates this geometry and without using proof by contradiction. However, Clairaut uses an algebra system to bypass this difficulty, something that, again, our system cannot do. For this reason, our system requires a much more extensive referential that includes those properties that we cannot generate to be able to solve the problems. This limitation adds another layer of complexity concerning the challenge of the level of granularity discussed in Font et al. (2018).

3.7.2 Possible with Some Fundamental Change

One of the biggest limitations of our solver is the necessity to provide, beforehand, the whole geometrical situation (the super-figure), including the set of elements that could be useful in a proof. This is very difficult to solve, as the construction of new elements in automated theorem proving in geometry is a whole research domain in itself. However, we envisioned a possible solution. Since QED-Tutrix is an online tool, and that the generation of the possible proofs can usually be done in a matter of seconds, it would be possible to wait for the student to construct a new element in the interface. Then, when constructed, QED-Tutrix sends it to the prover that attempts to infer new results from this new hypothesis. If we can reach the conclusion, then the hypothesis is indeed useful, and is added, in real time, to the HPDIC graph of the problem. This solution, however, would require a complete change of the interactions between QED-Tutrix and the solver and comes with its share of issues. A direct, less profound consequence of this limitation on the possible proofs is that it is impossible to discover, in the middle of a proof, that a line actually passes through a point. This is because it is necessary to provide the complete set of points the line passes through from the beginning when encoding the problem. This limitation could be solved, but it would require a drastic change to the way we represent elements in Prolog.

3.7.3 Intrinsically Difficult

A crucial issue comes from the floating-point precision of calculations in the machine. For example, let us assume that there is a right triangle ABC, whose dimensions we know. From these dimensions, we can calculate the cosine, sinus, and tangent of the angles. This calculation is done with a certain precision. This creates a first problem: what level of imprecision from the student is acceptable? If the cosine of the angle has a value of 0.7654321, is 0.8 acceptable? Is 0.77? Is 0.76? Still, while delicate, this question concerns the tutor aspect, and not the automated proving, and we will not discuss it further. However, there is another issue concerning approximations. In the previous example, now that we know the cosine of the angle, by using the reciprocal of the property, we can calculate the value of the sides of the triangle. This calculation is also done with a certain precision, and if the resulting value is even imperceptibly different from the already known one, then it is considered a new result that will lead to the calculation of a slightly different cosine, and so on. Hopefully, we were able to solve this issue by only considering the first result. When we already know the value of something, every inference that results in another measurement of that value does not do any calculation but uses the already known one. In this example, if we already know that the hypotenuse of the triangle is of length 5, then any inference which results in the measurement of that line segment will not even calculate it, but reuse the 5 that has been calculated previously. Lastly, the precision can also create problems when using an inference such as "two angles are equal if they have the same measure." Just checking the equality of their measures is not enough, since they could be very slightly different due to rounding. In that case, we allow for a difference of 1% between the two values. We chose that precision by trial-and-error, with the argument that, in high-school geometry problems, two different values that are supposed to be measured will differ by more than 1%. This is not a mathematically ideal solution, but for our purposes, it is enough.

4 Conclusion

In this chapter, we presented both the tutor software QED-Tutrix and the ancillary automated proof generator we developed to complete its set of available problems. QED-Tutrix provides a space for the students to solve high-school geometry proof problems. In accordance to the theory of mathematical working space, it helps the student in the three geneses of mathematical work, i.e., to learn and do mathematics. Furthermore, its interface allows the student to explore the problem, enter mathematical elements as they come to him or her, and redact a mathematically sound proof. Lastly, its virtual tutor is able to provide custom-tailored help to the student in case of an impasse in the resolution, by attempting to identify the very specific step of the proof that he or she is stuck on.

We then presented the four layers that compose the core of QED-Tutrix and that allow these functionalities. The first is the HPDIC graph, storing in an oriented graph the set of all possible proofs for a problem, which is useful to instantly assess if the sentence entered by the student is relevant to the problem resolution, and serves as a support for the next layers. Then comes the MIA, allowing a representation of the student's cognitive state by storing all the sentences he or she entered in the resolution. The third layer is the EDOI, aimed at quickly identifying the proof the student is most likely working on, a crucial step for both the redaction part of the resolution and the tutor system. Lastly, the GMD is a finite state machine that determines the series of help messages that are most suited to help the student in any given situation.

The first of these, the HPDIC graph, is a crucial part of the software. Furthermore, unlike the other three layers, it has to be generated beforehand for each geometry problem added to QED-Tutrix. To be able to quickly include new problems, and consequently to generate new HPDIC graph is, necessary for turning QED-Tutrix into a software truly relevant for mathematics education. Therefore, we dedicated a whole branch of the project to creating such an automated proof generator.

For this endeavor, logic programming was a natural choice, because of the closeness between the reasoning behind logic programming and the inference process used in mathematical proofs. Creating such a generator then became a task of implementing each mathematical property, definition and possible result in Prolog. That task was not without difficulties, since human reasoning is very different from a mechanical process of inference, and, despite the apparent rigor of mathematical reasoning, the reality of classes is that many assumptions and shortcuts are taken, with good reason. Therefore, we had to adapt the mechanical process of finding new inferences to the many subtleties of human reasoning. At the end, despite these difficulties, the resulting automated proof generator managed a very reasonable coverage of typical high-school problems.

Overall, while QED-Tutrix is by itself a very interesting tool for research purpose, the adjunction of an automated proof generator offers the possibility of encoding as many problems as is needed in the future, opening the possibility of QED-Tutrix becoming a truly useful software in high-school classrooms.

References

Aleven, V., Koedinger, K., Colleen Sinclair, H., & Snyder, J. (1998). Combatting shallow learning in a tutor for geometry problem solving. In *Intelligent tutoring systems* (pp. 364–373). Springer. https://doi.org/10.1007/3-540-68716-5_42.

Aleven, V., & Koedinger, K. R. (2000). Limitations of student control: Do students know when they need help? In *Intelligent tutoring systems* (Vol. 1839, pp. 292–303). Springer. https://doi.org/10.1007/3-540-45108-0_33.

Aleven, V., & Koedinger, K. R. (2013). Knowledge component (KC) approaches to learner modeling. *Design Recommendations for Intelligent Tutoring Systems, 1,* 165–182.

Aleven, V., Mclaren, B., Roll, I., & Koedinger, K. (2006). Toward meta-cognitive tutoring: A model of help seeking with a cognitive tutor. *International Journal of Artificial Intelligence in Education, 16*(2), 101–128.

Aleven, V. A., & Koedinger, K. R. (2002). An effective metacognitive strategy: Learning by doing and explaining with a computer-based cognitive tutor. *Cognitive Science, 26*(2), 147–179. https://doi.org/10.1207/s15516709cog2602_1.

Anderson, J. R. (1996). Act: A simple theory of complex cognition. *American Psychologist, 51*(4), 355. https://doi.org/10.1037/0003-066X.51.4.355.

Anderson, J. R., & Schunn, C. (2000). Implications of the ACT-R learning theory: No magic bullets. In *Educational design and cognitive science: Advances in instructional psychology* (pp. 1–33).

Arroyo, I., Beal, C., Murray, T., Walles, R., & Woolf, B. P. (2004). Web-based intelligent multimedia tutoring for high stakes achievement tests. In *Intelligent tutoring systems* (pp. 142–169). Springer. https://doi.org/10.1007/978-3-540-30139-4_44.

Balacheff, N., Caferra, R., Cerulli, M., Gaudin, N., Maracci, M., Mariotti, M. A., Muller, J. P., Nicaud, J. F., Occello, M., Olivero, F., et al. (2003). Baghera assessment project, designing an hybrid and emergent educational society.

Baulac, Y. (1990). *Un micromonde de géométrie, Cabri-géomètre.* Ph.D. Thesis, Université Joseph-Fourier-Grenoble I.

Botana, F., Hohenwarter, M., Janičić, P., Kovács, Z., Petrović, I., Recio, T., & Weitzhofer, S. (2015). Automated theorem proving in GeoGebra: Current achievements. *Journal of Automated Reasoning, 55*(1), 39–59. https://doi.org/10.1007/s10817-015-9326-4.

Braun, G., & Narboux, J. (2017). A synthetic proof of Pappus' theorem in Tarski's geometry. *Journal of Automated Reasoning, 58*(2), 209–230. https://doi.org/10.1007/s10817-016-9374-4.

Brousseau, G., & Balacheff, N. (1998). Théorie des situations didactiques: Didactique des mathématiques 1970–1990. La pensée sauvage Grenoble.

Buchberger, B. (1988). Applications of Gröbner bases in non-linear computational geometry. In *Trends in computer algebra* (pp. 52–80). Springer. https://doi.org/10.1007/3-540-18928-9_5.

Chou, S. C. (1988). An introduction to Wu's method for mechanical theorem proving in geometry. *Journal of Automated Reasoning, 4*(3), 237–267. https://doi.org/10.1007/BF00244942.

Chou, S. C., Gao, X. S., & Zhang, J. Z. (1994). *Machine proofs in geometry: Automated production of readable proofs for geometry theorems* (Vol. 6). World Scientific. https://doi.org/10.1142/9789812798152_0002.

Chou, S. C., Gao, X. S., & Zhang, J. Z. (1996). Automated generation of readable proofs with geometric invariants, I. Multiple and shortest proof generation. *Journal of Automated Reasoning, 17,* 325–347. https://doi.org/10.1007/BF00283134.

Corbeil, J. P., Gagnon, M., & Richard, P. R. (2020). Probabilistic approaches to detect blocking states in intelligent tutoring system. In *International Conference on Intelligent Tutoring Systems* (pp. 79–88). Springer.

Coutat, S., Laborde, C., & Richard, P. R. (2016). L'apprentissage instrumenté de propriétés en géométrie: Propédeutique à l'acquisition d'une compétence de démonstration. *Educational Studies in Mathematics, 93*(2), 195–221.

Coutat, S., & Richard, P. R. (2011). Les figures dynamiques dans un espace de travail mathématique pour l'apprentissage des propriétés géométriques. *Annales de Didactique et de Sciences Cognitives, 16,* 97–126.

Cyr, S. (2021). *Étude des référentiels de géométrie utilisés en classe de mathématiques au secondaire*. Master's thesis, Université de Montréal.

El-Khoury, S., Richard, P. R., Aïmeur, E., & Fortuny, J. M. (2005). Development of an intelligent tutorial system to enhance students' mathematical competence in problem solving. In *E-Learn: World Conference on E-Learning in Corporate, Government, Healthcare, and Higher Education* (pp. 2042–2049). Association for the Advancement of Computing in Education (AACE).

Falmagne, J. C., Cosyn, E., Doignon, J. P., & Thiéry, N. (2006). The assessment of knowledge, in theory and in practice. In *Formal concept analysis* (pp. 61–79). Springer. https://doi.org/10.1007/11671404_4.

Farid, O. (2020). *Extraction des connaissances en géométrie plane à partir d'énoncés de problèmes*. Master's thesis, Polytechnique Montréal.

Font, L., Cyr, S., Richard, P. R., & Gagnon, M. (2020). Automating the generation of high school geometry proofs using prolog in an educational context. arXiv:2002.12551.

Font, L., Richard, P. R., & Gagnon, M. (2018). Improving QED-Tutrix by automating the generation of proofs. arXiv:1803.01468.

Géométrix. (2021). http://geometrix.free.fr/site/.

Hohenwarter, M. (2013). GeoGebra 4.4–from desktops to tablets. *Indagatio Didactica, 5*(1).

Janičić, P., Narboux, J., & Quaresma, P. (2012). The area method: A recapitulation. *Journal of Automated Reasoning, 48*(4), 489–532. https://doi.org/10.1007/s10817-010-9209-7.

Jean-Daubias, S. (2000). *Pépite: un système d'assistance au diagnostic de compétences*. Ph.D. thesis, Université du Maine.

Kapur, D. (1986). Using Gröbner bases to reason about geometry problems. *Journal of Symbolic Computation, 2*(4), 399–408. https://doi.org/10.1016/S0747-7171(86)80007-4.

Koedinger, K. (1991). On the design of novel notations and actions to facilitate thinking and learning. In *Proceedings of the International Conference on the Learning Sciences* (pp. 266–273).

Koedinger, K. R., & Anderson, J. R. (1990). Abstract planning and perceptual chunks: Elements of expertise in geometry. *Cognitive Science, 14*(4), 511–550. https://doi.org/10.1207/s15516709cog1404_2.

Koedinger, K. R., & Anderson, J. R. (1993). Effective use of intelligent software in high school math classrooms.

Kordaki, M., & Mastrogiannis, A. (2006). The potential of multiple-solution tasks in e-learning environments: Exploiting the tools of Cabri geometry II. In *E-Learn: World Conference on E-Learning in Corporate, Government, Healthcare, and Higher Education* (pp. 97–104). Association for the Advancement of Computing in Education (AACE).

Kuzniak, A. (2011). L'espace de Travail Mathématique et ses genèses. *Annales de Didactique et de Sciences Cognitives, 16,* 9–24. https://halshs.archives-ouvertes.fr/halshs-01060043.

Kuzniak, A., Montoya-Delgadillo, E., & Richard, P. R. (2021). Mathematical work in educational context. *Mathematics Education in the Digital Era*. https://doi.org/10.1007/978-3-030-90850-8.

Kuzniak, A., & Richard, P. R. (2014). Espacios de trabajo matemático. puntos de vista y perspectivas. *Revista Latinoamericana de Investigación en Matemática Educativa, 17*(4). https://doi.org/10.12802/relime.13.1741a.

Leduc, N. (2016). *QED-Tutrix: système tutoriel intelligent pour l'accompagnement d'élèves en situation de résolution de problèmes de démonstration en géométrie plane*. Ph.D. thesis, École polytechnique de Montréal.

Luengo, V. (1997). *Cabri-euclide: Un micromonde de preuve intégrant la réfutation*. These de doctorat, INPG, France.

Luengo, V. (2005). Some didactical and epistemological considerations in the design of educational software: The Cabri-euclide example. *International Journal of Computers for Mathematical Learning, 10*(1), 1–29. https://doi.org/10.1007/s10758-005-4580-x.

Luengo, V., & Balacheff, N. (1998). Contraintes informatiques et environnements d'apprentissage de la démonstration en mathématiques. *Sciences et Techniques Educatives, 5,* 15–45.
Mathway | Math Problem Solver. (2021). https://www.mathway.com/Algebra.
Matsuda, N., & Vanlehn, K. (2004). Gramy: A geometry theorem prover capable of construction. *Journal of Automated Reasoning, 32*(1), 3–33. https://doi.org/10.1023/B:JARS.0000021960.39761.b7.
Matsuda, N., & VanLehn, K. (2005). Advanced geometry tutor: An intelligent tutor that teaches proof-writing with construction. In *AIED* (Vol. 125, pp. 443–450).
Melis, E., Goguadze, G., Libbrecht, P., & Ullrich, C. (2009). Activemath—A learning platform with semantic web features. In *The future of learning* (p. 159).
Narboux, J. (2006). Mechanical theorem proving in Tarski's geometry. In *International Workshop on Automated Deduction in Geometry* (pp. 139–156). Springer. https://doi.org/10.1007/978-3-540-77356-6_9.
Niss, M., & Højgaard, T. (2019). Mathematical competencies revisited. *Educational Studies in Mathematics, 102*(1), 9–28.
Papert, S. A. (2020). *Mindstorms: Children, computers, and powerful ideas.* Basic Books.
Py, D. (1994). Reconnaissance de plan pour la modélisation de l'élève. le projet Mentoniezh. *Recherches en Didactique des MathéMatiques, 14*(1/2), 113–138.
Py, D. (1996). Aide à la démonstration en géométrie: le projet Mentoniezh. *Sciences et Techniques Éducatives, 3*(2), 227–256.
Py, D. (2001). Environnements interactifs d'apprentissage et démonstration en géométrie.
Rabardel, P. (1995). *Les hommes et les technologies; Approche cognitive des instruments contemporains.* Armand Colin.
Richard, P. R., & Fortuny, J. M. (2007). Amélioration des compétences argumentatives à l'aide d'un système tutoriel en classe de mathématique au secondaire. In *Annales de didactique et de sciences cognitives* (Vol. 12, pp. 83–116).
Richard, P. R., Gagnon, M., & Fortuny, J. M. (2018). Connectedness of problems and impasse resolution in the solving process in geometry: A major educational challenge. In *International perspectives on the teaching and learning of geometry in secondary schools* (pp. 357–375). Springer.
Richard, P. R., Venant, F., & Gagnon, M. (2019). *Issues and challenges about instrumental proof.* Suisse: Springer.
Roll, I., Baker, R. S. D., Aleven, V., & Koedinger, K. R. (2014). On the benefits of seeking (and avoiding) help in online problem-solving environments. *Journal of the Learning Sciences, 23*(4), 537–560. https://doi.org/10.1016/S0360-1315(99)00030-5.
Tanswell, F. S., & Rittberg, C. J. (2020). Epistemic injustice in mathematics education. *ZDM, 52*(6), 1199–1210.
Tessier-Baillargeon, M., Leduc, N., Richard, P., & Gagnon, M. (2017). Étude comparative de systèmes tutoriels pour l'exercice de la démonstration en géométrie. *Annales de Didactique et de Sciences Cognitives, 22,* 91–117.
Trouche, L. (2003). Construction et conduite des instruments dans les apprentissages mathématiques: nécessité des orchestrations.
Wang, K., & Su, Z. (2015). Automated geometry theorem proving for human-readable proofs. In *Twenty-Fourth International Joint Conference on Artificial Intelligence.*
Webber, C., Bergia, L., Pesty, S., & Balacheff, N. (2001). The Baghera project: A multi-agent architecture for human learning. In *Workshop-multi-agent architectures for distributed learning environments.* In *Proceedings International Conference on AI and Education.* San Antonio, Texas.
Weber, G., & Brusilovsky, P. (2001). ELM-ART: An adaptive versatile system for web-based instruction. *International Journal of Artificial Intelligence in Education (IJAIED), 12,* 351–384.
Wu, H. (1979). An elementary method in the study of nonnegative curvature. *Acta Mathematica, 142*(1), 57–78. https://doi.org/10.1007/BF02395057.

Zhang, J., Yang, L., & Deng, M. (1990). The parallel numerical method of mechanical theorem proving. *Theoretical Computer Science, 74*(3), 253–271. https://doi.org/10.1016/0304-3975(90)90077-U.

A Decision Making Tool for Mathematics Curricula Formal Verification

Eugenio Roanes-Lozano and Angélica Martínez-Zarzuelo

1 Introduction

An appropriate organization of the educational contents taught and learned in today's classrooms is fundamental. Educational laws usually establish a possible organization of educational contents, making it very clear that it should be considered a brief general guide. In Martínez-Zarzuelo et al. (2017) we take advantage of the degree of freedom that these laws grant and we propose a grouping and organization of contents considering a criterion based on the meaningful learning theory (Ausubel, 1963; Ausubel & Barberán, 2002; Ausubel et al., 1976; Moreira, 2000).

The authors have been working on mathematics curricula organization for a long time. The basic idea of these investigations is to consider as a starting point a set of mathematical educational contents and to establish two binary relations among the contents: the relations "to be a prerequisite" and "to be an immediate prerequisite". More precisely, if Content_1 and Content_2 are two educational contents, we have used the following definitions (Martínez Zarzuelo, 2015; Martínez-Zarzuelo et al., 2016):

- Content_1 is a prerequisite of Content_2, denoted Content_1 ▶ Content_2, if and only if understanding Content_1 is required to understand Content_2.
- Content_1 is an immediate prerequisite of Content_2, denoted Content_1 ▷ Content_2, if and only if

---

E. Roanes-Lozano (✉) · A. Martínez-Zarzuelo
Depto. de Didáctica de Ciencias Experimentales, Sociales y Matemáticas, Facultad de Educación, Universidad Complutense de Madrid, Madrid, Spain
e-mail: eroanes@ucm.es

E. Roanes-Lozano
Instituto de Matemática Interdisciplinar, Universidad Complutense de Madrid, Madrid, Spain

© The Author(s), under exclusive license to Springer Nature Switzerland AG 2022
P. R. Richard et al. (eds.), *Mathematics Education in the Age of Artificial Intelligence*, Mathematics Education in the Digital Era 17,
https://doi.org/10.1007/978-3-030-86909-0_4

(i) understanding Content_1 is required to understand Content_2 (that is, Content_1 ▶ Content_2),
and
(ii) there is no Content_3 such that Content_1 ▶ Content_3 ▶ Content_2.

(from the formal point of view, it is smarter to define ▷ from ▶).

Therefore, the key idea is to address the different types of curricula as partially ordered sets (or directed graphs). This way it is possible to use a computer to perform calculations about the dependence among contents.

2 A Theoretical Proposal

In Roanes-Lozano et al. (2020b), a theoretical approach to the verification of a certain "official curriculum development" (in the sense that it matches a given "preprocessed official curriculum" set as reference) is developed.

Let us try to summarize that article. Three curricula are distinguished in it:

- There is an "official curriculum" (O), a very general small set of contents imposed by the education authorities.
- We suppose that a team of experts has detailed O and there exists a "preprocessed official curriculum" (C), that details O, and introduces the "immediate prerequisite" relation (▷) among contents. The transitive closure of ▷, that could be denoted "prerequisite" relation, is represented by a ▶. (C is not normally available, but must be prepared for our purposes from O.) Observe that, when facing the real situation, that is, from a constructive point of view, the "prerequisite" relation, ▶, is obtained as the transitive closure of the "immediate prerequisite" relation, ▷, defined by the team of experts that have prepared the "preprocessed official curriculum", C.
- Someone else proposes an "official curriculum development" (for instance, a textbook or a project-based learning proposal), with another "immediate prerequisite" relation (G,≻).

The following comprehensive verification process is proposed:

- Step 1a: contents soundness: all the contents in G appear in C.
- Step 1b: contents completeness: all the contents in O can be found in G.
- Step 2a: relation soundness: all "immediate prerequisite" relation ordered pairs of G appear in the transitive closure of the "immediate prerequisite" relation proposed by C (that is, ≻ ⊆ ▶).
- Step 2b: relation completeness: all "immediate prerequisite" relation ordered pairs of G appear as "immediate prerequisite" in C (≻ ⊆ ▷).
- Step 3: absence of cycles in G.

Let us clarify steps 2a and 2b. Step 2b is clearly a stronger condition than 2a, but their meaning in this context is different.

In Step 2a it has to be checked that all ordered pairs in $(G, \succ)$ are either explicitly included by the experts in $(C, \rhd)$ or can be inferred from the ordered pairs considered in $(C, \rhd)$. As $\blacktriangleright$ is the transitive closure of $\rhd$, this is equivalent to checking whether $\succ \subseteq \blacktriangleright$ or not. If it does not hold it can be an error of the experts that developed $(C, \rhd)$ or something wrong in $(G, \succ)$.

Meanwhile, if Step 2a has been passed, Step 2b checks if any intermediate content in C is bypassed in G (as G is an detailed extension of the brief O, this does not mean that G is wrong, but the absence of this ordered pair deserves a careful analysis).

3 Design and Implementation of the Theoretical Proposal

We consider that this theoretical proposal could be useful and have various applications in the educational context. If an "official curriculum" is provided, for instance, by the educational authorities, an "official curriculum development" could be proposed by authors of educational resources such as textbooks. Thus, these textbooks that are adjusted to the "official curriculum" could be evaluated with our proposal in a simple computational way once the mathematical contents and the "prerequisite" relation among them have been set. This would avoid manual checking and, above all, would provide a guarantee of the completeness and soundness of the educational resources based on what is approved by the academic authorities. Other examples of current educational interest are "official curriculum developments" corresponding to project-based learning proposals. With our idea, it could be automatically checked in a simple way if, for example, a project-based learning proposal complies with the educational contents planned for a certain educational period and if it is complete and sound.

We shall exemplify the approaches hereinafter with a real case taken from Martínez Zarzuelo (2015). For this, we will consider 112 different algebraic educational contents of the Spanish Compulsory Secondary Education (corresponding to Grades 7–10 in the K-12 system) and 261 ordered pairs of the "immediate prerequisite" relation. We focus on mathematical contents because the hierarchical structure of the mathematical discipline allows its concepts to be organized coherently according to a prerequisite relation, but the same ideas can also be applied to educational contents from other disciplines.

3.1 First Approach (Rule Based Expert System)

The inspiration to this work is the process used for Rule Based Expert Systems (RBES) knowledge extraction and verification. There are several different computational methods for this goal. We have considered one based on moving to an algebraic model of Boolean logic (Alonso & Briales, 1995; Chazarain et al., 1991; Hsiang, 1985; Kapur & Narendran, 1985; Roanes-Lozano et al., 1998):

$$A = (Z/2Z)[x_1, \ldots, x_n]/ < x_1^2 - x_1, \ldots, x_n^2 - x_n >$$

where $x_1,\ldots,x_n$ are polynomial variables, image of the propositional variables in the isomorphism between the Boolean algebra of logic and the polynomial Boolean algebra (depending on the operations considered, either a Boolean ring isomorphism or a Boolean algebra isomorphism can be considered).

The main result states that (Roanes-Lozano et al., 1998, 2010):

- The logical proposition Y is a tautological consequence of $\{Y_1,\ldots,Y_n\}$ if and only if $y + 1 \in < y_1 + 1,\ldots,y_n + 1 >$
- $\{Y_1,\ldots,Y_n\}$ is consistent if and only if the ideal $< y_1 + 1,\ldots,y_n + 1 >$ is not the whole ring (that is, $< y_1 + 1,\ldots,y_n + 1 > \neq < 1 >$)

(where $< y_1 + 1,\ldots,y_n + 1 >$ denotes the polynomial ideal generated by $y_1 + 1,\ldots,y_n + 1$ and the lowercase polynomial variables denote the image of the corresponding uppercase propositional variables in the isomorphism mentioned above). The ideal membership and the non-degeneracy of the ideal can be computed using "normal forms" and "Gröbner bases", respectively.

Note that if x is the polynomial translation of proposition X, $x + 1$ is the polynomial translation of the negation of Y. The reason for including the negations in the results above is that propositions are normally stated as "true", what corresponds in the algebraic model to stating that the value of their algebraic translation is 1, meanwhile what is convenient in algebra is to decide whether an expression vanishes, that is, it is equal to 0, or not.

If we denote $I = < x_1^2 - x_1,\ldots,x_n^2 - x_n >$, a RBES where the facts in a certain set are stated as true is modeled by:

$$A/(J + K) = (Z/2Z)[x_1, \ldots, x_n]/(I + J + K)$$

where J is the ideal generated by the negation of the rules and K is the ideal generated by the negation of the facts stated as true (see Roanes-Lozano et al. (2010) for details). A recent related paper is Alonso-Jiménez et al. (2018).

We could use the *Maple*[1] implementation used in the recent Roanes-Lozano et al. (2020a) for dealing with knowledge extraction and formal verification of RBES whose underlying logic is Boolean logic:

---

[1] *Maple* is a trademark of *Waterloo Maple Inc.*, Waterloo, ON, Canada.

A Decision Making Tool for Mathematics ... 81

```
> with(Groebner):
> with(Ore_algebra):
> SV:=x||(1..112);
> fu:=var->var^2-var:
> iI:=map(fu,[SV]);
> A:=poly_algebra(SV,characteristic=2):
> Orde:=MonomialOrder(A,'plex'(SV)):
> fu:=var->var^2-var:
> iI:=map(fu,[SV]):
> NEG:=(m::algebraic)->NormalForm(1+`m`,iI,Orde):
> `&AND`:=(m::algebraic,n::algebraic)->
>           NormalForm(expand(m*n),iI,Orde):
> `&OR`:=(m::algebraic,n::algebraic)->
>           NormalForm(expand(m+n+m*n),iI,Orde):
> `&IMP`:=(m::algebraic,n::algebraic)->
>           NormalForm(expand(1+m+m*n),iI,Orde):
> `&XOR`:=(m::algebraic,n::algebraic)->
>           (m &OR n) &AND NEG(m &AND n):
```

where the *Maple* functions NEG (prefix) and &AND, &OR, &IMP and &XOR (infix) are the algebraic translation of the logical connectives "negation", "conjunction", "disjunction", "implies" and "exclusive disjunction". Note that iI is the ideal of the squares of variables minus variables introduced in the algebraic model of RBES to force idempotency (denoted *I* above).

If we identify the "immediate prerequisite" relation with the logical implication, we can introduce the former in the following form (not all 261 rules are listed and an ellipsis is used for the sake of brevity):

```
> R1:=incognita &IMP exp_algebraica:
> R2:=incognita &IMP parte_literal_exp_algebraica:
  ...
> R261:=sistema_ecuaciones_grado_1_equivalente &IMP
          metodo_Gauss:
```

(the contents are in Spanish because they come from the study of the Spanish case Martínez Zarzuelo (2015), but most of them are very similar). They are stored in file Edges_Algebra_GB.txt.

We can use nicknames to shorten the names of the contents:

```
> incognita:=x1:
> exp_algebraica:=x2:
  ...
> ecuacion_explicita_recta:=x112:
```

(the nicknames are stored in file Vertices_Nicknames.txt).

So, after reading these two files from the *Maple* session:

```
> read(`c:/.../Vertices_Nicknames.txt`);
> read(`c:/.../Edges_Algebra_GB.txt`);
```

it is possible to define the ideal of rules, iJ:

```
> iJ:=[NEG(R1),NEG(R2),NEG(R3),...,NEG(R261)]:
```

We can now check if, for instance, content "exp_algebraica" (x_2) follows from content "suma_monomios" (x_6). iK denotes the ideal of what is stated as true:

```
> iK:=[NEG(x6)]:
> B:=Basis([op(iJ),op(iI),op(iK)],Orde):
> NormalForm(NEG(x2),B,Orde);
                    x2 + 1
```

The answer is not 0, so it does not follow. Once the Gröbner basis B is computed (what takes about 30 s on a standard laptop), each question takes very little (hundredths of a second).

It is now very easy, for instance, to exhaustively check all what follows from "suma_monomios" (x_6):

```
> W:=[]:
> for i from 1 to 112 do
            if NormalForm(NEG(x||i),B,Orde)=0
                then W:=[op(W),x||i]
            fi;
    od;
> W;
```

(the answer consists of 60 variables and is computed in slightly more than 3 s).

Reciprocally, "suma_monomios" (x_6) does follow from "exp_algebraica" (x_2):

```
> iK:=[NEG(x2)]:
> B:=Basis([op(iJ),op(iI),op(iK)], Orde):
> NormalForm(NEG(x6),B,Orde);
                    0
```

(and, as done above for x_6, it can be easily checked that 109 contents follow from x_2).

Nevertheless, the RBES approach, although is the inspiration for the present article and does work, does not take advantage of its potential, as it is designed to deal with RBES complex rules (that are logic propositions involving negations, disjunctions and conjunctions, that do not arise in this particular case, where all rules are of the form $Y_i \rightarrow Y_j$). Let us try another approach.

3.2 Second Approach (Graph Theory)

The idea of this second approach is based on using graph theory to model the educational contents and ordered pairs of the binary relation using a graph structure. More precisely, using a directed graph (also called digraph) structure.

Maple offers an efficient package for dealing with graphs named *GraphTheory*. We can approach the same questions of the previous subsection from this other approach. Now we have to begin by loading the package and the data of the graph:

```
> restart;
> with(GraphTheory):
> read(`c:/.../Vertices_Nicknames.txt`);
> read(`c:/.../Edges_Algebra_Networks.txt`);
```

As in the previous subsection, the first file introduces the nicknames of the vertices and the second one introduces the directed edges of the digraph, now as a set of ordered pairs:

```
> LC:={[incognita,exp_algebraica],
       [incognita,parte_literal_exp_algebraica],
       [exp_algebraica,termino_exp_algebraica],
       ...
       [sistema_ecuaciones_grado_1_equivalente,
        metodo_Gauss]}:
```

Let us suppose that this is the "preprocessed official curriculum" (*C*) that is set as reference. It is straightforward to define the corresponding graph in *Maple* and to plot it:

```
> C:=Digraph([x||(1..112)],LC);
  C := Graph 2: a directed unweighted graph with
              112 vertices and 261 arc(s)
> DrawGraph(C);
```

The output of this last line of code can be found in Fig. 1 (there are too many vertices to display their names). Surprisingly, two contents are clearly isolated.

We can easily look for them constructing the set $\{x_1, x_2,...,x_{112}\}$ and using the set difference operator minus and the command indets (that returns the variables in an expression):

```
> {x||(1..112)} minus indets(LC);
              {x69, x70}
```

Variables x_{69} and x_{70} are "funcion_valor_absoluto" (absolute value function) and "función trigonométrica" (trigonometric function), respectively. The experts that developed the "preprocessed official curriculum" should be contacted to confirm that their isolation is correct.

There is a very convenient command in the *GraphTheory* package, IsReachable, that checks whether a vertex is reachable from another one or

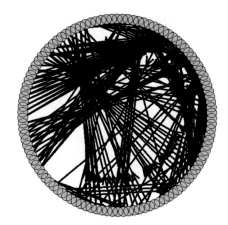

Fig. 1 Plot of the digraph "preprocessed official curriculum"

not. For instance, it can be used in this case to check whether x_2 is reachable from x_6 and vice versa:

```
> IsReachable(C,x6,x2);
                            false
> IsReachable(C,x2,x6);
                            true
```

These results are obtained in 0 s. This time is also obtained for computing which of the 112 contents follow from x_2:

```
> W:=[]:
> for i from 1 to 112 do
      if IsReachable(C,x2,x||i)
            then W:=[op(W),x||i]
      fi;
    od;
> W;
```

and W has, as computed in Sect. 3.2, 109 contents.

Therefore it is clear that this approach is much better in this case.

3.3 Case Study

Let us suppose that the list of vertices and the list of oriented edges (LC) used in Sect. 3.2 are those corresponding to the "preprocessed official curriculum" (the reference). The list of vertices is supposed to be known by the authors of educational resources, but the list of oriented edges (LC) is not.

Let us suppose that a textbook has been analyzed and the corresponding "official curriculum development" has been obtained and written in the same format as LC, and is denoted LG. And let us imagine that the author has reached almost the same graph as the set of experts but has forgotten to include the content "division_fraccion_algebraica" (x_{41}), and the corresponding ordered pairs. Let us proceed to see what happens when trying to verify this "official curriculum development".

Let us proceed as in Sect. 3.2, by also loading the file containing the set VO of vertices in the "official curriculum" (*O*) of Spanish Secondary Compulsory Education (ESO) and the file containing the set LG of edges corresponding to the "official curriculum development" proposal:

```
> restart;
> with(GraphTheory):
> read(`c:/.../Vertices_Nicknames.txt`);
> read(`c:/.../Edges_Algebra_Networks.txt`);
> read(`c:/.../Vertices_Algebra_RealDecretoESO.txt`);
> read(`c:/.../Edges_Algebra_Networks_Case.txt`);
```

We have to define two graphs, corresponding to the "preprocessed official curriculum" (*C*) and the "official curriculum development" (*G*):

```
> C:=Digraph([x||(1..112)],LC);
> G:=Digraph([x||(1..112)],LG);
```

```
> DrawGraph(G);
```
and we can easily plot the latter (Fig. 2):

We can now carry out the process proposed:

- STEP 1a: Check whether all the contents in the proposed "official curriculum development" (*G*) appear in the "preprocessed official curriculum" (*C*). The set difference of the vertices of *G* and *C* should be the empty set:

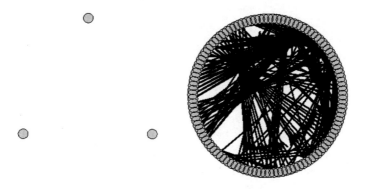

Fig. 2 Plot of the digraph "official curriculum development" proposal

```
> indets(LG) minus indets(LC);
                  {}
```
- STEP 1b: Check whether all the contents in the "official curriculum" (*O*) appear in the proposed "official curriculum development" (the set difference should be the empty set):

```
> VO minus indets(LG);
                  {x41}
```

In this case, the "official curriculum development" does not comply with the "official curriculum", as there is a content missing.

- STEP 2a: Do all the ordered pairs of the "immediate prerequisite" relation considered in the proposed "official curriculum development" (*G*) appear in the transitive closure of the "immediate prerequisite" relation of the "preprocessed official curriculum" (*C*)? Computing the transitive closure of the whole relation is computationally expensive, but to check whether certain directed edges are in the transitive closure of the relation is not (in fact IsReachable really checks that). The answer to the next lines of code (that stores in set H the edges that are not in the transitive closure of *C*) should be the empty set):

```
> H:={}:
> for i in LG do
        if not IsReachable(C,op(i))
               then H:={op(H),i}
        fi;
     od;
> H;
                  {}
```

- STEP 2b: Do all the ordered pairs of the "immediate prerequisite" relation considered in the proposed "official curriculum development" (*G*) appear as ordered pairs of the "immediate prerequisite" relation of the "preprocessed official curriculum" (*C*)? (the set difference should be the empty set):

```
> LG minus LC;
                  {}
```

- STEP 3: Are there cycles in the graph? There is a command in the *GraphTheory* that checks whether a graph is acyclic or not (so we expect a *true* answer):

```
> IsAcyclic(G);
                  true
```

We could even go further:

- STEP 1b (further): We could also check if all the contents in the "preprocessed official curriculum" (*C*) appear in the "official curriculum development" (*G*):

```
> indets(LC) minus indets(LG);
                  {x41}
```

In this case there are no content that the experts have detailed (added) to O (in C) that the author has not considered (in G). The author could be informed of this fact (if it arose), and the author could perhaps like to include it (although it would not be mandatory, as it complies with "official curriculum" (O)).
- STEP 2b (further): We could also check if all the ordered pairs in the "preprocessed official curriculum" (G) appear in the "official curriculum development" (G). In this case, at least all directed edges including x_{41} must arise:

```
> LC minus LG;
  {[x21, x41], [x24, x41], [x25, x41], [x41, x98]}
```

4 Conclusions

In RBES verification, what is checked is the logical coherence. For instance, in a medical RBES, a patient cannot be advised that he/she should take a drug that has a certain active ingredient, and, at the same time, that he/she should not take a drug that has that same active ingredient. What cannot be detected are errors or errata within the medical knowledge implemented. For instance, it cannot be detected in the verification process a (very serious) error such as "if the patient is dehydrated then do not give him any liquid". That is, we start from a supposedly correct knowledge and verify its logical coherence.

Similarly, in this case we suppose that a brief "official curriculum" (O) exists and that a team of experts has extended and detailed O in a "preprocessed official curriculum" (C), that will act as reference to check ulterior "official curriculum developments" (G). C is supposed to be correct. What can be checked is the correctness of G compared to C.

Developing the "preprocessed official curriculum" (C) from an "official curriculum" (O) is neither a short nor a trivial task. As said above, we had available the "preprocessed official curriculum" corresponding to the mathematics subjects of the Spanish Compulsory Secondary Education (Grades 7–10 in the K-12 system) developed by the second author for her Ph.D. Thesis (Martínez Zarzuelo, 2015). It is mainly based on different available "official curriculum developments" (textbooks) corresponding to that "official curriculum" and previous ones and required a long hard work.

The theoretical aspects of this proposal were discussed in Roanes-Lozano et al. (2020b). In this new article it has been proved that it is possible to implement the proposal. Obviously other implementations are possible (here, one closer to the inspiration of this work and other, much more efficient, have been detailed).

As in any computational field, an application is not design, developed and implemented if it is planned to be used just once. A proposal like this makes sense, for instance, in a country where textbooks (and perhaps project-based learning proposals) have to be checked and authorized. We believe that in such scenario our proposal is very convenient. It proposes a step forward, analogous to the one taken from manual RBES verification by a panel of experts, to computational verification of RBES.

Acknowledgements This work was partially supported by the Government of Spain under Grant PGC2018-096509-B-I00.

References

Alonso, J. A., & Briales, E. (1995). Lógicas Polivalentes y Bases de Gröbner. In C. Martin (Ed.), *Actas del V Congreso de Lenguajes Naturales y Lenguajes Formales* (pp. 307–315). Universidad de Sevilla.

Alonso-Jiménez, J. A., Aranda-Corral, G. A., Borrego-Díaz, J., Fernández-Lebrón, M. M., & Hidalgo-Doblado, M. J. (2018). A logic-algebraic tool for reasoning with Knowledge-Based Systems. *Journal of Logical and Algebraic Methods in Programming, 101*, 88–109.

Ausubel, D. P. (1963). *The psychology of meaningful verbal learning*. Grune & Stratton Inc.

Ausubel, D. P., & Barberán, G. S. (2002). *Adquisición y retención del conocimiento: Una perspectiva cognitive [Acquisition and retention of knowledge: A cognitive perspective]*. Paidós.

Ausubel, D. P., Novak, J. D., & Hanesian, H. (1976). *Psicología educativa: un punto de vista cognoscitivo* [Educational psychology: a cognitive point of view]. México: Trillas.

Chazarain, J., Riscos, A., Alonso, J. A., & Briales, E. (1991). Multi-valued logic and Gröbner Bases with applications to modal logic. *Journal of Symbolic Computation, 11*(3), 181–194.

Hsiang, J. (1985). Refutational theorem proving using term-rewriting systems. *Artificial Intelligence, 25*(3), 255–300.

Kapur, D., & Narendran, P. (1985). An Equational Approach to Theorem Proving in First-Order Predicate Calculus. In Proceedings of the *9th International Joint Conference on Artificial Intelligence (IJCAI-85)*, Vol 2, pp. 1146–1153. Los Angeles, CA, USA.

Martínez Zarzuelo, A. (2015). Selección, organización y secuenciación del conocimiento matemático mediante teoría de grafos (Ph.D: Thesis). Universidad Complutense de Madrid, Madrid.

Martínez-Zarzuelo, A., Roanes-Lozano, E., & Fernández-Díaz, M. J. (2017). Grouping Mathematical Contents Using Network Analysis Software. An Application to the Spanish Secondary Education Case. *International Journal for Technology in Mathematics Education, 24*(4).

Martínez-Zarzuelo, A., Roanes-Lozano, E., & Fernández-Díaz, M. J. (2016). A Computer Approach to Mathematics Curriculum Developments Debugging. *Eurasia Journal of Mathematics, Science & Technology Education., 12*(12), 2961–2974.

Moreira, M. A. (2000). *Aprendizaje significativo: Teoría y práctica [Meaningful learning: Theory and practice]*. Visor.

Roanes-Lozano, E., Casella, E. A., Sánchez, F., & Hernando, A. (2020a). Diagnosis in Tennis Serving Technique. *Algorithms, 13*(5), 106.

Roanes-Lozano, E., Martínez-Zarzuelo, A., & Fernández-Díaz, M. J. (2020b). An Application of Knowledge Engineering to Mathematics Curricula Organization and Formal Verification. *Mathematical Problems in Engineering, 3485846*, 1–12. https://doi.org/10.1155/2020/3485846

Roanes-Lozano, E., Laita, L. M., Hernando, A., & Roanes-Macías, E. (2010). An algebraic approach to rule Based Expert Systems. *RACSAM-Revista de la Real Academia de Ciencias Exactas, Fisicas y Naturales. Serie A. Matematicas, 104*(1), 19–40.

Roanes-Lozano, E., Laita, L. M., & Roanes-Macías, E. (1998). A polynomial model for multi-valued logics with a touch of algebraic geometry and computer algebra. *Mathematics and Computers in Simulation, 45*(1–2), 83–99.

A Classification of Artificial Intelligence Systems for Mathematics Education

Steven Van Vaerenbergh and Adrián Pérez-Suay

1 Introduction

Artificial Intelligence (AI) has a long history, starting from observations by the early philosophers that a reasoning mind works in some ways like a machine. For AI to become a formal science, however, several advances in the mathematical formalization of fields such as logic, computation, and probability theory were required (Russell & Norvig, 2009). Interestingly, the relationship between mathematics and AI is not unilateral, as AI, in turn, serves the field of mathematics in several ways. In particular, AI powers many computer-based tools that are used to enhance the learning and teaching of mathematics, several of which are the topic of discussion of this chapter.

The close relationship between AI and Mathematics Education (ME) dates back at least to the 1970s, and it has been discussed thoroughly in the scientific literature (Schoenfeld, 1985; Wenger, 1987; Balacheff, 1994). One could list several parallels between both fields, for instance, that they are both concerned with constructing sound *reasoning* based on the use of logic. Indeed, in ME, developing mathematical reasoning skills is an important educational goal, while many AI systems are designed to perform reasoning tasks in an automated manner. Also, modern AI techniques involve the concepts of *teaching* and *learning*, as some systems are required to learn

---

S. Van Vaerenbergh (✉)
Departamento de Matemáticas, Estadística y Computación, Universidad de Cantabria,
Avda. de Los Castros 48, 39005 Santander, Spain
e-mail: steven.vanvaerenbergh@unican.es

A. Pérez-Suay
Departament de Didàctica de la Matemàtica, Universitat de València,
Avda. Tarongers, 4, 46022 València, Spain
e-mail: adrian.perez@uv.es

© The Author(s), under exclusive license to Springer Nature Switzerland AG 2022
P. R. Richard et al. (eds.), *Mathematics Education in the Age of Artificial Intelligence*,
Mathematics Education in the Digital Era 17,
https://doi.org/10.1007/978-3-030-86909-0_5

models and concepts, either in an autonomous manner or supervised through some form of instruction.[1] Nevertheless, while these parallels exist, humans and machines clearly carry out these tasks in completely different ways. After all, as noted by Schoenfeld, "AI's perspective is severely distorted by the engineering perspective, and extrapolations to human performance can be dangerous" (Schoenfeld, 1985, p. 184).

Before continuing, we will take a closer look at what exactly AI is, including its subfield, machine learning.

1.1 Artificial Intelligence and Machine Learning

The literature contains many different definitions of AI, though they are all much related. Generally speaking, AI aims to create machines capable of solving problems that appear hard to the eyes of a human observer. Such problems may be related solely to thought processes and reasoning capabilities, or they may refer to exhibiting a certain behavior that strikes as intelligent (Russell & Norvig, 2009).

Historically, several simple AI programs have been designed using a set of predefined rules, which can often be represented internally as a decision tree model. For instance, a program for natural language understanding may look if certain words are present in a phrase, and combine the results through some set of fixed rules to determine the sentiment of a text. And in early computer vision systems, results were obtained by calculating hand-engineered *features* of pixels and their neighborhoods, after which these features were compared to the surrounding features using predefined rules. However, as soon as one tries to build a system with an advanced comprehension of natural language or photographic imagery, a very complex set of such internal rules is required, exceeding largely what can be manually designed by a human expert. To deal with this complexity, an *automated* design process for the set of internal rules is required, better known as *machine learning*.

Formally, machine learning (ML) is a subfield of AI that follows a paradigm known as *learning from examples*, in which a system is given practical examples of a concept or behavior to be learned, after which it develops an internal representation that allows its own output to be consistent with the set of given examples (Wenger, 1987). The concept of ML is perhaps best summarized by Tom Mitchell, who wrote it is "the field that is concerned with the question of how to construct computer programs that automatically improve with experience" (Mitchell, 1997). This definition highlights the three properties that any ML system should hold: (1) its learning is *automated*, as in a computer program that does not require human intervention; (2) the performance can be measured, which allows the system to measure *improvement*; and (3) learning is based on receiving examples (the *experience*). In summary, as more examples are processed, a well-designed ML system guarantees that its performance will improve.

---

[1] We will not enter into details regarding the similarities and differences in learning for AI and ME, as that discussion is slightly outside of the scope of this chapter.

The concept of computer programs that learn from examples has been around as long as AI itself, notably as early as Alan Turing's vision in the 1950s (Russell & Norvig, 2009). Nevertheless, it was not until the middle of the 1990s that a solid foundation for ML was established by Vladimir Vapnik in his seminal work on statistical learning theory (Vapnik, 1995). At the same time, the techniques of neural networks and support vector machines gained popularity as they were being applied in real-world applications, though only for simple tasks, as compared to today's standards. The latest chapter in the history of ML started around 2012, when it became clear that neural networks could solve tasks that were much more demanding than previously achieved, by making them larger and feeding them much more examples (Krizhevsky et al., 2012). These neural networks consist of multiple layers of neurons, leading to the name *deep learning*, where each layer contains thousands or even millions of parameters. They are responsible for many of the AI applications that people interact with at the time of writing, notably in image and voice recognition, and natural language understanding.

In this chapter, we aim to analyze the uses of AI, and in particular ML, in ME. The chapter is structured as follows. To fix ideas, in Sect. 2 we briefly discuss some contemporary AI-based tools that are being used by mathematical learners. Based on these examples, we introduce a high-level taxonomy of AI systems that serve as building blocks for tools in ME, in Sect. 3. We then show, in Sect. 4, how the proposed taxonomy allows us to provide an in-depth analysis of the different AI systems currently used in some ME tools. We finish with a discussion on future research required to build complete student modeling systems, in Sect. 5, and conclusions, in Sect. 6.

2 A Glimpse of the Present

The following are two examples of AI-based tools for mathematics education. These examples feature some representative AI techniques that will be at the basis of the taxonomy we propose later, after which we will revisit them for a more in-depth analysis.

2.1 A New Breed of Calculators

In 2014, a mobile phone application called Photomath[2] was released that quickly became very popular among schoolchildren (and less so among teachers) (Webel & Otten, 2015). It allows the user to point a phone camera at any equation in a textbook and instantly obtain the solution, including the detailed steps of reasoning. If the user wishes so, she can even request an alternative sequence of steps for the

---

[2] https://photomath.app/.

solution. While the first version of the software would only work on pictures of clean, textbook equations, the technology was later upgraded to recognize handwriting as well. Several other apps have followed suit since, notably Google's Socratic[3] and Microsoft Math Solver.[4]

The emergence of these tools, known as "camera calculators", can be mainly attributed to advances in image recognition technology, and in particular to optical character recognition (OCR) algorithms based on deep learning. Once the OCR algorithm has translated the picture into a mathematical equation, standard equation solvers can be used to obtain a solution. The third and last stage of the application consists in explaining the solution to the user, for instance, by means of a sequence of steps. In the case of the solution of a linear equation, this stage is resolved algorithmically, requiring little AI. Nevertheless, it is often possible to find a shorter or more intuitive solution by thinking strategically (see, for instance, the example in Webel & Otten, 2015, p. 4). For an AI to do so, it would require the capabilities of mimicking human intuition or exploring creative strategies. We will discuss these capabilities later, in Sect. 4.2.2.

Interestingly, these apps have reignited the discussion on the appropriate use of tools in mathematics education, reminiscent of the controversy on the use of pocket calculators that started several decades ago (Webel & Otten, 2015). Indeed, they could be seen as examples of a new generation of "smarter" calculators, that limit the number of actions and calculations that the user must perform to reach a solution. Such new, smarter calculators are not limited to equation solvers only, as they can be found in other fields as well. For instance, as of version 5, the popular dynamic geometry software (DGS) GeoGebra[5] includes a set of automated reasoning tools that allow the rigorous mathematical verification and automatic discovery of general propositions about Euclidean geometry figures built by the user (Kovács et al., 2020). Rather than merely automating calculations, as common digital calculators do, these tools allow to *automate the reasoning*, to a certain extent.

2.2 Blueprint of a Data-Driven Intelligent Tutoring System

Our second example concerns an interactive tool for learning and teaching mathematics. In particular, in the following, we describe a hypothetical interaction between a student and an intelligent tutoring system (ITS).

Hypatia, a student, logs onto the system through her laptop and she starts reading a challenge proposed by the system. This time, the challenge consists in solving an integral equation. Hypatia is not sure how to start, and she spends a few minutes scratching calculations on her notepad. The system, after checking her profile in the database, infers that she needs help, and offers a hint on screen. Hypatia now knows

---

[3] https://socratic.org/.
[4] https://math.microsoft.com/.
[5] https://www.geogebra.org/.

how to proceed and advances a few steps toward the solution. However, some steps later, she makes a mistake in a substitution. The system immediately notices the mistake and identifies it as a common error (a "bug"). Through the visual interface, the system tells Hypatia to check if there were any mistakes in the last step. She reviews her calculations and quickly corrects the mistake. The system encourages her for spotting the error, and she continues to solve the exercise successfully. At this point, the system shows her a summary of the solution and reminds her of the hints she was given. She can then choose to review any of the steps and their explanation, or continue to the next problem. If she chooses to continue, the system will present her a problem that has been designed specifically to advance along her personalized learning path.

Before Hypatia started using this ITS, the system's database already contained the interactions of many other students. By using data mining techniques, it was able to identify a number of "stereotype" student profiles. The first time Hypatia interacted with the ITS, the system's AI analyzed her initial actions to build an initial profile for her, based on one of the stereotype profiles. As she now performs different problem solving sessions, the system adds more of her interactions to its database, which allows it to identify behavioral patterns and build a more refined student model for her. This, in turn, allows the system to personalize her learning path and to offer her more relevant feedback when she encounters difficulties.

Certainly, the above-described example is not purely hypothetical, but based on real ITS that are used in practice today. We will return to this example later on.

3 A Taxonomy of AI Techniques for Mathematics Education

We now propose a taxonomy of AI techniques that are used in digital tools for ME. The taxonomy consists of four categories that span the entire range of such AI systems. While each of the categories is motivated by some aspect of the previous examples, we include a more comprehensive list of particular cases from the literature for each of them. Furthermore, we shed some light on the current technological capabilities of these AI systems.

3.1 Information Extractors

We use the term *information extractors* to refer to AI technologies that take observations from the real world and translate them into a mathematical representation (Fig. 1). A classic example in this category consists in parsing the text of algebraic word problems into equations (Koncel-Kedziorski et al., 2015). More advanced information extractors can operate on digitized data from a sensor, such as a camera or a microphone, to which they apply an AI algorithm to extract computer-interpretable mathematical information.

information extractor

Fig. 1 Representation of an information extractor. The globe represents observations from the real world, and the summation sign represents mathematical information

An example of information extractors that operate on visual data was given in Sect. 2.1, where the initial stage of the described camera calculator translates a picture into a mathematical equation. The AI required to perform OCR in these information extractors operates in two steps: First, it employs a convolutional neural network (CNN) to recognize individual objects in an image. In essence, a CNN is a particular type of artificial neural network that is capable of processing spatial information present in neighborhoods of pixels by applying (and learning) digital filters (Krizhevsky et al., 2012). Then, the individually recognized objects are transformed into a sequence, which was traditionally performed by techniques such as Hidden Markov Models (Rabiner & Juang, 1986), but is now implemented as neural network based techniques such as Long Short-Term Memory networks (Hochreiter & Schmidhuber, 1997) and transformers (Vaswani et al., 2017).

Visual information extractors are not only used to digitize algebraic equations, but can also be used to extract other types of mathematical information from the real world. For instance, in the MonuMAI project (Lamas et al., 2021), extractors based on CNN are used to obtain geometrical information from pictures of monuments. And some camera calculators, such as Socratic, allow to take pictures of word problems, which are transformed to text, interpreted, and converted into a mathematical representation.

Finally, sensor data from a student may be used to extract information for an ITS (see Sect. 4.2). In particular, these systems may require information about the student's state of mind during the resolution of a mathematical problem. In this category we encounter AI techniques for facial expression recognition (Li & Deng, 2020), speech emotion recognition (Fayek et al., 2017), and mood sensing through electrodermal activity (Kajasilta et al., 2019).

3.2 Reasoning Engines

In software engineering, a *reasoning engine* is a computer program that is capable of inferring logical consequences from a set of axioms found in a knowledge base, by

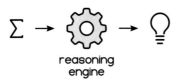

Fig. 2 Representation of a reasoning engine. It receives a mathematical problem as an input and outputs the corresponding solution

following a set of predefined rules (Furht, 2008). For the current context of mathematics education, we employ a broader definition of reasoning engines that includes all software systems that are capable of automatically solving a mathematically formulated problem (Fig. 2). A very simple such system consists of an equation solver, whose action is limited to transforming the (set of) equations into their canonical form and applying the formula or the algorithm to solve them (Arnau et al., 2013). Several types of more sophisticated reasoning engines exist in the mathematical research literature, for instance, *automated theorem provers* (ATP), whose aim is to verify and generate proofs of mathematical theorems (Loveland, 1978). While proof verification is a simple mechanical process that only requires checking the correctness of each individual step, proof generation is a much harder problem, as it requires searching through a combinatorial explosion of possible steps in the proof sequence.

A novel contribution in the development of reasoning engines lies in the use of ML techniques, which has been fueled by the success of deep learning in pattern matching problems (Krizhevsky et al., 2012). These techniques follow the standard ML paradigm that requires a set of training examples: The ML algorithm, typically a deep neural network, learns a model in order to explain as much of the training examples as possible. The learned model is completely data driven, without any hard rules or logic programmed into it.

ML algorithms could improve current ATP techniques by encoding human provers' intuitions and predicting the best next step in a proof (Gauthier & Kaliszyk, 2015; Loos et al., 2017; Schon et al., 2019). Furthermore, neural networks for natural language processing are being used to train machines to solve word problems and to perform symbolic reasoning, yielding currently some limited but promising results. For instance, Saxton et al. (2019) generated a data set of two million example problems from different areas of mathematics and their respective solutions. Several neural network models were trained on these data and, in general, a moderate performance was obtained, depending on the problem type. Deep learning is also being used to solve differential equations (Arabshahi et al., 2018; Lample & Charton, 2019), perform symbolic reasoning (Lee et al., 2020), and solve word problems (Wang et al., 2017; Wang et al., 2018). Note that these methods typically operate on text data and they perform the action of the information extractor and the reasoning engine using a single AI. Finally, in the ML community there is a growing interest in automating abstract reasoning. Research in this area currently focuses on solving visual IQ tests, such as variants of Raven's Progressive Matrices (Barrett et al., 2018; Chollet, 2019), and causal inference, which deals with explaining cause-effect relations, for instance from a statistical point of view (Pearl et al., 2019).

3.3 Explainers

While reasoning engines can solve mathematical problems and generate correct proofs, they do not necessarily produce results that can be read by a human. Sometimes this is simply not needed, for instance, when an ATP is used in research to verify a theorem that requires a long and complex proof, prone to human errors. But in a different context, for instance that of the mathematical learner, it becomes important to have proofs that are understandable (Ganesalingam & Gowers, 2017).

In the AI community, interest in explainable methods has recently surged. Part of this interest is due to legal reasons, as some administrations demand that decisions taken by an AI model on personal data be accompanied by a human-understandable explanation (Meng-Leong, 2019). While some early AI systems generated models that could easily be interpreted, modern AI techniques, especially deep learning systems, involve opaque decision systems. These algorithms operate in enormous parametric spaces with millions of parameters, rendering them effectively black-box methods whose decisions cannot be interpreted. To solve this issue, the research field of *explainable AI* is concerned with developing AI methods that produce interpretable models and interpretable decisions (Adadi & Berrada, 2018; Molnar, 2019; Arrieta et al., 2020). We will refer to AI methods that produce understandable explanations as *explainers* (Fig. 3).

From a technical point of view, there exist two types of explainers. The first type is modules that can be added onto existing, opaque AI systems. They perform what is called *post-hoc explainability*, and may do so for instance by approximating the complex model with a simpler, interpretable one. Some different post-hoc explainability approaches are illustrated in Arrieta et al. (2020, Fig. 4). The second type of explainable AI consists of models that are interpretable by design. Under our terminology, these correspond to reasoning engines that are restricted to only producing interpretable solutions. Of these two types, the former has the advantage that it does not require replacing the entire reasoning engine, which is usually hard to design and train in the first place.

In the field of ME, explainers have been built principally for solving math equations step by step, for instance, in the open source project mathsteps.[6] In ATP, on the other hand, explainability is a fairly new research line. In order to apply a post-hoc explainer onto an ATP, it might be necessary to construct an ATP based purely on logic, though, as (Fu et al., 2019) notes, "while logic methods proposed have always been the dream of mankind, their applications are limited due to the massive search space". One case in point is found in DGS, where geometric automated theorem provers (GATP) are now being integrated (Quaresma, 2020). State-of-the-art

---

[6] https://github.com/google/mathsteps.

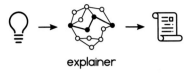

Fig. 3 Representation of a post-hoc explainer. It translates a machine-code solution into a sequence of logical, human-readable steps

GATP are based on algebraic methods, and their results cannot be translated into human-readable proofs (Quaresma, 2020; Kovács & Recio, 2020). For this reason, explainability is to be introduced in ATP by designing ATP techniques that are transparent by design (Ganesalingam & Gowers, 2017; Meng-Leong, 2019). In the case of GATP, this approach is currently very limited, as discussed in Font et al. (2018).

3.4 Data-Driven Modeling

Up till this point, we have described several techniques and scenarios in which substantial amounts of data are generated: the extraction techniques from Sect. 3.1 distill real world and sensor observations into numerical information and mathematical representations; and, in Sect. 2.2, Hypatia interacts with an ITS that relies on a database of student information and completed student tasks, which increases each time a student uses the system. In modern AI systems, data mining and machine learning techniques are used to analyze these data and to convert them into insights and practical models. These techniques, which we will refer to as *data-driven modeling*, cover a broad area and make up the final class of AI in ME (Fig. 4).

Data-driven modeling is employed for several reasons in ME. First, it may allow building models that are used to improve specific aspects of the learning process of individual students. These include AI models to predict a student's performance (Cortez & Silva, 2008; Smith et al., 2015; Asif et al., 2017), to determine at what specific problem step a student learns a concept (Baker et al., 2010), or even to detect that a student tries to game an ITS (Baker et al., 2008).[7] The ML techniques that are used to construct these models are mainly regression techniques (to obtain predictors that produce numerical values) and classification algorithms (to predict categorical or qualitative variables).

Second, data-driven modeling techniques can be used on large collections of student data, in a big-data fashion. A classic application in this category consists in analyzing completed student tasks in order to build a database of common errors, or "bugs" (Wenger, 1987; Chrysafiadi & Virvou, 2013), which is an important component of an ITS. A different application consists in modeling complete student populations, which can be useful to group students into different "stereotypes", and

---

[7] We discuss student modeling in more detail in Sect. 4.2.1.

Fig. 4 Representation of a data-driven model. After receiving data from different sources, it infers a model for the data that can be used to produce predictions

is typically performed by clustering algorithms. Another application is the large-scale analysis of student profiles and completed tasks to improve the personalization of learning paths in an ITS. In this case, recommendation algorithms can be employed (Chrysafiadi & Virvou, 2013). Finally, while studies in this field are mostly restricted to single schools or data from single ITS, it is easy to imagine that data-driven modeling can be applied to larger populations of students, for instance, on a national level, where they could be used to make statistical assessments about the effectiveness of specific aspects of a curriculum.

4 The Present, Revisited

Armed with the taxonomy introduced in Sect. 3, we can now revisit the examples from Sect. 2 and analyze their AI and ML techniques in more detail, pointing out some capabilities that may be added in the future.

4.1 AI-Based Calculators

The "camera calculator", described earlier, operates as shown in Fig. 5: First, the user captures a problem, for instance, by taking a picture, which is translated by an information extractor into its mathematical formulation. Second, a reasoning engine solves the problem and produces a solution, in machine code. Third, an explainer translates the machine code into a human-readable reasoning sequence. In the case of simple problems, the reasoning engine and explainer could be replaced by a single module. Finally, the complete solution is presented to the user, who may request additional information on each of the steps.

The described workflow is valid for a wide range of calculators: The extractor could operate on different types of data, such as text from word problems, or voice commands, which are transcribed to text. As for the reasoning engine, many ATP and advanced computational engines are available, including the solvers Mathemat-

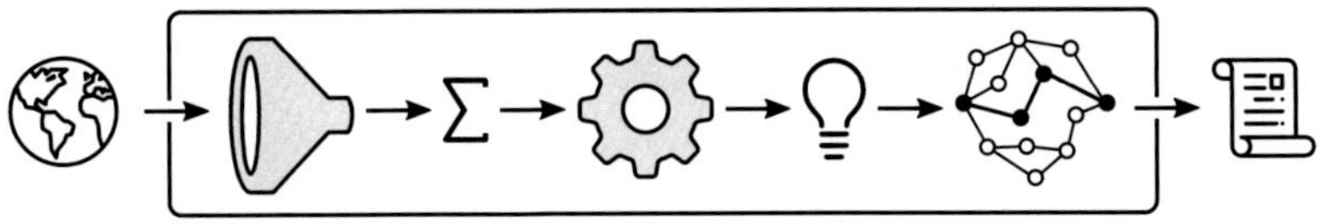

Fig. 5 The workflow of a camera calculator

ica[8] and Maple.[9] In this context, a pioneering role in the development of AI-based calculators is played by the WolframAlpha computational engine,[10] which operates on written queries and combines database look-ups with the computational power of Mathematica. It includes some *explainer* capabilities as well, as it provide feedback on the solution and links to related educational resources. WolframAlpha was launched in 2009, making it a forerunner of current AI-based calculator such as the ones included in personal digital assistants, notably Siri,[11] Cortana,[12] Alexa[13] and Google Assistant.[14]

Another type of AI-based calculator is found in DGS with reasoning capabilities, which take geometric constructions as an input. While these tools contain a reasoning engine in the form of their GATP, they do not have explainer capabilities, as mentioned before, since the GATP they include produce proofs that cannot be translated to human-readable reasoning (Quaresma, 2020; Kovács & Recio, 2020).

Presently, it is not clear how these new tools should fit in current mathematics curricula. If they are allowed without restrictions, some opponents claim that they will keep students from learning. Others recognize that the role of these tools must be debated in the educational community. Some proponents point out that the availability of these tools produces a shift in the desired objectives of ME (Kovács & Recio, 2020).

4.2 Intelligent Tutoring Systems

An ITS is a computer-based learning tool that makes use of AI to create adaptive educational environments that respond both to the learner's level and needs, and to the instructional agenda (Graesser et al., 2012). While an ITS may share some underlying technologies with the AI-based calculators we described, they are much more complex tools, and they are fundamentally interactive. Here, we will review and discuss some relevant ITS that have been proposed in the literature.

---

[8] https://www.wolfram.com/mathematica/.
[9] https://www.maplesoft.com/.
[10] https://www.wolframalpha.com/.
[11] https://www.apple.com/siri/.
[12] https://www.microsoft.com/cortana/.
[13] https://developer.amazon.com/alexa/.
[14] https://assistant.google.com/.

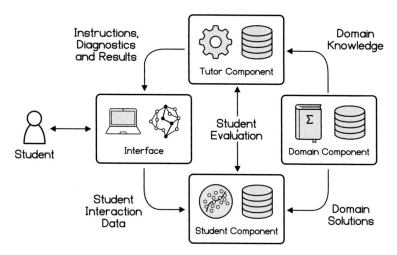

Fig. 6 Components of an ITS, in relationship to the introduced taxonomy

Typically, an ITS involves four different components (Wenger, 1987; de Souza & Varella Ferreira, 2002; Shute & Psotka, 1996), as represented in Fig. 6: (i) a domain component to encode the expert knowledge, (ii) a student component to represent student knowledge and behavior, (iii) a tutor component to select the best pedagogical action, and (iv) an interface to interact with the student. The domain component includes, among others, expert knowledge, databases of tasks, and databases of bugs. The student component includes student models and student data, such as the detailed history of completed student tasks. After each interaction, the student actions are analyzed and the models are updated to reflect the new data. During the interaction with the student, the tutor uses a reasoning engine to track the reasoning of the student. It uses data from the student and domain components to spot bugs, offer feedback and personalize the learning path.

In some ITS, the tutor relies on expert knowledge with exact inference rules, which allow it to know a priori all possible solution paths to a problem. Examples include the Hypergraph Based Problem Solver (HBPS) ITS, which deals with word problems (Arnau et al., 2013), and the QED-Tutrix ITS, used for solving high-school level geometric proof solving problems (Leduc, 2016). In general though, much of the information available in an ITS is incomplete or uncertain. Modeling the student, for instance, involves making inferences about the student's knowledge and behavior. Hence, probabilistic and approximate techniques such as Bayesian networks or fuzzy modeling are needed. An example is found in the TIDES ITS, proposed in Danine et al. (2006), which uses a Bayesian network to model student behavior based on the bugs that the student commits.

Currently, large parts of ITS that are being used in practice are designed and adapted using data-driven approaches, as described in Sect. 3.4. For instance, (Kurvinen et al., 2020) describes the ViLLe ITS, whose commercial version uses AI techniques to improve the learning experience, based on the data of millions of student interactions.

4.2.1 Narrow Student Modeling

The literature on student modeling is vast, and currently there exist dozens of student models that are used in practical ITS (Chrysafiadi & Virvou, 2013; Sani et al., 2016; Abyaa et al., 2019). Nevertheless, the majority of student models focus only on one specific aspects of the student, which is why we will refer to them as specific or "narrow" models. For instance, a student model may be constructed solely to predict student performance, and a different model may represent their competences in mathematics. A comprehensive student model, as envisioned decades ago, should include both a *complete* model of the student's knowledge as well as model of his behavior (Balacheff, 1993). This requires a more advanced AI, which we will discuss briefly in Sect. 5.

4.2.2 Some Notes on Exploration, Creativity and Randomness

The interaction with an ITS guarantees that a specific learning occurs and that a target performance is reached. Nevertheless, if the ITS is to provide a rich experience in which it can "determine the nature of the underlying meaning", it should contain environments that allow the student to freely explore problem situations (Balacheff & Kaput, 1996). This is obtained by "guided discovery learning", in which the system can shift between a tutor-like behavior for some situations and an open, exploratory environment for others.

In the AI literature, exploration is a prevailing theme. It is especially used in the subfield of ML known as "reinforcement learning", in which an agent explores an environment to determine how to maximize some reward over time. Successful applications of this field include robotics software, where the AI has to learn how to interact with the physical world, and strategy games, where the system must figure out how to beat the game using its custom set of rules.

AI is usually not associated with creativity, or the capability to come up with creative solutions. What is more, popular belief has it that AI is suited only to provide "mechanical" solutions, while creativity is reserved for humans. Nonetheless, it is precisely the AI techniques that require exploration that show strong indications that AI is capable of showing creative behavior. A striking example was seen in the 2015 tournament of the game Go between world champion Lee Sedol and AlphaGo, an AI-based Go program developed by Google DeepMind (Silver et al., 2017). The AI was first trained on records of human Go games, and then set out to battle clones of itself in order to continue improving. Interestingly, this approach led it to discover strategies that were previously unknown to human players.

In general, exploration and creativity require a certain component of "randomness". Several studies have been performed on this topic in the AI literature. For instance, randomness can be used to initiate the exploration of the solution space in neural networks. A comprehensive introduction to this topic can be found in Scardapane and Wang (2017).

5 Modeling the Mathematical Learner: a Most Ambitious Goal

While the advances in AI and ML over the past decade have been impressive, it is important to put them in context. In particular, the recent well-known "breakthroughs" in AI are all techniques that are very good at solving a very specific problem, such as recognizing objects in pictures or translating text into a different language. If the specific setting is changed, they may not function properly. For instance, if a model is trained on recognizing animals in *pictures*, it may not return a correct answer when given a *drawing* of an animal. This capability of transferring a learned concept from one situation to a new context is known as "generalization", and humans are particularly good at it. It is also one of the traits expected from Artificial General Intelligence (AGI), as opposed to the described narrow AI systems. A discussion on generalization capabilities in AI can be found in Kansky et al. (2017).

In the previous sections, we have discussed several examples of student modeling, most of which are narrow modeling techniques, as they only cover one specific aspect of the student's knowledge or behavior. In order to design a complete student model, we believe the AI needed is akin to AGI, in that it should thoroughly understand several fields and it should be able to generalize. The following is a non-exhaustive list of properties: First, it should master the understanding of physics, which is being researched in robotics AI. In a mathematics learning setting, physics are often needed to interpret word problems, and they are indispensable to describe what is happening in photographic imagery. Second, it should feature strong natural language understanding. This is currently a very active field in ML, with the best results being obtained by large neural networks. One such noteworthy system is the Generative Pre-trained Transformer 3 (GPT-3), a neural network built by OpenAI that contains 175 billion parameters and was trained on 45 TB of text data (Brown et al., 2020). It shows remarkable text-analysis capabilities and can correctly answer many text queries by producing arbitrarily long human-like text passages. Third, the AI should have reasoning capabilities that allow it to solve mathematical and other problems. As briefly touched throughout this chapter, this would require cognitive abilities in mathematical, logical, and abstract reasoning. Its generalization capabilities would furthermore allow it to relate knowledge from different fields. Finally, it would require knowledge from cognitive and developmental psychology to understand the student's actions and general behavior. This aspect is perhaps the most complex to model, and ML research in this area is currently very limited.

6 Conclusions and Discussion

In this chapter, we have presented an overview of contemporary AI techniques that are being used in digital ME tools. To provide a framework for this analysis, we have established a taxonomy of four different classes that cover each of these techniques: Information extractors, which convert data from the real world into a mathematical representation; Reasoning engines, which are solvers for mathematical problems; Explainers, which translate machine reasoning into human-interpretable steps; and data-driven modeling techniques, which are used to distill useful information and models from the data generated by students, for instance, in ITS. We have also given more an in-depth analysis of AI-based calculator apps, which we consider to be the next generation of pocket calculators, and we have related the proposed taxonomy to the different components in a modern data-driven ITS.

We leave the reader with some ideas on AI-based tools for ME that we may see in the near future. First, progress in AI is currently dominated by ML-based techniques. The influence of these techniques is also noticeable in the experimental tools for ME that we have discussed. For instance, ML is used to automate perception in information extractors, to encode human intuition for searching in large solution spaces, and to analyze large volumes of data that are being generated in online education platforms. This trend will likely continue, with advances in ML being used to improve digital ME tools.

Second, a recurrent theme throughout this chapter is the existence of parallels between research in AI and research in ME. For one, many of the questions asked in the design of digital ME tools can be found in AI research as well. And, as such, many state-of-the-art techniques that are developed in AI can solve problems that are encountered while building ME tools, in particular in ITS. Nevertheless, it must be noted that the mentioned AI techniques were developed in fields other than ME, with goals other than ME in mind, after which they were "borrowed" to be used in ME tools. This imbalance is certainly fueled by the massive interest that exists nowadays in the AI space, and one could wonder what incentives in research and industry would be required to start a new generation of ME-first AI techniques. A related observation is that the field of AI has advanced greatly over the past decade, partly due to the habit of publishing novel algorithms as open source software. In ME, digital tools are currently difficult to access: Most of them are either private (and closed-source) initiatives or academic prototypes that are not maintained after their research project finishes. A notable exception is found in DGS such as GeoGebra.

Finally, we can reflect on the transformation that occurred around five decades ago with the advent of pocket calculators. At that time, there existed a large experimental space in which many different ideas were tried out, after which some standard tools emerged that are still in use today. Currently, AI-based ME tools are in a similar experimental phase, and although it may take some years or decades, we do expect to see a similar appearance of a set of standard AI-based applications for ME.

References

Abyaa, A., Idrissi, M. K., & Bennani, S. (2019). Learner modelling: Systematic review of the literature from the last 5 years. *Educational Technology Research and Development, 67*(5), 1105–1143.

Adadi, A., & Berrada, M. (2018). Peeking inside the black-box: A survey on explainable artificial intelligence (XAI). *IEEE Access, 6,* 52138–52160.

Arabshahi, F., Singh, S., & Anandkumar, A. (2018). Towards solving differential equations through neural programming. In *ICML Workshop on Neural Abstract Machines and Program Induction (NAMPI)*.

Arnau, D., Arevalillo-Herráez, M., Puig, L., & González-Calero, J. A. (2013). Fundamentals of the design and the operation of an intelligent tutoring system for the learning of the arithmetical and algebraic way of solving word problems. *Computers & Education, 63,* 119–130.

Arrieta, A. B., Díaz-Rodríguez, N., Del Ser, J., Bennetot, A., Tabik, S., Barbado, A., et al. (2020). Explainable artificial intelligence (XAI): Concepts, taxonomies, opportunities and challenges toward responsible AI. *Information Fusion, 58,* 82–115.

Asif, R., Merceron, A., Abbas Ali, S., & Haider, N. G. (2017). Analyzing undergraduate students' performance using educational data mining. *Computers & Education, 113,* 177–194.

Baker, R. S. J. d., Goldstein, A. B., & Heffernan, N. T. (2010). Detecting the moment of learning. In *International Conference on Intelligent Tutoring Systems* (pp. 25–34). Springer.

Baker, R., Walonoski, J., Heffernan, N., Roll, I., Corbett, A., & Koedinger, K. (2008). Why students engage in gaming the system behavior in interactive learning environments. *Journal of Interactive Learning Research, 19*(2), 185–224.

Balacheff, N. (1993). Artificial intelligence and mathematics education: Expectations and questions. In T. Herrington (Ed.), *14th Biennal of the Australian Association of Mathematics Teachers* (pp. 1–24), Perth, Australia, 1993. Curtin University.

Balacheff, N. (1994). Didactique et intelligence artificielle. *Recherches en Didactique des Mathématiques, 14,* 9–42.

Balacheff, N., & Kaput, J. J. (1996). *Computer-based learning environments in mathematics* (pp. 469–501). Netherlands, Dordrecht: Springer.

Barrett, D., Hill, F., Santoro, A., Morcos, A., & Lillicrap, T. (2018). Measuring abstract reasoning in neural networks. In J. Dy & A. Krause (Eds.), *Proceedings of the 35th International Conference on Machine Learning.* Proceedings of Machine Learning Research (Vol. 80, pp. 511–520). PMLR, 10–15 Jul 2018.

Brown, T., Mann, B., Ryder, N., Subbiah, M., Kaplan, J. D., Dhariwal, P., Neelakantan, A., Shyam, P., Sastry, G., Askell, A., Agarwal, S., Herbert-Voss, A., Krueger, G., Henighan, T., Child, R., Ramesh, A., Ziegler, D., Wu, J., Winter, C., Hesse, C., Chen, M., Sigler, E., Litwin, M., Gray, S., Chess, B., Clark, J., Berner, C., McCandlish, S., Radford, A., Sutskever, I., & Amodei, D. (2020). Language models are few-shot learners. In H. Larochelle, M. Ranzato, R. Hadsell, M. F. Balcan & H. Lin (Eds.), *Advances in neural information processing systems* (Vol. 33, pp. 1877–1901). Curran Associates, Inc.

Chollet, F. (2019). On the measure of intelligence. arXiv:1911.01547.

Chrysafiadi, K., & Virvou, M. (2013). Student modeling approaches: A literature review for the last decade. *Expert Systems with Applications, 40*(11), 4715–4729.

Cortez, P., & Silva, A. M. G. (2008). Using data mining to predict secondary school student performance. In A. Brito & J. Teixeira (Eds.), *Proceedings of 5th Future Business Technology Conference* (pp. 5–12). EUROSIS-ETI, April 2008.

Danine, A., Lefebvre, B., & Mayers, A. (2006). Tides-using Bayesian networks for student modeling. In *Sixth IEEE International Conference on Advanced Learning Technologies (ICALT'06)* (pp. 1002–1007). IEEE.

de Souza, M. A. F., & Varella Ferreira, M. A. G. (2002). Designing reusable rule-based architectures with design patterns. *Expert Systems with Applications, 23*(4), 395–403.

Fayek, H. M., Lech, M., & Cavedon, L. (2017). Evaluating deep learning architectures for speech emotion recognition. *Neural Networks, 92,* 60–68.

Font, L., Richard, P. R., & Gagnon, M. (2018). Improving QED-Tutrix by automating the generation of proofs. In P. Quaresma & W. Neuper (Eds.), *Proceedings 6th International Workshop on Theorem proving components for Educational software (ThEdu'17).* Electronic Proceedings in Theoretical Computer Science (Vol. 267, pp. 38–58). Open Publishing Association.

Fu, H., Zhang, J., Zhong, X., Zha, M., & Liu, L. (2019). Robot for mathematics college entrance examination. In *Electronic Proceedings of the 24th Asian Technology Conference in Mathematics, Mathematics and Technology, LLC.*

Furht, B. (2008). *Encyclopedia of multimedia.* Springer Science & Business Media.

Ganesalingam, M., & Gowers, W. T. (2017). A fully automatic theorem prover with human-style output. *Journal of Automated Reasoning, 58*(2), 253–291.

Gauthier, T., & Kaliszyk, C. (2015). Premise selection and external provers for HOL4. In *Proceedings of the 2015 Conference on Certified Programs and Proofs* (pp. 49–57). ACM.

Graesser, A. C., Conley, M. W., & Olney, A. (2012). Intelligent tutoring systems. In *APA educational psychology handbook.* Application to Learning and Teaching (Vol. 3, pp. 451–473). American Psychological Association.

Hochreiter, S., & Schmidhuber, J. (1997). Long short-term memory. *Neural Computation, 9*(8), 1735–1780.

Kajasilta, H., Apiola, M.-V., Lokkila, E., Veerasamy, A., & Laakso, M.-J. (2019). Measuring students' stress with mood sensors: First findings. In *International Conference on Web-Based Learning* (pp. 92–99). Springer.

Kansky, K., Silver, T., Mély, D. A., Eldawy, M., Lázaro-Gredilla, M., Lou, X., Dorfman, N., Sidor, S., Phoenix, S., & George, D. (2017). Schema networks: Zero-shot transfer with a generative causal model of intuitive physics. In *International Conference on Machine Learning* (pp. 1809–1818). PMLR.

Koncel-Kedziorski, R., Hajishirzi, H., Sabharwal, A., Etzioni, O., & Dumas Ang, S. (2015). Parsing algebraic word problems into equations. *Transactions of the Association for Computational Linguistics, 3,* 585–597.

Kovács, Z., & Recio, T. (2020). GeoGebra reasoning tools for humans and for automatons. In *Proceedings of the 25th Asian Technology Conference in Mathematics.*

Kovács, Z., Recio, T., Richard, P. R., Van Vaerenbergh, S., & Pilar Vélez, M. (2020). Towards an ecosystem for computer-supported geometric reasoning. *International Journal of Mathematical Education in Science and Technology.*

Krizhevsky, A., Sutskever, I., & Hinton, G. E. (2012). ImageNet classification with deep convolutional neural networks. In F. Pereira, C. J. C. Burges, L. Bottou & K. Q. Weinberger (Eds.), *Advances in neural information processing systems* (Vol. 25). Curran Associates, Inc.

Kurvinen, E., Kaila, E., Laakso, M.-J., & Salakoski, T. (2020). Long term effects on technology enhanced learning: The use of weekly digital lessons in mathematics. *Informatics in Education, 19*(1), 51–75.

Lamas, A., Tabik, S., Cruz, P., Montes, R., Martínez-Sevilla, Á., Cruz, T., & Herrera, F. (2021). MonuMAI: Dataset, deep learning pipeline and citizen science based app for monumental heritage taxonomy and classification. *Neurocomputing, 420,* 266–280.

Lample, G., & Charton, F. (2019). Deep learning for symbolic mathematics. In *Proceedings of ICLR.*

Leduc, N. (2016). *QED-Tutrix: Système tutoriel intelligent pour l'accompagnement des élèves en situation de résolution de problèmes de démonstration en géométrie plane.* Ph.D. thesis, École Polytechnique de Montréal.

Lee, D., Szegedy, C., Rabe, M. N., Loos, S. M., & Bansal, K. (2020). Mathematical reasoning in latent space. In *Proceedings of ICLR.*

Li, S., & Deng, W. (2020). Deep facial expression recognition: A survey. *IEEE Transactions on Affective Computing.*

Loos, S., Irving, G., Szegedy, C., & Kaliszyk, C. (2017). Deep network guided proof search. *LPAR-21. 21st International Conference on Logic for Programming, Artificial Intelligence and Reasoning.*

Loveland, D. W. (1978). *Automated theorem proving: A logical basis.* Elsevier.

Meng-Leong, H. (2019, July). Future-ready strategic oversight of multiple artificial superintelligence-enabled adaptive learning systems via human-centric explainable AI-empowered predictive optimizations of educational outcomes. *Big Data and Cognitive Computing,3*(3), 46.

Mitchell, T. M. (1997). *Machine learning.* McGraw-Hill.

Molnar, C. (2019). *Interpretable machine learning.* Lulu Publishing.

Pearl, J., Glymour, M., & Jewell, N. P. (2019). *Causal inference in statistics: A primer.* Wiley.

Quaresma, P. (2020). Automated deduction and knowledge management in geometry. *Mathematics in Computer Science, 14*(4), 673–692.

Rabiner, L. R., & Juang, B. H. (1986). An introduction to hidden Markov models. *IEEE ASSP Magazine, 3*(1), 4–16.

Russell, S., & Norvig, P. (2009). *Artificial intelligence: A modern approach* (3rd ed.). USA: Prentice Hall Press.

Sani, S. M., Bichi, A. B., & Ayuba, S. (2016). Artificial intelligence approaches in student modeling: Half decade review (2010–2015). *IJCSN-International Journal of Computer Science and Network, 5*(5).

Saxton, D., Grefenstette, E., Hill, F., & Kohli, P. (2019). Analysing mathematical reasoning abilities of neural models. In *Proceedings of ICLR.*

Scardapane, S., & Wang, D. (2017). Randomness in neural networks: An overview. *Wiley Interdisciplinary Reviews: Data Mining and Knowledge Discovery, 7*(2), e1200.

Schoenfeld, A. H. (1985). Artificial intelligence and mathematics education: A discussion of Rissland's paper. In E. Silver (Ed.), *Teaching and learning mathematical problem solving: Multiple research perspectives* (pp. 177–187). Hillsdale, NJ: Lawrence Erlbaum Associates.

Schon, C., Siebert, S., & Stolzenburg, F. (2019). Using ConceptNet to teach common sense to an automated theorem prover. In *Proceedings ARCADE 2019.*

Shute, V. J., & Psotka, J. (1996). Intelligent tutoring systems: Past, present, and future. In D. Jonassen (Ed.), *Handbook of research on educational communications and technology* (pp. 570–600). New York: Macmillan.

Silver, D., Schrittwieser, J., Simonyan, K., Antonoglou, I., Huang, A., Guez, A., et al. (2017). Mastering the game of go without human knowledge. *Nature, 550*(7676), 354–359.

Smith, A., Min, W., Mott, B. W., & Lester, J. C. (2015). Diagrammatic student models: Modeling student drawing performance with deep learning. In *International Conference on User Modeling, Adaptation, and Personalization* (pp. 216–227). Springer.

Vapnik, V. N. (1995). *The nature of statistical learning theory.* Berlin, Heidelberg: Springer-Verlag.

Vaswani, A., Shazeer, N., Parmar, N., Uszkoreit, J., Jones, L., Gomez, A. N., Kaiser, Ł., & Polosukhin, I. (2017). Attention is all you need. In I. Guyon, U. V. Luxburg, S. Bengio, H. Wallach, R. Fergus, S. Vishwanathan & R. Garnett (Eds.), *Advances in neural information processing systems* (Vol. 30, pp. 5998–6008). Curran Associates, Inc.

Wang, Y., Liu, X., & Shi, S. (2017). Deep neural solver for math word problems. In *Proceedings of the 2017 Conference on Empirical Methods in Natural Language Processing* (pp. 845–854).

Wang, L., Zhang, D., Gao, L., Song, J., Guo, L., & Tao Shen, H. (2018). MathDQN: Solving arithmetic word problems via deep reinforcement learning. In *Thirty-Second AAAI Conference on Artificial Intelligence.*

Webel, C., & Otten, S. (2015). Teaching in a world with PhotoMath. *The Mathematics Teacher, 109*(5), 368–373.

Wenger, E. (1987). *Artificial intelligence and tutoring systems: Computational and cognitive approaches to the communication of knowledge.* Morgan Kaufmann.

AI and Mathematics Interaction for a New Learning Paradigm on Monumental Heritage

Álvaro Martínez-Sevilla and Sergio Alonso

1 Introduction

The purpose of this chapter is to show how AI can help, from fields such as computer vision or deep learning, to understand and explore mathematical concepts. We present two AI-based systems that incorporate interaction with reality in an environment that eases the use of mathematical concepts and applications in a highly motivational field for its cultural and historical load: the monumental heritage. To do so, in the rest of this section we present how digital technologies have been introduced into mathematics education and how the intrinsic mathematics in monuments conform an interesting tool to explain and reinforce mathematical concepts. In Sect. 2, we present the main forms of mathematical reasoning and how AI offers a new way of reasoning worthy of investigation. In Sect. 3, we present MonuMAI, an AI-driven environment designed to help the general public to do monumental analysis. Not only do we present the tool but also we describe some of the citizen science methodology that has been developed with it and the problems that arose during the development and use of the system. In Sect. 4, we present a novel approach to construct an automatic geometrical model of architectural façades. To do so, we have to overcome some issues as the rectification of images and photographs in order to get an image without perspective distortions that can be later automatically analyzed to discover geometrical properties as symmetries. In Sect. 5, we briefly describe some of the educational experiences

Á. Martínez-Sevilla (✉)
Research Institute on Data Science and Computational Intelligence, Universidad de Granada, Granada, Spain
e-mail: asevilla@ugr.es

S. Alonso
Software Engineering Department, Research Institute on Data Science and Computational Intelligence, Universidad de Granada, Granada, Spain
e-mail: zerjioi@ugr.es

that we have carried out using the presented tools. Finally, in Sect. 6 we provide some indications on the future works that we expect to develop.

1.1 Mathematics Education: Reality, Models, Computation and Solutions

Mathematics education has been a unique and exceptionally powerful way of teaching our young people. It usually takes problems from reality and models them by determining objects and relationships between them. Finally, it develops a rigorous method to combine these objects and relationships in order to obtain answers to the problems posed, making the solutions explicit in the terms in which we had initially taken them from our real context. To address a set of problems expressible under the same model, it develops theory and methods, in which finally mathematics education will have to be able to transmit the model and the general solution methodology. Of course, it is also important to have the ability to recognize reality and fit it into one of the given models (or a variant). Inspiration from reality is a central source of motivation. The *example* is the most evident proof that our mathematical and model development works on concrete cases and deepens the appreciation of reality in a systematic and clarifying way.

Therefore, the connection with reality is an inalienable way of teaching mathematics. This occurs more intensely at the first levels, in primary and secondary education. As mathematics education advances at higher levels, this connection becomes less necessary. At university levels, it provides students with methods and developments, which they have to be able to put into practice in contexts outside mathematics itself.

Perhaps the most time-consuming part of mathematics education is the one which focuses on managing the model to obtain the solutions. That is, developing correct computational mechanics that proceeds without inappropriately mixing objects or relationships between them. Fortunately, the algorithmic nature of the mathematical reasoning process has led to automating many of these processes with the help of digital technology in recent decades. This has allowed us to free ourselves from having to calculate on each occasion in very repetitive processes which are not really interesting or enriching for the understanding of the world that we intend for our students.

This allows us to focus on the concepts, on reality and its conversion into a model, on the identification of its elements and on the correct interpretation of the obtained solution. Mathematics education must be able to devote adequate time to each part of the mathematical learning process, giving perhaps greater importance to concepts and reality, and perhaps less to calculation and the use of digital support tools. In fact, one of the objectives of technology has been to simplify these processes and to make them increasingly intuitive and less dependent on specific programming skills, except in very specific and specialized cases of problem-solving at a high educational level or in specialized training.

1.2 Computational Support in Mathematics Education

The first portable electronic calculators, with low power consumption chips, appeared in 1970, and they represented a revolution, a great leap from the mechanical machines of arithmetic calculation that were highly developed toward the middle of the last century. With only a disposable battery and an integrated circuit, calculators are capable of simple and direct computations. The low cost and portability made them become a usual element of the school portfolio in the following years (particularly since 1976). They have been incorporating more and more calculation functions, such as trigonometric or statistics. However, some discussion persisted among different educational systems on the desirability of keeping in teaching the ability to perform the accounts with pencil and paper or with a calculator.

This first generation school machines reach what we could call the second generation with the incorporation of symbolic abilities. In 1987, HP launched a model with symbolic equation solving capabilities. In the following years, improvements and the incorporation of more symbolic capabilities followed, running in parallel with improvements in hardware, where PDAs and Pocket PCs were added to the repertoire of calculation devices. Also, software has improved in order to process text and spreadsheets. They are also connectable devices, via cables, infrared beams and other technologies, which makes them more versatile.

Parallel to these advances, countless symbolic calculus packages for personal computers emerged. However, the rigidity and complexity of their programming made them a tool for professionals or high educational levels, and they were generally rejected by mathematics teachers in primary or secondary schools.

We could say that the third generation starts with the presentation of GeoGebra (2021) by Markus Hohenwater in 2002. It is a software program that aims to combine dynamic geometry with symbolic computation, allowing intuitive and easy handling of the provided functions. GeoGebra is widely accepted within the educational community and has propitiated the creation of a large-scale network of collaborators who create applications with a didactic approach. In recent years, GeoGebra has improved its functionalities and even created specialized versions for the web and smartphone devices. Thus, it is becoming a very used tool for mathematics education due to its ease of use, flexibility and pedagogical adequacy, maintaining the possibility of understanding mathematical concepts, while facilitating their calculation.

The devices that we have called first and second generation have a common characteristic: they make calculations, either numerical or symbolic, but they do not interact with reality. This interaction appears with GeoGebra, with the ability to load any images and construct dynamic geometry on them. This leads to an innumerable source of didactic applications on concrete problems, taken from immediate everyday life. GeoGebra intends to expand these capabilities with the introduction in 2018 of the GeoGebra AR version, which runs on a smartphone and uses the device's camera video stream to build mathematical objects over it. Although still limited in the ability to load arbitrary Geogebra-like files, it is a big step in the integration of the computing device—here mainly to plot mathematical objects—with the interaction with the world.

However, we believe that the potential of GeoGebra in its interaction with reality is largely unexplored. In Botana et al. (2020), we propose the creation of an extension for GeoGebra that is able to interact in a deeper way with reality, through the use of a wide range of sensors of a smartphone: camera, accelerometer, compass, and positioning system (currently GPS), in such a way that allows measuring distances, angles, speeds, geographical positions, directions and so on. GeoGebra would thus take advantage of the large number of sensors in a smartphone to integrate the reality-model interface in the calculation device, something that started motivating mathematics education and which is still difficult to achieve by computer systems.

The fourth generation of these digital devices to facilitate mathematics education are Artificial Intelligence (AI) systems. They can use the computational capacity and connectivity of smartphones while integrating them with tasks such as cloud computing, Internet of Things, and access to databases, and thus constitute a whole universe of possibilities for interactive calculation against reality. The coming years will surely bring us an exciting path in the development of this integration to perform increasingly complex tasks in a simpler and more intuitive way. The use of expert knowledge to approach and integrate other scientific or humanistic subjects, analyzing solutions and proposing application scenarios or intensifying the use of technology are some of the desirable features in what we call STEAM (science, technology, engineering, arts and math) education. These features will undoubtedly come hand in hand with the creation of AI-based systems. Particularly, in the following sections we will describe two of these systems that incorporate interaction with reality.

1.3 The Mathematics in Monuments: A Unique Form of Mathematics Education

One way to bring students closer to mathematics is by connecting it to everyday life. Not only does it allow a surprising and close application scenario but it is also very accessible for special sessions and projects in the classroom and outside. In fact, some technological applications have been developed in this sense specifically dedicated to supporting the educational actions with mathematics in the scenario of a city.

From this point of view, mathematics linked to the monuments have a special place: to the rich mathematics contained in monuments, we must add their interdisciplinary possibilities and the motivation they generate. Their mathematical properties can be presented together with their historical-artistic aspects as well as other scientific aspects linked to monuments. They offer great opportunities to study through technology, making this field one of the most promising in mathematics of STEAM projects and orientations.

In this sense, we have already developed a methodology (Martíez-Sevilla, 2017a) and mathematical content resources (Martíez-Sevilla, 2017b, 2020) as well as technological ones (Botana et al., 2020) that allow an approach to the field of teaching mathematics based on monuments. Beyond simple contents, mathematics in monuments offers the rich complexity of the intricacies of the relationships in which artists, architects and geometricians have come together to design the appearance of buildings to last for the future and to transmit a symbol to future generations. These symbolic aspects are perhaps the most attractive and motivating in the use of monuments in mathematics education. It is so that on occasions, they have even made it possible to provide a new sense of historical-artistic interpretation mediated by new research based on mathematics (Martínez-Sevilla & Cruz Cabrera, 2021). Moreover, the use of mathematics just for functional or decorative aspects, where calculus and geometry usually play an important role, has allowed numerous teaching resources based on them.

Naturally, this scenario could not be unaware of the great development of Artificial Intelligence in the last decade (Fiorucci et al., 2020). Both as a tool for monumental analysis and as a proposal for an an interdisciplinary educational approach which manages resources, technologies and modes of reasoning typical of a recent discipline, but with increasing depth in current teaching. Within this framework, we have developed the MonuMAI tool (Lamas et al., 2021) as an integrated and growing system to approach the monumental environment through AI and mathematics.

2 Forms of Mathematical Reasoning: Deduction, Induction, and Abduction

In this section, we introduce the classical forms of mathematical reasoning (deduction and induction) and present how AI offers a different (but interesting) way of reasoning (abduction).

2.1 Classical Forms of Reasoning

Among the most defined forms of mathematical reasoning, there are two that are usually applied and taught in mathematics: **deduction** and **induction**.

In **deductive reasoning**, a conclusion is obtained from a finite number of previously established hypotheses or premises. A series of rules, called inference rules, intervene in its definition, which we will use in our reasoning. Thus, we would have, as a more formal definition, that *a deduction is a finite set of steps, in which in each one we use either one of the hypotheses (or axioms added to these) or a formula obtained by applying one of the inference rules to two of the previous steps*. In this way, the first two steps must necessarily be hypotheses or axioms. The last step is the so-called conclusion, deduced from the set of hypotheses. From this mode of reasoning, science has elaborated the *hypothetical-deductive reasoning*, where the observation of reality and the experimentation play their role to verify the system of hypotheses and conclusions.

The other mode of reasoning, that is, **inductive reasoning**, works in a very different fashion. In general, in inductive reasoning we go from the particular to the general. Induction studies the properties that a certain object or sequence fulfills in order to extend or generalize this property to a wider family in which essentially the same norms are fulfilled. It is a usual way of learning in children and even in the daily reasoning of our life. As an example, we use this kind of reasoning if we observe on multiple occasions that after dark clouds rain comes. We will then tend to deduce that rain comes from the presence of dark clouds when this may not really be the case: the consequence of dark clouds may also be snow, or no rain due to other meteorological variables.

However, in classical mathematics, violations of a general formulated rule are not tolerable. That is why they use the most far-reaching format of induction reasoning, the so-called *Complete Induction*, which in its strongest form can be formulated as follows.

If $A \subset N$ and

1. $0 \in A$
2. for any $n > 0$, $\{0, 1, \ldots, n-1\} \subseteq A$, then $n \in A$

then $A = N$.

This is the so-called **Second Principle of Induction**. There is another, with a weaker formulation, known as the **First Principle of Induction**. Both principles are equivalent and constitute an axiom of natural numbers. In fact, they are valid for any set that is a *Well Order*, that is, in which each subset has a minimum element. Moreover, the demonstration of this property can be deduced from the Principle of Induction.

Therefore, this is a form of mathematical reasoning that allows to extend a property to the whole set of natural numbers as long as that property has the first element that fulfills it and that given an arbitrary number that fulfills it, we can verify that the following one also fulfills it (weak formulation).

Inductions can also be applied to two or more variables. All of them are ways to generalize a demonstration in an automatic way to a larger set than the starting one. They form the basis of the logical work with some interesting mathematical objects as recurrences or successions.

2.2 AI and Its Forms of Reasoning

However, it is not easy to fulfill the deduction and induction requirements in practical applications. Classical mathematics, which is the one that is mainly taught in primary and secondary schools, has established a type of universal truth based on binary logic, whether it is fulfilled or not, and with no other options to consider. Usually, however, we are presented with situations that are not so clear-cut.

For example, reasonings in which we can use the word *probable*. Estimates about the occurrence of some event, which may trigger a deduction, but only in a probabilistic way. Even the application of some rules may be affected by such a measure of probability. This type of reasoning is frequently found in daily life and in scientific applications, but it is not usually trained in mathematics lectures (Batanero et al., 2005, 2016). On many occasions, we talk about probability, and we estimate it with measurements, but usually disconnected from the world of logic, without incorporating it within a context of inference rules. Artificial Intelligence basically works with this kind of reasoning and that is why familiarity with it should be the first step to acquire this argumentative ability.

Therefore, a new mathematics education is necessary, not only to provide validity to the statements obtained by deduction or induction, but also to those obtained by **abduction**: the method of reasoning in which probability is incorporated in the premises or the rules of inference and also in the conclusion.

There are several types of logic that incorporate the management of uncertainty in different forms and for different purposes: *modal logic*, *probabilistic logic* or *fuzzy logic*. In them, not only probabilities are quantified for a premise or inference rule but also imprecise concepts are handled to indicate different types of quantities in everyday life. For example, in modal logic we can use statements such as "it is necessary that" or "it is possible that", in probabilistic logic, we can use "probably (with a crisp measure)" and in fuzzy logic, we can use linguistic terms (qualitative quantifiers) such as "many" or "surely".

Mathematics education should make an effort to incorporate these types of logics that are capable of managing the uncertainty of an answer in a generalized way. Although modal logic can be understood as a formal and didactic step in that management and fuzzy and probabilistic logics as technical tools that allow reasoning in environments with partial information and uncertainty, it is important to incorporate them with the appropriate depth to each educational stage as well as to provide examples and applications that practice with them. Today's world is already working through AI, and we can find some examples of the interest that it brings forward to mathematics education (Chassignol et al., 2018; Gadanidis, 2017). It is used as a base and therefore an adequate understanding of its possibilities and results will only be possible with an education that incorporates adequate conceptual models of reasoning.

One of the AI-based technologies that can be successfully used to illustrate the power of AI and its way of reasoning is pattern recognition, a set of techniques aimed to recognize in an automated way patterns and other regularities in data. Please note that pattern recognition differs from pattern matching in the sense that the latter is aimed to find *exact* coincidences in the data while the former pretends to find softer regularities in the data (not exact coincidences but with a certain degree of flexibility). As an analogy, we can say that pattern matching is as strict as inductive and deductive reasoning processes, while pattern recognition follows a more abduction-like form of reasoning.

Particularly, visual pattern recognition takes as input images and tries to find characteristics in those images to classify them (or to classify sub-elements in the images). Due to its visual nature, it makes for an appropriate tool to use with scholars.

3 MonuMAI: An AI-Driven Environment for Monumental Analysis

As we have mentioned before, monuments can concentrate a large part of the mathematical knowledge of their time, as well as valuable historical and artistic information. This is particularly intense on monumental façades, so we will focus mainly on them. The analysis of a façade requires a lot of expert knowledge:

- different styles usually share some similar architectural elements,
- monuments usually have a mixture of styles and periods, and
- some are not executed under canonical criteria, but with singularities.

All these factors, with the addition of the deterioration and alterations of the monuments as time passes, make this kind of analysis a quite difficult task.

In order to facilitate it and, at the same time, to create a teaching support tool that can serve in an interdisciplinary way in subjects of mathematics, art and history, we have created MonuMAI (Monuments with Mathematics and Artificial Intelligence). MonuMAI is an interdisciplinary project of research, education and scientific dissemination, joining researchers from Math, AI, general Computer Science, and History of Art, altogether scientific communicators and educational advisors. It is a STEAM project which has as its main objective to deal with science and art by means of technology.

MonuMAI, born in September 2018, is a project of the Research Institute on Data Science and Computational Intelligence[1] of the University of Granada[2] and the Descubre Foundation.[3] You can check out more information and details about it at http://monumai.ugr.es/ (Fig. 1). The related app can be downloaded from both the AppStore (iOS) and Play Store (Android), or directly from the links on the project page.

3.1 MonuMAI Dissection: How to Classify Monuments Using Deep Learning

MonuMAI consists of three main blocks:

---

[1] https://dasci.es/.
[2] https://www.ugr.es/.
[3] https://www.fundaciondescubre.es/.

AI and Mathematics Interaction for a New Learning Paradigm ... 115

Fig. 1 The front page of the MomuMAI web page and the splash screen of its app

1. A monuments dataset;
2. A deep learning pipeline;
3. A mobile app to integrate the previous ones.

In the following we describe the three parts:

Monuments dataset: An extensive dataset (selected and annotated by experts in Art History) has allowed us to train the algorithm for the recognition of up to 15 key architectural elements, easily recognizable by their geometry (Fig. 2). These in turn will allow us to distinguish up to 4 from the most important artistic styles in Europe: Renaissance, Baroque, Gothic and Hispano-Muslim. This dataset is a database of labeled images of monumental façades. Such a database did not exist to date. The MonuMAI dataset includes 6650 tagged images of the 15 key elements. The dataset is the basic component, a result of expert knowledge, on which MonuMAI bases its learning through its pipeline.

Deep learning pipeline: The MonuMAI deep learning pipeline's main purpose is the detection of key elements in an image using the MonuMAI Key Element Detection (MonuMAI-KED) model and also the classification of artistic styles.

MonuMAI-KED is based on a novel taxonomy of monumental heritage (MonuNet), specifically created for this task (Fig. 3). This taxonomy incorporates a classification of styles according to the recognition of the previous 15 key architectural elements. The classes cover arch, structural support objects and decoration, horseshoe arch, lobed arch, flat arch, pointed arch, ogee arch, trefoil arch, serliana, triangular pediment (or pointed pediment), segmental pediment, gothic pinnacle, rounded arch, lintelled doorway, porthole, solomonic column and broken pediment (Fig. 2).

So, how does MonuMAI classify the style of the monuments? The 15 architectural elements allow us to establish some identification criteria for styles. For example, the Gothic style is characterized by the use of ogee arches, pointed arches, gothic pinnacles and trilobed arches. On the other hand, the Hispano-Muslim style is associated with horseshoe or lobed arches while the Baroque style uses broken pediments or solomonic columns. In this way, a pseudo tree appears, to classify styles: the 4 child nodes of the root are determined based on the descendant nodes that are reached. We

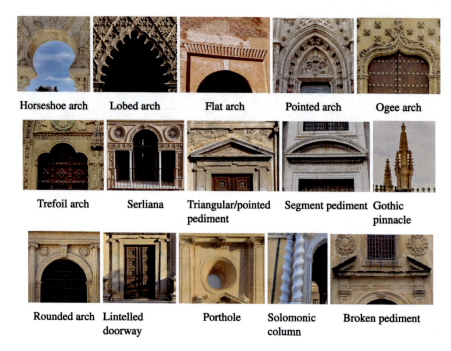

Fig. 2 The 15 architectural elements for MonuMAI-KED

would like to remark that MonuNet is not really a tree, but actually a graph with a few cycles. This is so because some styles share elements (descendants nodes), such as the rounded arch or the porthole for the Renaissance and Baroque styles.

This is also an interesting matter, since in education and some simplified scenarios binary or n-ary taxonomies are used but usually in the form of a tree, where there are no cross characteristics between several predecessor nodes. However, reality is not so perfectly separable, and usually establishing a realistic taxonomy implies having to classify the same object in several categories and having to go to secondary categories to make a proper classification. In fact, we can find this issue in numerous everyday examples and it appears in the field of science as well. For example, although Darwin's well-known "tree of life" drawing is really a tree-like graph, we know that the complexity of life in many species requires a continuous modification of its taxonomy tree labels and branches. This is done to rearrange the new knowledge, and sometimes this connects leaves (or nodes) in some biological species (Dale, 2017).

Thus, it is interesting to note that MonuMAI's classification may sometimes output a double-class as Renaissance/Baroque. The only possibility to discern those two styles is finding in the façade other recognized key elements which are typical of only one of the styles. If this is not possible, the uncertainty in the classification will remain. This is not a bug in the application, but a singularity of the artistic classification. Moreover, the experts themselves are sometimes unable to distinguish

AI and Mathematics Interaction for a New Learning Paradigm … 117

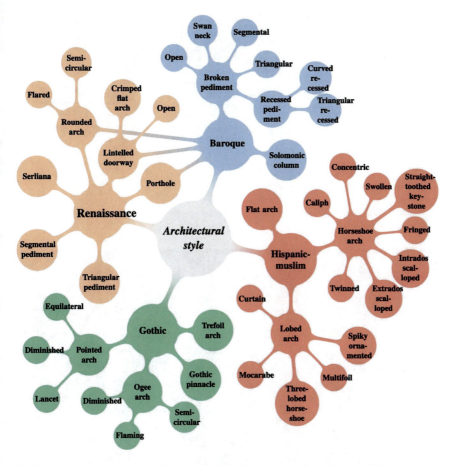

Fig. 3 MonuNet: Taxonomy in the form of a rooted pseudo tree for style recognition

one style from another except when using added information such as the date of construction or the author of the design. Classification in artistic-historical styles is a matter that is far from being completely finished in the state of the art of this social science.

Mobile app: The MonuMAI app uses the two previous elements to operate in real-life conditions. Our tests have shown that the MonuNet architecture and the detection model provide excellent results even in harsh real-world conditions such as side shot photographies, noise or blur in the image.

The app offers the interaction between the user and our image classification process. The user, once registered, can take an image of a monumental façade and select a region of interest on his mobile device. Once this is done, the image is sent to our server where it is processed by MonuMAI-KED which uses a Convolutional Neural Network (CNN) to analyze positive identifications (with enough probability) of the

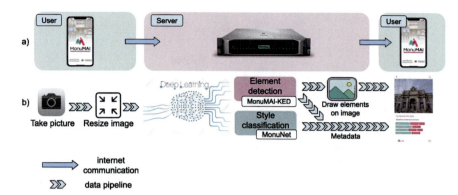

Fig. 4 MonuMAI deep learning pipeline. **a** Sequenced communications flow. **b** Image processing status in each stage

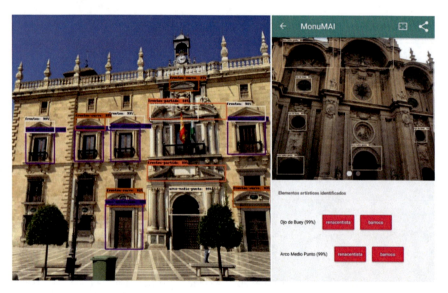

Fig. 5 Application examples in the user app: Left: Royal Court of Justice of Granada (with framed detected elements); Right: Granada Cathedral (identification and labeling of elements)

key elements detected in the image. I will also draw a frame around each of the analyzed regions for each key element on the image. Based on the detected key elements, MonuNet will offer a style classification proposal—probabilistically quantified—as metadata added to the image. Finally, the annotated image will be returned to the user. The whole process is shown in Fig. 4.

The output resulting from applying MonuMAI-KED and MonuNet to an image is then displayed in the user's app (Fig. 5).

For the operation of MonuMAI-KED, we use a two-step strategy. First, a selective search method that is implemented as a fast region-based CNN makes a proposal

AI and Mathematics Interaction for a New Learning Paradigm ... 119

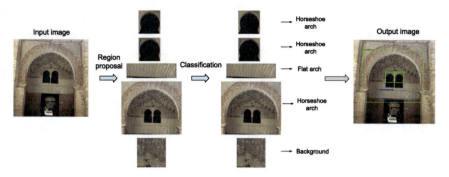

Fig. 6 MonuMAI's process for recognizing key elements in an image

of candidate regions in each input image. The search for candidate regions is done through regression learning, considering the visual features and region shape of the selected elements. In the second step, each region is classified by a CNN as an architectural element or background. For the regions classified as key elements, the architectural information is added to the monumental image. The process is shown in Fig. 6.

3.2 The Citizen Science Methodology on MonuMAI

One of the novel points in the methodology is that the user actively participates in the improvement of the dataset through their added knowledge. It is done applying the Citizen Science methodology: citizens participate in the project and contribute with their collaboration to achieve scientific purposes. In our setting, the user can contribute in the improvement of the scientific results offered by MonuMAI. After classifying and returning the image to the user, it is asked for their opinion regarding the classification offered, to which they can respond with "I agree", "I don't know" or "I don't agree". MonuMAI will value these opinions to add them to the dataset in cases where it has confirmed the offered classification, or it will lower the probability in cases of disagreement. The dataset therefore improves with its use, thus appreciably increasing the precision in the responses of the MonuMAI system.

Citizen Science has shown itself as a powerful tool for doing science in recent years. The European Union even has a Research Program based on it: Science With And For Society.[4] Moreover, in terms of scientific possibilities, not only for participation but also for the addition of knowledge distributed in society, the case of Deep Learning and Citizen Science is a special pairing. Where scientists cannot reach, Citizen Science can reach.

---

[4] https://ec.europa.eu/programmes/horizon2020/en/h2020-section/science-and-society.

Until now, previous existing models were based on machine learning and computer vision techniques, which require a high degree of supervision and manual intervention. These techniques produce models with little scalability, that is, the possibility to add new elements for the detection or artistic styles. Furthermore, their operation in real environments is usually difficult due to the presence of noise and spurious elements in the image. However, they usually offer satisfactory results. Our approach in which we use machine learning through deep learning techniques, supported by a Citizen Science methodology. is a better way to increase and improve the dataset, and thus recognition.

3.3 Learning and Mistakes

How does MonuMAI learn? The implemented learning model based on deep learning technologies is not the one employed in usual mathematical reasoning, where deductive or inductive processes are applied to achieve the final decision. Deep learning works as a computational brain, and its learning model is similar to the one employed by children: it reasons by analogy based on the accumulation of cases. It takes into account the context and infers knowledge—in our case probabilistically— to be able to decide among the different architectural styles after the inspection and detection of key elements. From the dataset of annotated images, it learns the differential characteristics of each architectural element. The algorithm then is able to search over every new image and process all the characteristics it has learnt and thus to recognize elements that have similar properties, even if it has never seen them. In this sense, MonuMAI's learning process is intelligent behavior.

At this point, we would like to point out that even intelligent behavior makes mistakes: For example, MonuMAI did learn that in the context of horseshoe arches, bricks are usually associated in their appearance, forming part of it, or in the decoration of its flat lintel (whenever it exists). Thus, if it found a rounded arch from the Mudejar style (post-fifteenth century style that takes some of the earlier Islamic construction techniques, very frequent in Spain), then MonuMAI used this context association and erroneously determined that the arch was a horseshoe arch and thus concluded that the architectural style was Hispanic-Muslim, when in reality it corresponded to a Christian church.

3.4 A Critical Look at MonuMAI: How to Troll the System

To test the abilities of MonuMAI and to check the kind of errors that it can produce, we propose a challenge to the students who use it (from an educational point of view): Can MonuMAI be trolled? Can the user input images that clearly are not related to architectural monuments and make its inner algorithms to fail and recognize elements and even determine a particular architectural design?

AI and Mathematics Interaction for a New Learning Paradigm … 121

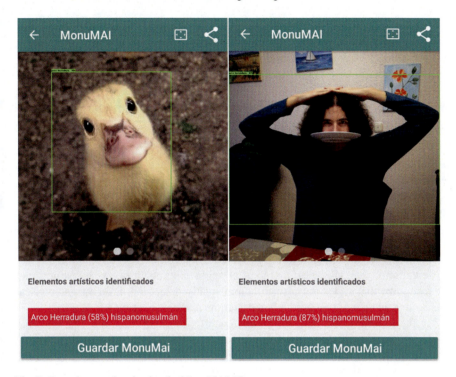

Fig. 7 Some images that deceive the MonuMAI AI

The students were pretty fast responding to the challenge. They found some example images in which MonuMAI only recognizes "shapes in their immediate context" and not the overall appearance of the image. For example, in Fig. 7 (left) the image depicts a duck. However, for MonuMAI it is not a duck but a horseshoe arch. The head with a notch on the neck, and the surrounding darker brown color (similar to horseshoe arches) makes MonuMAI infer a small probability (58%) for that architectonic feature. In Fig. 7 (right), a student tried to fool the system (and succeeded): MonuMAI produced a high probability (87%) of the image depicting a horseshoe arch.

These funny examples demonstrate that the creativity of a motivated student knows no limits. Moreover, these examples offer a great opportunity to discuss with the students how a particular AI (like the MonuMAI one) may be very good in a particular task (recognize architectural styles) but it may miserably fail when faced with examples that are completely out of its scope: it is important to remark that the MonuMAI AI has never faced an image of an animal or person and therefore it cannot recognize and discriminate against them. On the other hand, the human brain has a much more general intelligence that has previously faced much more images and concepts and therefore is much more capable of differentiating against those malicious examples. However, not all humans are able to distinguish among different architectural styles. To summarize, MonuMAI is good at classifying architectural elements and style over photographies of architectural façades, but it can say nothing correct about photographs of other different subjects.

4 Toward the Construction of an Automatic Geometric Model (AGM) of Architectural Façades

Architectural façades are a great example to understand how geometry has been extensively used in different arts and monuments. The use of particular proportions, symmetries, repeating patterns and so on can be discovered in a great variety of building façades.

Moreover, AI techniques can be used to ease the analysis of those mathematical properties and even to automatically detect the geometrical properties of the different elements in a façade. We have already covered how an AI-based mobile app can detect different art styles. However, that kind of application could be expanded not only to recognize that styles but also to automatically detect the different geometrical constructs that have been used when designing and constructing a façade.

In the following, we describe an approach to develop such a tool that analyzes and constructs geometrical models of architectural façades based in different AI techniques and computer vision technologies. This approach assumes that all the analysis must be done over a single photograph (probably taken with a mobile phone or typical camera) and without any other particular knowledge about the target. This assumption is a quite restrictive one as many advanced technologies as photogrammetry (Taddia et al., 2020) (construction of a 3D model from several pictures of the subject taken from different points of view), 3D scanning (Wojtkowska et al., 2021) and so on could provide much more precise results. However, we do not take into account these techniques and tools as they are not usually available for the general public.

4.1 The Problem of Perspective in Photographies

One of the first problems in order to automatically analyze a façade that has been photographed is the effects of the perspective (Soycan & Soycan, 2019). These perspective effects produce that the proportions and angles presented in the façade are not maintained in the photograph (which is a 2D representation of a 3D object). Here, it is important to remark that the human brain has evolved to be able to overcome those effects and can usually infer some geometrical properties in the photograph. Let's take as an example Fig. 8 where 2 buildings are depicted in a synthetic render. It is obvious that from the perspective chosen some of the elements in the façades are not represented at the same scale (the green circles in Fig. 9) and are deformed (the green circles are depicted as ellipses) and some parallel lines (the red ones) are not really parallel in the photograph. However, our brain can easily overcome those

AI and Mathematics Interaction for a New Learning Paradigm ... 123

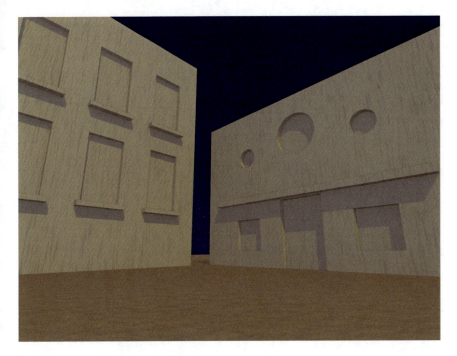

Fig. 8 Synthetic render of two buildings

difficulties and can instantly recognize that the ellipses represent circles of the same size in the façade and that the red lines are parallel ones in the real world. We think that eliciting these issues regarding perspective effects in traditional perspective drawing and photography is a very interesting topic as they can remain unnoticed by the public due to the good job that our brain does interpreting drawings and photographs.

In fact, to partially avoid those perspective effects, it would be necessary to make the photograph in a very particular scenario: we need to use specialized perspective corrected lenses, shoot at an appropriate angle in which the sensor of the camera is parallel to the façade plane and as close to the center of the façade both in the horizontal and vertical axes (which is usually not possible as the photographer cannot usually put the camera at the appropriate height). In Fig. 10, we depict the previous façade simulating those ideal conditions and we can see that the effects of the perspective are minimized.

4.2 Correcting Perspective Issues in Façades

Fortunately, today we can use computers in order to apply transformations on photographs to rectify the perspective-related issues. Image rectification consists of a

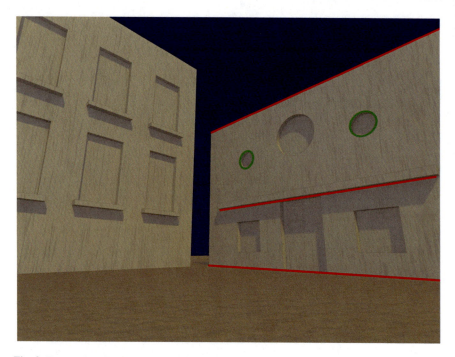

Fig. 9 Perspective distortions over elements in the façade: green circles are depicted as ellipses with different sizes, and parallel lines (in red) are depicted as not parallel

transformation process in which the original image is projected onto a different plane (in our case, a plane that makes the façade plane and the sensor of the camera to be parallel). In order to do this rectification, a projective transformation must be applied. If we know the camera angles with regard to the façade plane it is possible to compute the precise projective transformation that should be applied Soycan and Soycan (2019). These camera angles are usually called *yaw* (rotation around the vertical axis), *pitch* (rotation around the side-to-side axis) and *roll* (rotation around the front to back axis). For example, in Fig. 8 the angles of the camera were set as $yaw = 45°$, $pitch = 15°$ and $roll = 0°$.

In our approach, we assume that we do not know those angles. However, we must note that many cameras nowadays (specially the ones integrated into mobile phones) have built-in sensors to measure that angles. Having a reasonable approximation of those rotations could be very helpful in the rectification process.

An approach to compute those angles (and even different parameters of the lens and camera) is offered in some panoramic creation software such as Panorama Tools (2007). One of the modules of this software (PTOptimizer) can take as an input a photograph and a list of lines which are identified as vertical and horizontal lines of the façade that we want to use as the rectification plane. From this information, the program executes several optimization algorithms and as a result

Fig. 10 Using a particular lens configuration and carefully choosing the angle and camera position, the perspective effects are minimized

provides an estimation of the yaw, pitch and roll and also the camera lenses' focal length.

4.3 The Hough Transform to Identify Basic Geometries in an Image

Therefore, in order to obtain the camera orientation we need to identify vertical and horizontal lines in the façade that we want to rectify. To do so, we propose the use of the Hough transform (Duda & Hart, 1972; Ballard, 1981). This transform allows detecting analytically defined shapes. In our case, the shapes that we want to detect are straight lines and thus, to do so, the image in which the transform is going to be applied is usually filtered by means of a Canny filter (Canny, 1986) which emphasizes all edges (abrupt changes in illumination) in the original image.

Once those edges have been emphasized, each pixel in the image "votes" in a parameter space matrix. In the case of straight lines, the parameter space is a two-dimensional one (the angle θ of the normal of the line and its algebraic distance ρ from the origin. Once the voting has finished, the algorithm just searches for local

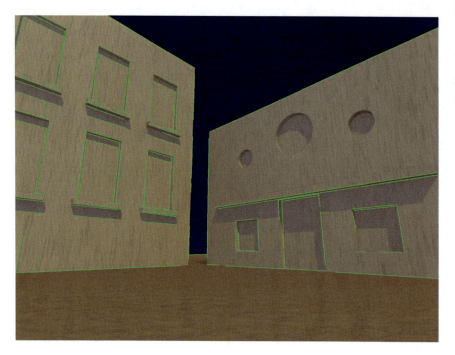

Fig. 11 After the application of the Hough transform, we obtain the straight lines in the image

maximums in the voting matrix determining the parameters of the detected lines. In Fig. 11, we show the detected straight lines after applying the Hough transform on our synthetic render.

4.4 Straight Line Classification and Optimization of the Camera Angles

As a final step before applying the optimization algorithm to detect the orientation of the camera and to rectify the image, we need to sort the lines into different groups: The vertical and horizontal lines in the façade that we want to rectify and lines corresponding to other façades or elements in the image. To do so, we propose the use of a simple clustering algorithm that takes into account the θ angle of the lines (with respect to the bottom side of the image). We can also compute the vanishing points of the clustered lines (which are supposed to be parallel in the real world if they are horizontal and vertical lines in the façade in order to detect outliers that do not belong to the clusters we need).

In Fig. 12, we can see the different clusters of lines that we can obtain: In green, we have the horizontal lines in the façade of interest. In blue, we have marked the

AI and Mathematics Interaction for a New Learning Paradigm ...

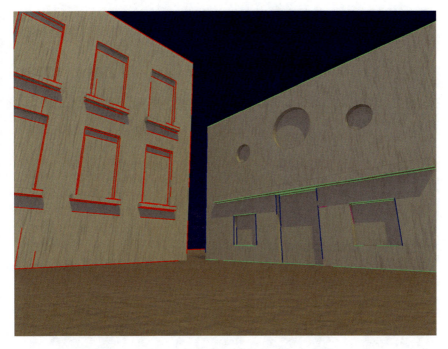

Fig. 12 After clustering the lines, they are classified into the horizontal (green) and vertical (blue) lines in the façade that we want to rectify. Other lines correspond to the other façade (red) or even to different features in the image as shadows or textures (purple)

vertical lines for the same façade. In red, we have the lines that correspond to a different façade and finally in purple, we have some lines in the façade of interest that correspond to other features in the image (shadows in this case).

At this point, we can run the optimization algorithm with the selected vertical and horizontal lines in order to rectify the image. For the test image, the optimization algorithm gives the values $yaw = 44.9°$, $pitch = 15°$ and $roll = 0°$ which are very close to the values that we used to generate the initial synthetic image. Finally, we can apply the appropriate projective transformation to get the rectified image. In Fig. 13, we show the result of applying the projective transformation. As it can be appreciated, the façade plane is now parallel to the camera sensor, and therefore the proportions of the elements are maintained and the vertical and horizontal lines in the façade are now parallel.

We want to remark that this mechanism to rectify images can be used as a good example of how optimization algorithms work and to show the kind of results that can be achieved with them. Playing with the input straight lines and the possible optimization parameters (not only yaw, pitch and roll but also focal length, lens distortions and so on) is a very graphical example of the strengths of these algorithms.

A final example of the good results that can be achieved to rectify photographs of façades in an automatic manner is shown in Fig. 14. There, we deal with a real photo-

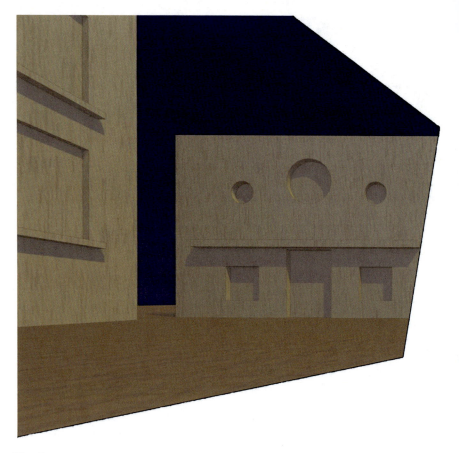

Fig. 13 Rectified image. Now the façade maintains the proportions among its elements and the vertical and horizontal lines are parallel

graph of the façade of the Granada Cathedral. In the same photograph, the façades of the adjoining buildings have been captured. Applying the described techniques, we have been able to reconstruct not only the main cathedral façade but also the other façades in the photograph.

4.5 Constructing an Automatic Geometrical Model for the Façade

Once we have a rectified Image, we can apply different techniques in order to create a geometrical model based on the characteristics of the façade. To do so, we can extend the Hough transform to detect more geometrical features as, for example,

AI and Mathematics Interaction for a New Learning Paradigm ...

Fig. 14 Real example of the rectification of a photograph of the façade of the Granada Cathedral. **a** Original image, **b** rectified cathedral façade, **c** rectified left façade and **d** rectified right façade

circles, ellipses, squares and rectangles, which are typically used in architecture, and determine their properties (positions, proportions and so on). Moreover, we can also run some pattern recognition algorithms (similar to the ones implemented in MonuMAI) to detect more complex features. As an example in Fig. 15, we have applied an extended Hough transform to detect circles, squares and rectangles. We have successfully detected (in green) the most important features in the façade (windows, door, ledge and round insets). However, the Hough transform has also detected some artifacts due to the effects of the shadows, perspective, textures or even the overall structure of the façade. Please note that the position of each feature and its dimensions and proportions would never be detected with total accuracy: for example, the two square windows may not have been detected with exactly equal size and perfectly aligned.

Once the detection is finished, we'll have a collection of features with their relative positions in the façade and we can study the geometrical relations among them. We can apply different techniques (from AI techniques as neural networks and deep learning to a brute force algorithm) to identify those relations. We would start from some basic relations as "similarity": for example, we could detect that some windows in the façade (rectangles) are similar elements (with approximate sizes and

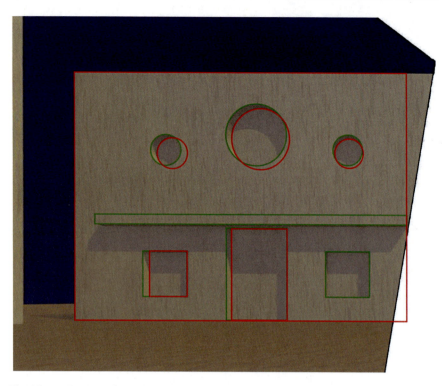

Fig. 15 Application of an extended Hough transform in order to detect circles, squares and rectangles. Green: real detected features in the façade. Red: artifacts detected by effects of perspective, shadows or the overall façade structure

proportions). Later we will try to identify more complex relations as translations, symmetries (along an axis or points) or even more complex geometrical patterns.

In Fig. 16, we show the results of the computation of some geometrical properties for the façade by analyzing the features obtained with the extended Hough transform: In blue, we have detected a vertical axis of symmetry. To do so for each possible vertical axis in the image, we have computed the relative positions of each of the detected features and we have mirrored and matched them (with a certain tolerance for imprecision). We have selected the axis in which the matching of the mirrored features was maximized. In orange and purple, we have depicted some alignment horizontal lines that can be found in the images: the center of the windows and door and the bottom of the circular insets. To obtain those lines, we have followed a similar approach: we have tested all possible horizontal alignment lines in the image against the centers and boundary of the detected features (with some allowed tolerance).

Finally, we want to emphasize that the aims of the proposed method to recreate the geometrical model of façades are multiple:

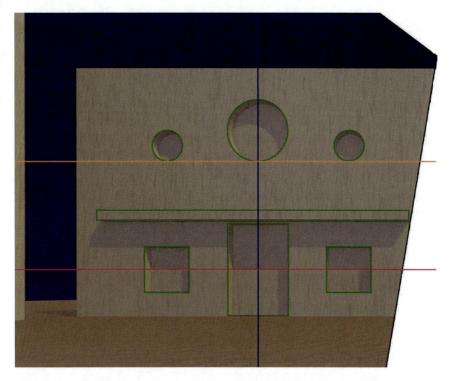

Fig. 16 Detection of some geometrical properties of the façade: a vertical axis of symmetry (blue) and some horizontal alignment lines: center of windows and door (purple) and bottom of the round insets (orange)

- The rectification process is a very good example of how the human brain is very well suited to deal with problems that are difficult to solve by a machine (the effects of the perspective). Therefore, students can better understand how pre-processing the data we work with (in this case an image) is a fundamental step in order to later implement more powerful AI techniques.
- An automated method that "discovers" the geometrical properties of a façade can discharge some of the most tedious work for the person studying it. For example, if the algorithm detects some symmetries among the elements in the façade, the scholars can put all their effort into the interpretation of that geometrical characteristic.
- If the method is applied to a sufficient number of façades, it may be possible to determine which geometrical constructs are used in different architectural styles, thus enriching the mathematical interpretations of each style. For example, presumably in the Renaissance style, we can detect a higher use of the golden ratio than in other styles.

5 Experiences in Education

Our team has built MonuMAI as a tool to be used in education and also in mathematical dissemination. In fact, we have already started teaching it in official maths courses in secondary education as well as in unique activities like science fairs and other educational activities.

In this regard, MonuMAI is part of some Applied Mathematics courses in regular secondary education (15 years). These courses have been carried out, designing a competency scheme for joint use with the subject of Geography and History, with 4 phases:

1. Finding information from buildings in the city of particular historical and cultural interest, highlighting the architectural style of the façade. This affected the *Consciousness and Cultural Expressions* competence.
2. Comparing and verifying the information provided by MonuMAI about the buildings with the information previously collected: *Science and Technology* competence and *Digital* competence.
3. Identifying mathematical ratios on the façade of the visited buildings (MonuMAI Lab): *Mathematical* competence and *Digital* competence.
4. Working in a group in an organized and effective way to complete the global task of gathering and approving the information. *Sense of Initiative and Entrepreneurship* competence and *Learn to Learn* competence.

The activity evaluations for the students were carried out through rubrics, resulting in very satisfactory evaluations in most cases. They were incorporated into the personal evaluation of each student in these subjects.

With respect to the second use in unique educational activities, in addition to lectures, we have done workshops and monumental street walks to show students the possibilities of its application and emphasize the logic with which it reasons, through its use in real cases. Also, we have presented it in various educational and scientific conferences (Descubre Foundation, 2019).

In the lectures and workshops, a set of photographs of especially significant monuments or whose determination leads to cases of interest have been used. Given that almost all students in secondary education own a smartphone and the app is free, this allows each student to do their own experimentation with the software. Among the images provided, there are some with unequivocal identification and others with insufficient identification, for which completion of it is necessary to obtain other shots of the same building (Fig. 17). In these workshops, students have also been encouraged to troll MonuMAI as has been previously mentioned. In the monumental walks, these practices have been carried out live on monuments in different cities, discussing and explaining the outings offered by MonuMAI. Thus, students can see how the probability of determination changes according to light conditions, perspective or distance.

We have also developed a children's teaching kit (MonuMAI_Toy), through the use of simple school games of construction wooden pieces. We have added to Mon-

AI and Mathematics Interaction for a New Learning Paradigm ...					133

Fig. 17 Use of MonuMAI in education. Left: Example of manual proportions adjustment with MonuMAI Lab. Right: A workshop with some printed images

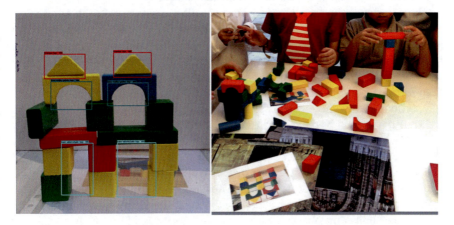

Fig. 18 MonuMAI for children. Left: recognition of wooden building blocks. Right: Playing and learning with `MonuMAI_toy`

uMAI the ability to recognize those pieces with their corresponding detected architectural elements (Fig. 18). Children can assemble their wooden building and then apply MonuMAI themselves to detect the assembled elements. The result of the use of this kit has always been a full commitment of the students and children developing a natural curiosity on how MonuMAI carries out such a classification.

Moreover, MonuMAI has also been integrated as a tool within the Educational Innovation Program "Living and Feeling the Heritage" of the Ministry of Education of the Regional Government of Andalucía (Spain) with specified didactic objectives and achieving a high compliance with them, according to the tutors of these projects.

Regarding the application in the education of the AGM, this will be carried out during the present year 2021. So, at the moment of writing this chapter, we still do not have an assessment in this regard.

6 What's Next for the Future?

We have shown some of MonuMAI's strengths and weaknesses in interdisciplinary work with mathematics, AI and art history. MonuMAI is already in its second version, but we hope soon to be able to present the third one with more functionalities and tools, like the previously presented AGM. It is clear that the functionalities provided by the latter will make the recognition of architectural elements and classification of styles much more precise, by being able to rectify the images and make frontal shots of the façades. In addition, the AGM will be possible to be used as a basis for educational work in mathematics on monumental façades.

The algorithmic approach that MonuMAI will use will be twofold: On the one hand, Computer Vision techniques, through the angular use of the Hough transform and its variants, as explained in Sect. 4 and, on the other hand, with the use of Deep Learning through CNN. Both approaches differ greatly, although in our system they complement each other. The AGM will enhance the deep learning algorithm while deep learning will offer us the direction of what kind of mathematical constructs (proportions, geometric elements, etc.) to look for in the monument through the AGM. But what will be the performance of both in purely mathematical tasks?

In some applications, the AGM could clearly be advantageous: for example, to automatically detect proportions in rectangular spans. Since in the model we must already have the segments of the lines that form it, detecting a rectangular proportion only requires making a search in the transformed domain. Two perpendicular segments that join at a point with an added condition: the quotient between their lengths must be the number that marks the proportion. This search, which involves measurement and is carried out directly, seems more efficient and faster to do it through the AGM, than by any other method.

However, other purposes can test both approaches. Like the one proposed in this question: is it possible to automatically learn to detect a symmetry (axial or central) in a façade? And a turn, a translation, or other isometries of the plane?

The most immediate answer is that with the AGM tool, it would be a relatively easy task, at least from a conceptual point of view: Detecting an (axial) symmetry from the AGM would consist of an exhaustive search for possible axes parallel to the vertical or horizontal segments (1 search variable). It also needs a soft verification step for each of these possible axes, of a symmetrical coincidence of the model, or part of it within a precision range. For a translation, the complexity will rise to two variables, always performing the search in the transformed domain. But this conceptual simplicity is tarnished, however, by the possible high computational cost that such tasks would have.

And how would MonuMAI's deep learning approach this task? MonuMAI can be extended to recognize new architectural elements. To do so, we have to expand the dataset with a sufficiently large and representative set of images of the new element to be recognized. As an example, more than 230 images of ogee arches were supplied to be able to recognize this type of arch. However, all these arches share a common geometry: a symmetrical arch that begins with a convex circumferential arc and then passes to another concave arc at an inflection point, joining the two parts at a cuspidal point. Therefore, it is reasonable to think that MonuMAI can be trained with an extended dataset from artificial geometric images to increase the performance of visual recognition. This is a line in which we hope to reach conclusions in the near future.

However, tackling more abstract concepts such as "symmetry" is a clearly much difficult task since it cannot be easily captured in a homogeneous set of images. Each example of symmetry can have a different geometric shape, sometimes so different that it will have very little to do with the other examples, except for the abstract fact of the symmetry itself. Will a deep learning-based AI be able to detect this kind of abstraction? Although we have our own opinions, the answer is, however, still unknown.

In this direction, other possibilities can be the use of more advanced systems based on deep learning or computer vision. These developments may be incorporated in the future into systems focused on teaching mathematics. For example, the creation of applications that are able to detect mathematical objects that are expressed visually. In this group, we can include graphs, lattices, tessellations or even curve characteristics, such as continuity or differentiability. The future presents some exciting possibilities for systems that work with mathematical concepts by means of AI. Systems that could detect properties, and what could be better, offer the possibility of a discussion about the concepts involved. But this will only be useful, and even possible, if today's mathematics education begins a transition from the old paradigms of finalist calculation to the new ones of learning based on logic and the concepts that Artificial Intelligence incorporates.

References

Ballard, D. (1981). Generalizing the hough transform to detect arbitrary shapes. *Pattern Recognition, 13*(2), 111–122.
Batanero, C., Chernoff, E. J., Engel, J., Lee, H. S., & Sánchez, E. (2016). *Research on teaching and learning probability*. Springer.
Batanero, C., Henry, M., & Parzysz, B. (2005). The nature of chance and probability. In G. Jones (Ed.), *Exploring probability in school, mathematics education library* (pp. 15–37). Springer.
Botana, F., Martínez-Sevilla, A., Kovács, Z., & Recio, T. (2020). *Automatically augmented reality with GeoGebra* (pp. 347–368). Brill, Leiden, The Netherlands. https://doi.org/10.1163/9789004408845_015.
Canny, J. (1986). A computational approach to edge detection. *IEEE Transactions on Pattern Analysis and Machine Intelligence, PAMI-8*(6), 679–698. https://doi.org/10.1109/TPAMI.1986.4767851.

Chassignol, M., Khoroshavin, A., Klimova, A., & Bilyatdinova, A. (2018). Artificial intelligence trends in education: A narrative overview. *Procedia Computer Science, 136*, 16–24 . *7th International Young Scientists Conference on Computational Science*, YSC2018, 02–06 July2018, Heraklion, Greece.

Dale, M. R. (2017). Shapes of graphs: Trees to triangles. In *Applying graph theory in ecological research* (pp. 37–53). Cambridge University Press. https://doi.org/10.1017/9781316105450.003.

Dersch, H. (2007). Panorama tools, open source software for immersive imaging. In: *International VR Photography Conference 2007 in Berkeley, Keynote Address*.

Descubre Foundation. (2019). Monumai citizen science. In *EUSEA Annual Conference of the European Science Engagement Association*. https://eusea.info/activity/monumai-citizen-science/.

Duda, R. O., & Hart, P. E. (1972). Use of the hough transformation to detect lines and curves in pictures. *Communications of the ACM, 15*(1), 11–15. https://doi.org/10.1145/361237.361242.

Fiorucci, M., Khoroshiltseva, M., Pontil, M., Traviglia, A., Del Bue, A., & James, S. (2020). Machine learning for cultural heritage: A survey. *Pattern Recognition Letters, 133,* 102–108. https://doi.org/10.1016/j.patrec.2020.02.017.

Gadanidis, G. (2017). Artificial intelligence, computational thinking, and mathematics education. *International Journal of Information and Learning Technology, 34*(2), 133–139.

Geogebra web page. Retrieved May 03, 2021, from https://www.geogebra.org/.

Lamas, A., Tabik, S., Cruz, P., Montes, R., Martínez-Sevilla, A., Cruz, T., & Herrera, F. (2021). MonuMAI: Dataset, deep learning pipeline and citizen science based app for monumental heritage taxonomy and classification. *Neurocomputing, 420,* 266–280.

Martíez-Sevilla, A. (2017a). Artistic heritage meets GeoGebra: A closer view to research on history of art and mathematics dissemination. In *Communication in the Geogebra Global Gathering (GGG-17)*. Johannes Kepler University, Linz (Austria). https://www.geogebra.org/m/wXrYA38J.

Martíez-Sevilla, A. (Ed.) (2017b). Paseos Matemáticos por Granada. Un estudio entre arte, ciencia e historia. Editorial UGR, Granada, Spain.

Martíez-Sevilla, A. (2020). Matemáticas, tecnología y arte. una propuesta educativa STEAM. *Uno: Revista de didáctica de las matemáticas, 87,* 15–21.

Martínez-Sevilla, A., & Cruz Cabrera, P. (2021). El módulo constructivo y la orientación del Palacio de Carlos V de Granada: Arte, Geometrímbolo. *Arte, Individuo y Sociedad (ARIS), 33*(1), 29–47. https://doi.org/10.5209/aris.67029.

Soycan, A., & Soycan, M. (2019). Perspective correction of building facade images for architectural applications. *Engineering Science and Technology - An International Journal, 22*(3), 697–705. https://doi.org/10.1016/j.jestch.2018.12.012.

Taddia, Y., Gonzalez-Garcia, L., Zambello, E., & Pellegrinelli, A. (2020). Quality assessment of photogrammetric models for facade and building reconstruction using DJI phantom 4 RTK. *Remote Sensing, 12*(19).

Wojtkowska, M., Kedzierski, M., & Delis, P. (2021). Validation of terrestrial laser scanning and artificial intelligence for measuring deformations of cultural heritage structures. *Measurement, 167*. https://doi.org/10.1016/j.measurement.2020.108291.

AI-Supported Learning of Mathematics

Introduction to Section 2 by the Coordinator Jana Trgalová

Artificial intelligence (AI) has entered mathematics education a few decades ago with software such as dynamic geometry (DG) or computer algebra systems (CAS). Feedback provided by this technology is recognized as its essential feature that supports learners' meaning making when they interact with the software (Balacheff, 1993). Recent technological developments result in new features added to these systems allowing the automatic discovery and proving of geometric properties (Kovács et al., 2018) or providing tutorial interventions supporting learners in solving mathematical problems (Richard et al., 2019). These developments open new possibilities for the interactions between the learner and the AI-supported *milieu* (Brousseau, 1997) conducting to new ways of knowledge construction.

Chapters in the Section 2 of the book focus on the complicit interactions, at the human initiative, between learners and AI-supported *milieus* for mathematical activity. This echoes a recent report on the place and impact of AI in our society (Villani, 2018) that calls for thinking about the modes of complementarity between humans and intelligent systems:

Que ce soit au niveau individuel ou collectif, cette complémentarité peut prendre plusieurs formes et peut être aliénante comme libératrice. Au cœur du développement de l'IA doit résider la nécessité de mettre en œuvre une complémentarité qui soit capacitante, en ce qu'elle permet de désautomatiser les tâches humaines (p. 12)[1].

> The issue of human system complementarity that empowers learners or teachers, addressed in the six chapters of this Section is tackled from four perspectives: (1) that of the design of intelligent systems (upstream from their use by teachers or learners)—*Grugeon-Allys et al.*, (2) that of the impact of AI-supported systems on mathematical activity and on the resulting learning—*Narboux & Durand-Guerrier*, (3) that of the interactions between learners and systems in solving mathematical activities—*Blažek & Pech, Hašek*, and *Dana-Picard & Kovács* and (4) that of knowledge and competencies required for empowering humans in their relationships with the systems—*Betteridge et al.*

---

[1] Whether at the individual or collective level, this complementarity can take many forms and can be alienating or liberating. At the heart of the development of AI must reside the need to implement a complementarity that is empowering, in that it makes it possible to de-automate human tasks. (Our translation)

AI applied to education opens ways to the development of systems supporting personalized/adaptive learning (Xie et al., 2019; Villani, 2018). In their chapter, *Grugeon-Allys, Chenevotot-Quentin and Pilet* report about a research conducted since two decades taking up this challenge. The authors provide a detailed account of the elaboration of didactic models of elementary algebra and geometry knowledge domains and learners enabling to design, develop and implement an automated diagnostic tool. This tool aims at describing students' profiles, building groups of students according to these and identifying their learning needs. Collaboration with researchers in computer science allowed for the design and implementation of adaptive learning paths—exercises and feedback—tailored to the students' needs. These personalized learning solutions, articulating knowledge and learner didactic models and AI techniques, can significantly help teachers to deploy differentiated methods of mathematics teaching.

Narboux and Durand-Guerrier address the issue of "reification of mathematical knowledge" (Balacheff, 1993, p. 3) related to proof in Coq, an AI-supported proof assistant. The authors conduct epistemological analyses of proofs of two well-known theorems—the sum of angles of a triangle being equal to two right angles and the Varignon theorem—in different mathematical settings on the one hand, and with the Coq assistant on the other hand. The formalization of the proofs in Coq highlights several issues, such as a number of implicit steps in proofs in a non-digital setting that need to be made explicit within Coq, or the impossibility in Coq to deduce facts from a drawing reflecting a *didactic contract* (Brousseau, 1997) between a teacher and the students. The authors discuss didactic implications of their analyses and open perspectives to exploit them in prospective mathematics teacher education.

According to Wilf (2005), computers have significantly impacted mathematical research, as he claims:

It begins with wondering what a particular situation looks like in detail; it continues with some computer experiments to show the structure of that situation for a selection of small values of the parameters of the problem; and then comes the human part: the mathematician gazes at the computer output, attempting to see and to codify some patterns. If this seems fruitful then the final step requires the mathematician to prove that the pattern she thinks she sees is in fact the truth, rather than a shimmering mirage above the desert sands.

Examples reported in chapters by *Blažek & Pech*, *Hašek* and *Dana-Picard & Kovács* resonate with this idea transposed to mathematics learning through problem-solving. *Blažek & Pech* explore how a dynamic geometry environment (DGE) can support learners' experiments, which leads the authors to suggest a model theorizing ways a DGE can help finding "empirical facts" leading progressively to problem solution. The operationality of the model is illustrated on two examples taken from the upper secondary school mathematics. Likewise, *Hašek* explores the interactions of learners (prospective mathematics teachers) with computer algebra and DG systems equipped with automated theorem proving features to solve two historical geometry problems. Experiences described by the author tend to show the potential of the environment to foster creative approaches to solving traditional problems. In their chapter, *Dana-Picard & Kovács* show how affordances offered by exploiting "networking between technologies", namely DG and CAS, can offer a new approach to studying traditional mathematical topics (isoptics of plane curves). Based on the reported human-driven experiments, the authors discuss the possibilities of automation of technology networking, thus opening avenues to considering AI techniques.

Last but not least, the rapid development of AI raises the issue of education towards the development of the skills to understand the ongoing transformations of our society and to adapt to a constantly changing world, and therefore to accompany us in the construction of an empowering complementarity with the machines. The school mission becomes therefore giving everyone a general knowledge of AI and algorithms (Villani, 2018). Following this line of thought, *Betteridge et al.* share their experiences with teaching programming in mathematics courses, opening a discussion of important issues related to AI, such as the changes of mathematics curricula that seem unavoidable in the AI age or the ethics underlying the tools for mathematics education.

References

Balacheff, N. (1993). Artificial Intelligence and Mathematics Education: Expectations and Questions. *14th Biennal of the Australian Association of Mathematics Teachers* (pp. 1–24). Perth, Australia.

Brousseau, G. (1997). *Theory of didactical situations in mathematics.* Dordrecht: Kluwer Academic Publishers.

Kovács, Z., Recio, T., & Vélez, M. P. (2018). Using Automated Reasoning Tools in GeoGebra in the Teaching and Learning of Proving in Geometry. *International Journal of Technology in Mathematics Education 25*(2), 33–50.

Richard, P. R., Venant, F., & Gagnon, M. (2019). Issues and Challenges in Instrumental Proof. In G. Hanna et al. (Eds.), *Proof Technology in Mathematics Research and Teaching* (pp. 139–172). Mathematics Education in the Digital Era, Vol. 14. Cham: Springer.

Villani, C. (2018). *Donner un sens à l'intelligence artificielle. Pour une stratégie nationale et européenne.* Rapport d'une mission parlementaire. https://www.aiforhumanity.fr/pdfs/9782111457089_Rapport_Villani_accessible.pdf

Wilf, H. S. (2005). Mathematics: An Experimental Science. In W. T. Gowers (Ed.), *Princeton Companion to Mathematics.* Princeton: Princeton University Press.

Xie, H., Chu, H. C., Hwang, G. J., & Wang, C.C. (2019). Trends and development in technology-enhanced adaptive/personalized learning: A systematic review of journal publications from 2007 to 2017. *Computers & Education*, 140.

Using Didactic Models to Design Adaptive Pathways to Meet Students' Learning Needs in an Online Learning Environment

Brigitte Grugeon-Allys, Françoise Chenevotot-Quentin, and Julia Pilet

In many education systems, diagnostic assessment to plan paths for students is a complex issue. We discuss the didactic conditions in order to design learning paths tailored to the students' learning needs in an intelligent technological environment for a given mathematical domain.

What are the conditions for defining the didactic models of knowledge, student, tasks, paths and the ontology which formalizes them? We will elaborate a methodology based on epistemological, institutional and cognitive approaches that underlie the analysis of students' learning needs in relation to the knowledge to be taught at a given grade level.

We will illustrate this on the French online learning environments *Pépite* and *MindMath*.

1 Introduction

This chapter concerns Sect. 2 "AI-supported learning of mathematics". It focuses on the design of learning environments for teaching and learning mathematics for students in middle/lower secondary grades (12–16-year olds). The research presented herein is based on results obtained from the *Pépite* project (Grugeon, 1997; Delozanne

B. Grugeon-Allys (✉)
Université Paris Est-Créteil, LDAR, Créteil, France
e-mail: brigitte.grugeon@orange.fr

F. Chenevotot-Quentin
Institut national supérieur du professorat et de l'éducation Lille Haut de France, LDAR, Villeneuve-d'Ascq, France

J. Pilet
Université Paris Est-Créteil, LDAR,, Créteil, France

© The Author(s), under exclusive license to Springer Nature Switzerland AG 2022
P. R. Richard et al. (eds.), *Mathematics Education in the Age of Artificial Intelligence*, Mathematics Education in the Digital Era 17,
https://doi.org/10.1007/978-3-030-86909-0_7

et al., 2010; Grugeon-Allys et al., 2012) and *MindMath* (Jolivet, Lesnes-Cuisiniez & Grugeon-Allys, 2021), in two mathematics fields (algebra and geometry), which resulted in two learning environments with the same names. Situated at the crossroads of the didactics of mathematics and computer science, gradually integrating adaptive learning and learning analytics, this research adopts an interdisciplinary approach that enables it to fully integrate the field of Interactive Learning Environments (ILE).

In France, *Pépite* learning environment concerns the assessment and regulation of learning in algebra for students in middle/lower secondary grades (12–16-year olds), in the classroom. Usually, assessment results are generated from standardized and psychometric models and many online exercises do not interpret the procedures and reasoning used by students to solve exercises. Such an approach has shown some strengths and limitations for making instructional decisions (Kettelin-Geller & Yovanoff, 2009). One may rightfully ask oneself how a didactic approach would allow to identify the features of appropriate digital diagnostic assessment of, for and as learning. Since the 1990s, this research based on a close collaboration between researchers in didactics of mathematics (LDAR)[1] and researchers in computer science for ILE (LIP6 MOCAH team),[2] has developed multidisciplinary projects (Delozanne et al., 2010; Grugeon-Allys et al., 2012) concerning the design, development and implementation of a digital diagnostic assessment tool *Pépite*. It aims to describe students' profiles to automatically build groups of students according to their profiles and to identify the learning needs of the students in elementary algebra. *Pépite* can be used by teachers to plan differentiated instruction for defined groups of students. *Pépite* was disseminated on platforms[3] that are largely used by teachers and students. Exercises offered in differentiated instruction have not been digitalized in the online platform but stored as pdf files in a database.

In France, the *MindMath* project was initiated to overcome the limitations of the *Pépite* environment. *MindMath* learning environment provides students in middle/lower secondary grades (12–16 years old) with adaptive paths in algebra and geometry, both in and out of the classroom. *MindMath* results of the collaboration between teams of researchers in computer science (LIP6 MOCAH team) and in didactics of mathematics (LDAR) as well as companies (Tralalère, Cabrilog, Domoscio, Breakfirst). First, the diagnostic assessment of student knowledge aims to describe and update the student profile in order to identify their learning needs and exercises to work on. Then, to do this, the learning environment must offer learners a list of exercises from the curriculum and relevant feedback to the learners' mathematical activities. Unlike *Pépite*, exercises allocation and feedback decisions are automatically calculated, and adaptive learning paths are fully automated through adaptive learning and learning analytics.

---

[1] LDAR: André Revuz Didactics Laboratory, Universities of Cergy-Paris, Lille, Paris, Paris-Est-Créteil, Rouen. https://www.ldar.website/.

[2] LIP6: Computer Science Laboratory, Paris-Sorbonne University. https://www.lip6.fr/recherche/team.php?acronyme=MOCAH.

[3] *Pépite* tools are available on *LaboMep* platform (developed by *Sésamath*, a French maths' teachers association https://labomep.sesamath.net/) until 2012 and then on *WIMS* environment (an online learning environment spanning learning from primary school to university not only in mathematics).

This chapter focuses on the potential of contributions from the didactics of mathematics, combined with the theories and expertise developed in AI, to model exercise paths within digital systems for students learning. This chapter does not deal with the development of online learning environments, their uses, or the analysis of experiments. More specifically, the articulation between the didactics of mathematics and AI can perform tasks that normally require human intelligence, including the automation of certain tasks, and activities associated with human reasoning such as decision-making, learning, and problem-solving.

With regard to Balacheff's work (1994), this chapter addresses the following questions. How to support the didactic and computational transpositions necessary for the representation of human knowledge in an online learning environment? This research is anchored in a didactic rather than computer-based approach to address issues relating to the didactic modelling of a field of knowledge and of the learner. With regard to the didactic modelling of a field of knowledge, how can we best model knowledge in order to establish links with a computable symbolic model? The challenge is to ensure the conditions for the epistemological validity of these interactive learning environments, particularly from the perspective of the field of tasks accessible to learners, the characteristics of how interactions are managed, and the coherence and consistency of the system in terms of knowledge representation in the interface. Indeed, these elements have an important impact on the conceptualization of mathematical objects and on the mathematical activity developed by the students. Concerning modelling of the learner, Balacheff (1994, p. 24) distinguishes two levels. On the one hand, behavioral modelling aims to organize and build up a corpus of observables enabling us to "*account for the student's behaviors*" (ibidem, p. 24). On the other hand, the epistemic modelling consists of "*attributing meaning to these behaviors*" (ibidem, p. 24) based on didactic analyses. According to Balacheff, these two levels are necessarily articulated in the interactive learning environment, but only the level of epistemic modelling allows a diagnostic function (ibidem, p. 26). This is why we are mainly interested in the epistemic modelling of the learner and in its potential to carry out a diagnosis followed by appropriate pathways as we do in the *Pépite* and *MindMath* learning environments.

This chapter explores the following questions:

- What are didactic conditions on the representation of knowledge that allow the design of exercises paths accessible to learners, epistemologically valid, and computable in an online learning environment integrating AI?
- What are didactic conditions that allow to describe the student's profile and to automate the selection of tasks (for diagnosis and for pathways) so that they are epistemologically valid to accompany the student in an online learning environment integrating AI?

This chapter is organized as follows. Using an example, we specify the issue and multidisciplinary theoretical approaches. Then we describe didactic models carried out, first for *Pépite* and then for *Mindmath* online learning environments.

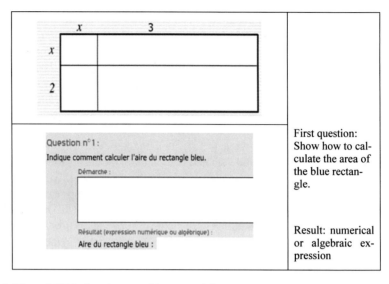

Fig. 1 The task "Calculate the area of the rectangle"

2 Theoretical Elements, Foundations of Didactic Modelling

2.1 An Example to Address the Roles of the Knowledge Model and of the Learner Model

Let us consider the tasks[4] of the algebraic domain and, in particular, the following task[5]: "Calculate the area of the rectangle" presented in Fig. 1.

Our challenge is to motivate the role of the didactic analysis to characterize a task and to allow the automatic calculation of the next task to propose to a student according to his answer. What mathematical knowledge is involved? At what grade level is it offered in the school curriculum? What is/are the solutions(s) expected at this school level? How to interpret the student's mathematical activity on the task? Which tasks should be proposed next?

To answer these questions, we must combine several approaches:

- Epistemological approach: to identify the type of task and position it among those of the algebraic domain;
- Institutional approach: to situate the solutions in relation to those expected according to the curriculum of the educational institution where the student is learning;

---

[4] We now use the term task, with reference to the Anthropological Theory of Didactics (ATD) to refer to a question in an exercise.

[5] Task from *Pépite* environment.

- Cognitive approach: to identify the knowledge and errors of the student solving the task and characterize his/her learning needs thanks to prior positioning of the tasks according to their complexity and the difficulties encountered by the student.

From an epistemological perspective, this task is a modelling task that involves the notion of the area of a rectangle and the translation between two registers of representation, on the one hand that of figures and on the other hand that of algebraic writings with the use of brackets. Other types of tasks are constitutive of the algebraic domain, for example, proving the equivalence of two expressions, expanding or factoring an algebraic expression. From an institutional perspective, this task is consistent with the 9th-grade[6] curriculum on algebraic expressions in France. The expected responses are as follows: $(x + 2) \times (x + 3)$ or $x^2 + 2x + 3x + 6$. The second response can be obtained by considering the property of area additivity on separate domains or the property of (double and simple) distributivity to perform the product. From a cognitive perspective, other tasks of the same type can prove more or less complex, depending on the adaptations left to the student (Robert, 1998, 2010). Another task, "Calculate the area of the rectangle" where the rectangle is a combined square with side x and rectangle of respective sides x and 3, is less complex than the previous one because it leads to the product of a monomial by a sum $x (x + 3)$.

Let us consider the following solutions provided by two 9th-grade students: $5x^2$ for student A, $x + 2 \times x + 3$ for student B. We make several hypotheses. Student B's response indicates a relevant modelling process in relation to the model of the rectangle area, but a difficulty relating to the translation between registers of representation. While student A's response indicates a difficulty in the modelling process or calculation relating to the variable status, and an aborted negotiation of the epistemological gap between arithmetic and algebra (Kieran, 2007). Students A and B have not built the same relationship to algebra during their schooling, nor did they achieve the same conceptual development in relation to algebraic expressions: an arithmetic relationship to algebra for student A, a relationship to algebra in the process of being built with regard to the rules of translation for student B. This highlights the need to use a learner model at the epistemic level to consider the student's actual activity.

It is the a priori analysis of the task that allows us to characterize it with respect to the model of knowledge (algebraic domain), to interpret the students' responses, to code them with respect to the learner model and to make hypotheses on the tasks to be proposed afterwards. The aim of this research is, therefore, to computerize this analysis on which the epistemological validity of online learning environments is based.

We now specify the theoretical framework for a didactic modelling of knowledge and of a learner that ensures the conditions for an epistemological validity of online learning environments integrating AI. We show the links with the generation of mathematical domain tasks accessible to learners and pathways adapted to the learning needs of students.

---

[6] 3ème in France.

2.2 Praxeological Model of Knowledge

We rely on the Anthropological Theory of Didactics—ATD (Chevallard, 1992)[7] to model a domain of knowledge (Grugeon, 1997). Mathematical knowledge is a product of human activity developed in a given institution (Chevallard, 1999). Modelling knowledge of a mathematical domain, for a given institution, consists in describing it by means of praxeologies that structure this domain (Chevallard, 1999) in order to relate them to those of the school curriculum. A praxeology is structured as a quadruplet: on the one hand, the type of tasks and the technique(s) used to solve the tasks of a given type, and on the other hand, the technology developed to justify the techniques (properties, rules, logical arguments) and the theory that justifies the technology. To avoid confusion between technology in the praxeological sense and in the computer technology sense, we use the term justification from now on as Taranto et al. (2020). Praxeologies are not isolated but structured in relation to each other: punctual praxeologies (PMO) aggregate into local praxeologies (LMO) around a justification, then in regional praxeologies (RMO) around a theory, and finally global praxeologies (GMO) around several theories for a field of study.

We use a Praxeological Reference Model (PRM)[8] related to a mathematical domain (Bosch & Gascon, 2005) as an analysis tool to characterize praxeologies. We distinguish praxeologies to be taught developed in the school institutions curriculum, those taught by teachers but also praxeologies learned by the students and try to put them in relation to each other to characterize the praxeologies to be worked on by the students during the course in the online learning environment. A PRM aims to characterize the epistemological aspects of the pieces of knowledge of this field and to describe the appropriate praxeologies associated with the targeted knowledge.

2.3 Praxeological Model of Algebraic Knowledge

Modelling the domain of elementary algebra (Grugeon, 1997) in *Pépite* and *Mind-Math* is based on a summary of research on didactics of algebra (Chevallard, 1985, 1989; Kieran, 2007) and falls within the scope of the secondary school curriculum. The Praxeological Reference Model of the algebraic domain is structured on the basis of two regional praxeologies (RMO), one relating to algebraic expressions (Pilet, 2015) and the other relating to equations (Sirejacob et al., 2018). Each RMO is structured into four local praxeologies (LMO): *modelling, representation, proof* and *calculation* (Gantois & Schneider, 2012) as presented in Fig. 2.

---

[7] Before 1997, Chevallard introduces the notion of relationship to knowledge and distinguishes between institutional and personal relationships to knowledge.

[8] Bosch and her team (Ruiz-Munzon et al., 2020) also use the term "epistemological model of reference", emphasizing its description in terms of praxeologies.

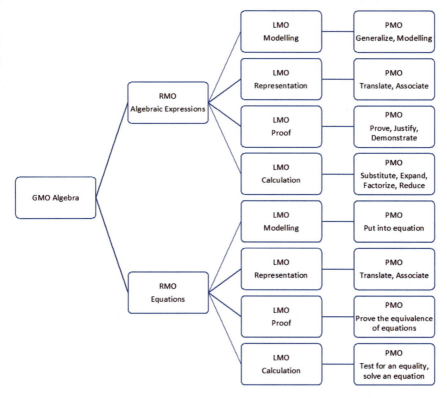

Fig. 2 Modelling algebraic domain knowledge using a Praxeological Reference Model

2.4 Didactic Modelling of Tasks

In order to make this model of knowledge operational in an online learning environment to automatically generate tasks, we define a didactic task model by the following elements, as brought in the previous example and illustrated for *Pépite* and *MindMath*:

- Generate (a) task(s) (from a category of a given type of task or from a pathway);
- Identify the knowledge involved for each task, from the regional to the specific level, to characterize the a priori role(s) of the task in learning and thus to position the tasks in relation to each other;
- Determine the educational level of a task;
- Characterize the complexity of a task;
- Indicate techniques allowing to solve it and the associated justifications (potential error if erroneous justifications).

The operationalization through the selected task variables is different according to *Pépite* (Chenevotot-Quentin et al., 2016) and *MindMath* environments. In *MindMath*, we have refined these variables by integrating them into the T4TEL framework developed by Chaachoua (2018).

2.5 Didactic Modelling of the Learner

During the education received, the student pursues the process of conceptualization of notions (Vergnaud, 1990), which he or she encounters from one institution to another where the praxeologies taught may vary (Wozniak, 2012). If the epistemological breaks are not sufficiently negotiated or the work of justification is not sufficiently developed, the student may continue to use reasoning, which may be appropriate at one point in his schooling, but which may no longer be appropriate in a new institution or may lead to erroneous conceptions and procedures. This can contribute to the construction of an unsuitable or even erroneous personal relationship with the knowledge at stake, as illustrated by the modelling techniques of students A and B.

The didactic model of the learner is conceived as an intelligible model of the student's personal relationship to knowledge in a given institution, relating to the mathematical domain worked on (Grugeon, 1997). Its aim is to describe the main features of the student's actual (sometimes erroneous) mathematical activity during the resolution of tasks in the domain. For each task, considering a description of the student's activity in terms of a technique at the microscopic level makes it difficult to achieve a meaningful and operative synthesis of the student's actual activities on the praxeologies of the domain of knowledge worked on. The learner's model is, therefore, based on a macroscopic level of description of the student's activity, considering the most frequently used justifications for the praxeologies worked on. We distinguish three a priori modes of justification: an expected mode of justification (noted M1), an old mode of justification (M3), and a mode of justification under construction but containing erroneous rules or properties (M2). Error classes are associated with the erroneous rules or properties. This epistemic model of the learner makes it possible to calculate and automatically update the student's profile, to identify his or her learning needs and then to automatically propose tasks to be worked on in pathways.

In the case of the algebraic domain, the student profile includes the modes of justification according to the local praxeologies of *modelling, representation, proof* and *calculation* and, in addition, the successes in the exercises. For the *modelling* praxeology, M1 corresponds to algebraic modelling, M2 to erroneous symbolic modelling linked to a lack of knowledge of the magnitudes involved or to erroneous translation, and M3 to arithmetic modelling. Student A using arithmetical approaches falls into M3 and student B into M2.

2.6 In Conclusion: An Ontology to Establish the Links Between the Knowledge Model and the Learner Model

The cross-referencing of epistemological, institutional and cognitive approaches, at different scale levels, allows to define:

- a praxeological model of knowledge to model a domain of knowledge and tasks, at different school levels;
- a model of the learner to describe and update the student's profile.

The question of linking them is crucial to enable a digital system to update the student's profile and to make decisions on the automatic generation of new tasks to be proposed. This link is achieved by an ontology *"(which) defines a set of elementary representations with which to model a domain of knowledge or discourse. Elementary representations are typically classes, attributes (or properties) and relations (or links between instances of classes)"* (Gruber, 2009).

As we successively illustrate for *Pépite* and *MindMath* online learning environments, by showing the potentialities and limits of the ontology, the one built for *Pépite* is only partial compared to the one defined in *MindMath*. These ontologies make it possible to automate the decision-making and the calculation of pathways composed of epistemologically valid tasks, meeting the conditions of a computational transposition, to accompany students in the evolution of their learning in an ILE.

3 *Pépite* Online Learning Environment Around Assessment and Regulation of Algebra in the Classroom

Pépite online learning environment consists in assessing and regulating learning of algebra in secondary school for work in the classroom (12–16 years old students). We present the diagnostic test that allows the elaboration of the student's profile and then the didactic models developed to support their computer representation and the design of pathways.

3.1 Knowledge Model and Diagnostic Tasks

The design of the diagnostic tasks and the analysis of the students' responses to the test are based on the knowledge model of elementary algebra presented above. We rely on the 9th-grade level test for 14-year-old students to explain the structure of *Pépite* test and to describe the responses analysis.

Table 1 Values of the variables for the third task "Calculating the area of the rectangle"

Variables	8th	9th	10th
Structure of the expression	First degree	Second degree	Second degree
Magnitudes involved	Perimeter	Area	Area
Number of variables	1	1	2
Example of expression	$4x + 7$	$(x + 2)(x + 3)$	$(a + 3)(a + b)$
Registers	Figures to algebraic expressions	Figures to algebraic expressions	Figures to algebraic expressions
Transformation of the algebraic expression	Reduce	Develop and reduce	Develop and reduce
Coefficients	{2; 3; 4; 5; 6; 7; 8; 9}	{2; 3; 4; 5; 6; 7; 8; 9}	{2; 3; 4; 5; 6; 7; 8; 9}
Name of the variables	x, y	x, y	{a, b, c, u, m, n, x, y, u, v}

The diagnostic test is composed of ten diagnostic tasks (27 individual items[9]) covering the four praxeologies of the PRM of the algebraic domain (Fig. 2): *Modelling* (formulas, putting into the equation) (5 items), *Representation* (relationships between registers of representation) (16 items), *Proof* (generalization by algebraic expressions and proof) (8 items), *Calculation* (expanding algebraic expressions, solving equations) (4 items) (Grugeon-Allys et al., 2018). The tasks may be multiple-choice items or open-ended items as illustrated in (Ibidem, Figs. 1a–c and 2, p. 249). We defined three versions of *Pépite* diagnostic test according to the school level of 8th-, 9th- and 10th-grades students. These versions are automatically generated and composed of similar tasks (same type of tasks and same proportion of tasks of a given type covering the domain) but instantiated by different values of the variables characterizing the tasks, according to the school level considered: type of tasks, structure of the algebraic objects and properties involved, expected justification, representation registers, complexity of the task. We illustrate this in Table 1 on the task "Calculating the area of the rectangle" already presented (Fig. 1) which is the third task of the test for 9th-grade students

For example, the equivalent of the third task for the 8th grade[10] is to produce the expression $4x + 7$ for the perimeter of the polygon (Fig. 3).

3.2 Analysis of Test Responses

An initial analysis of the answers is carried out for each task on the basis of an a priori didactic analysis. For each of the students' possible answers, the analysis

---

[9] The total of the four categories of items is 33; whereas, the total of individual items is 27 because some items appear in several categories.

[10] 4ème in France.

Fig. 3 Figure accompanying the task "Calculate the perimeter of the polygon" in the 8th-grade test

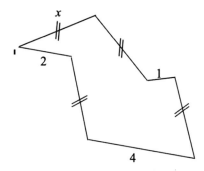

characterizes not only their validity (correct or not) but also the techniques and justifications to which they belong, associates them with classes of errors and situates the justification they involve in relation to the institutional expectations. Thus, student's consistency in algebraic activity is not analysed only in terms of validity (V) but in four other dimensions: the use of letters as variables (L), the translations from a semiotic register to another (T), the algebraic rationality (J) and the algebraic writing produced during symbolic transformations (EA), described in Table 2.

So, the diagnostic system provides a set of codes that characterize the response according to anticipated responses. As shown in Table 2, students' responses are coded with assessment criteria depending on knowledge and justification involved in the techniques. Contrary to usual assessment practices, we do not attribute different assessment criteria for each technique for each task. This would lead to a multiplicity of criteria on various tasks and would be unusable for a cross-analysis on all the tasks of the test.

We illustrate the task analysis with the task "Calculating the area of the rectangle" (Fig. 1). The different responses presented in the example are thus coded[11] V1 L1 EA1 T1 for $(x + 2) \times (x + 3)$ or $x^2 + 2x + 3x + 6$, V3 L4 EA4 T4 for $5x^2$ and V3 L3 EA3 T3 for $x + 2 \times x + 3$.

3.3 Student's Profile

The student's cognitive profile is supported by the learner model defined below and characterizes each student's mode of justification according to the local praxeologies of *calculation, proof, modelling* and *representation*. Due to the small number of tasks on the use of the algebraic tool for modelling and proof for the lowest grade levels, the *proof* and *modelling* praxeologies have been grouped together. We, therefore, distinguish three main praxeologies: Algebraic Calculation (CA) relied on *calculation* praxeology, Use of Algebra (UA) relied on *proof* and *modelling* praxeologies and Algebraic Translation (TA) relied on *representation* praxeology. The

---

[11] The algebraic rationality (J) is not involved for this task because no proof is expected.

Table 2 Assessment criteria according to five dimensions

Assessment dimensions	Assessment criteria
Validity of the response (V)	V0: No answer V1: Valid and optimal answer V2: Valid but not optimal answer V3: Invalid answer Vx: Unidentified answer
Use of letters as variables during *Modelling* or *Proof* (L)	L1: Correct and optimal use of letters L2: Correct but not optimal use of letters L3: Letters are used with incorrect rules L4: Letters are used as labels or abbreviations L5: No use of letters Lx: No interpretation
Translation from a semiotic register to another during *Representation* (T)	T1: Correct and optimal translation T2: Correct but not optimal translation T3: Incorrect translation taking into account the relationships T4: Incorrect translation without taking into account the relationships Tx: No interpretation
Algebraic rationality during *Proof* (J)	J1: Correct algebraic reasoning J2: Arithmetic reasoning J3: Algebraic reasoning but using incorrect rules Jx: No interpretation
Algebraic writing produced during symbolic transformations during *Calculation* (EA)	EA1: Reasoned and controlled algebraic calculation EA2: Correct algebraic calculation but without arguments EA3: Incorrect calculation based on syntactic rules (without taking into account the equivalence of expressions) EA4: Incorrect calculation based on arithmetic rules EAx: No interpretation

student's profile is described by a triplet (CA_i, UA_j, TA_k) with i, j and k ∈ {1,2,3} that locates the learned praxeologies relative to algebraic knowledge (Table 3). To compute it, the diagnostic system goes through all the tasks of the test and collects similar codes on different tasks. The algorithm for calculating the modes for UA is given in Appendix. The diagnostic also indicates the success rates and personal features (relative strengths and limitations, false rules and correct rules).

Figure 4 shows the individual global diagnosis for a 9th-grade student with CA3-UA3-TA3 on *LaboMep* platform[12] (wording intended for teachers). This student gives not much sense to algebraic activity and doesn't use it as a tool for solving problems. His personal features enlighten his strengths and weaknesses.

---

[12] The *LaboMep* interface is no longer operational due to the evolution of the implemented technologies.

Table 3 Justification modes according to *calculation, proof-modelling* and *representation* praxeologies

Justification modes	Level	Description
Algebraic Calculation CA (*calculation* praxeology)	CA1	Reasoned and controlled calculation taking into account the equivalence of expressions
	CA2	Calculation based on syntactic rules without taking into account the equivalence of expressions
	CA3	Calculation with arithmetic strategies and without operating priorities
Use of Algebra UA (*proof* and *modelling* praxeologies)	UA1a	Algebraic tool mastered
	UA1b	Algebraic tool adapted in some types of problems
	UA2	Algebraic tool used but without sense for letters
	UA3	Low because arithmetic reasoning
Algebraic Translation TA (*representation* praxeology)	TA1	Controlled translation
	TA2	Translation without support on the reformulation
	TA3	Translation as to schematise

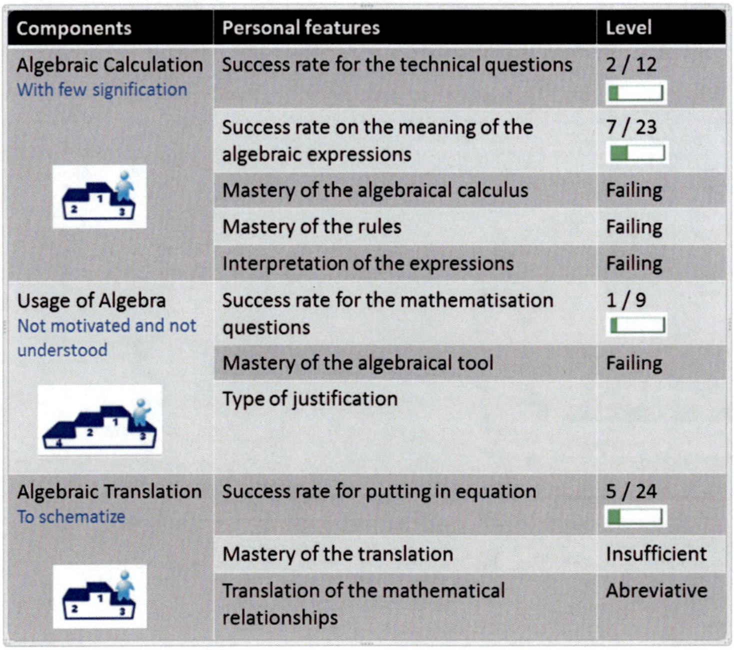

Components	Personal features	Level
Algebraic Calculation With few signification	Success rate for the technical questions	2 / 12
	Success rate on the meaning of the algebraic expressions	7 / 23
	Mastery of the algebraical calculus	Failing
	Mastery of the rules	Failing
	Interpretation of the expressions	Failing
Usage of Algebra Not motivated and not understood	Success rate for the mathematisation questions	1 / 9
	Mastery of the algebraical tool	Failing
	Type of justification	
Algebraic Translation To schematize	Success rate for putting in equation	5 / 24
	Mastery of the translation	Insufficient
	Translation of the mathematical relationships	Abreviative

Fig. 4 An overview of the individual global diagnosis for a 9th-grade student

3.4 Groups and Exercises Path

After the diagnostic test is passed in the classroom, the system automatically sets three groups of students who have close cognitive profiles, i.e. similar modes of justification. Figure 5 shows three groups built on *LaboMep* platform in a 9th-grade class: (1) Group A (0 out of 23) students making sense of algebra and developing an intelligent and controlled use of algebraic calculation, (2) Group B (7 out of 23) students making sense of algebraic calculation and beginning to develop an intelligent and controlled practice of algebraic calculation, (3) Group C (16 out of 23) students practicing algebraic calculation with little control, often blindly, using false rules more or less frequently.

Each group of students is given a pathway consisting of a set of tasks corresponding to the praxeologies to be worked on, according to the learning objectives targeted by the teacher (Grugeon-Allys et al., 2012; Pilet et al., 2013; Pilet, 2015). These tasks aim to ask old or erroneous techniques or justifications identified by the diagnosis or to continue the construction of the praxeologies involved. The tasks are not automatically generated by the system, which only gives the link to resources in a mathematical database. The tasks are indexed by the modes of justification (UA, TA, CA) and the variables defined in the didactic model of the tasks. The pathways can be proposed by teachers to these groups of students at different moments of the study, in particular when introducing new knowledge or resuming their study (Pilet, 2015).

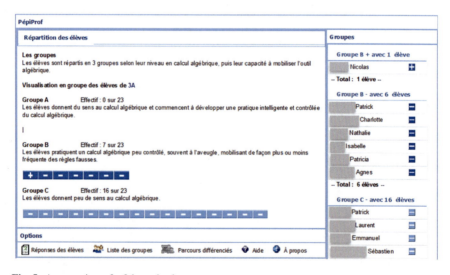

Fig. 5 An overview of a 9th grade class

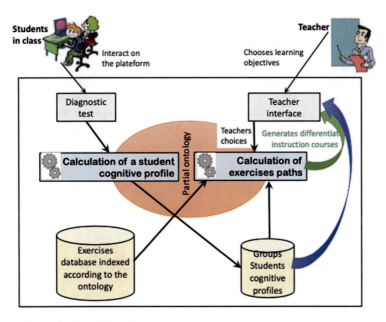

Fig. 6 *Pépite* online learning environment

3.5 Computer Model and Partial Ontology

After the students passed *Pépite* assessment, the environment (Fig. 6) automatically calculates an individual student's profile, as well as profiles for groups of students. Thanks to an ontology taking into account the algebraic praxeologies according to the school level, we have indexed assessment diagnostic tasks with variables presented above and sets of tasks. Then, the teacher can choose a teaching objective and the system automatically selects differentiated instruction sessions for identified groups of students of the class.

3.6 Limits

The calculation of justification modes and student groups is handled by a didactic algorithm that does not allow them to be updated as the student solves new tasks because it does not articulate didactic models and AI techniques. However, making a course dynamic for the students presents challenges: adapting the tasks to the student's responses during the course, managing the individual support of the students through appropriate feedback. This is the challenge of the following project entitled *MindMath*.

4 The *MindMath* Online Learning Environment Around Adaptive Paths in Algebra and Geometry

The general aim of the *MindMath* project is to produce an online learning environment to support the independent work of secondary school students (12–16-year olds) in their learning of algebra and geometry by offering them adaptive courses (Lesnes-Cuisiniez & Grugeon-Allys, 2019; Jolivet, Lesnes-Cuisiniez, & Grugeon-Allys, 2021). As discussed above, the knowledge model for the algebraic domain is based on the PRM for elementary algebra. The knowledge model for the field of figures of plane geometry is developed by Lesnes-Cuisiniez in his Ph.D. We focus on the domain of algebra and show the exploitation of the knowledge and learner models in the design of the *MindMath* learning environment.

4.1 Relations Between Praxeologies in the Didactic Model of Knowledge

In the PRM, the praxeologies are linked to each other, which characterizes the links between the knowledge. For the algebraic domain, the GMO contains the RMOs of algebraic expressions and equations the latter integrating the LMOs of *modelling, representation, proof* and *calculation*. The *modelling* LMO mobilizes the *representation* LMO while the *proof* LMO mobilizes the other three (Fig. 7). These relationships are considered in the *MindMath* ontology.

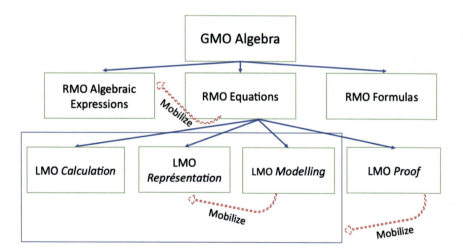

Fig. 7 Examples of relationships between praxeologies at different levels of inclusion

4.2 Didactic Model of Task Families

We present the operationalization in *MindMath* of the didactic task model, defined in the theoretical part. To model punctual praxeologies, Chaachoua (2018) developed the T4TEL framework and introduced task-type generators. We adapt the use of task-type generators by distinguishing between task-type variables (Chaachoua et al., 2019) and task variables (Grugeon-Allys et al., 2018; Lesnes-Cuisiniez & Grugeon-Allys, 2019). Choosing the values of the task-type variables allows defining different generators of task types, considering the different properties at stake in the LMO according to the structure of the mathematical objects. The task variables and their values have a double function: on the one hand, to characterize the scope of certain techniques, and on the other hand, to characterize the complexity of tasks. We, therefore, define task-type generators by instantiating task-type values and, for a given task type, task families by instantiating task variables. A task family generator is thus defined by an action verb, a fixed complement, a set of task-type variables and a set of task variables with their associated values. The tasks of a task family are considered to be similar except for the randomness of generation. This operationalization of the task model makes it possible, on the one hand, to produce task families by type of task, to motivate the need to move from one justification to another by playing on the scope of the techniques, and, on the other hand, to produce task families whose resolution makes it possible to further develop the techniques and justifications by playing on the complexity of the tasks.

For instance, the punctual praxeology "Solve first degree equation" is about the GMO of elementary algebra, related to the RMO of equations and in the LMO *calculation*. One of the task-type generators in the LMO is defined by the action verb "Solve", the fixed complement "first degree equation" and the following variables:

- Two task-type variables: (VT1) structure of the equation and (VT2) number of solutions of the equation,
- Three task variables: (Vt_P1) nature of solutions, (Vt_C1) nature of coefficients and (Vt_C2) complexity of rewriting.

The different values of VT1 specify the first-degree equation structures worked on in secondary school and distinguish between: equations of the form $ax + b = c$ (a $\neq 0$) or of the form $ax + b = cx + d$ (a-c $\neq 0$) or that $P(x) = Q(x)$ (with $P(x)$ and $Q(x)$ expressions of the first degree) reducing to $ax + b = cx + d$, equations of the form $P(x) = Q(x)$ (with $P(x)$ and $Q(x)$ expressions of the second degree and degree (P-Q) = 1). The variable Vt_P1 characterizes the families of tasks solved exclusively by non-arithmetic techniques, which is a way to make arithmetic techniques inoperative. Two values are associated to Vt_P1 according to the set of numbers to which the solution of $ax + b = cx + d$ belongs to $\mathbb{D}$ (decimal numbers) or $\mathbb{Q} \setminus \mathbb{D}$ (non-decimal rational numbers). The two variables Vt_C1 and Vt_C2 allow to characterize the complexity of the equation to be solved. Vt_C1 concerns the nature of the coefficients (integer, relative or rational). Vt_C2 concerns the complexity of the rewriting according to

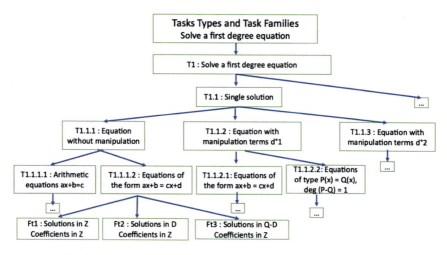

Fig. 8 Task families

whether it calls upon the punctual praxeologies *develop* and *reduce* an algebraic expression.

We first define generators of task types to structure the set of first-degree equations (see values of VT1 and VT2), relevant to secondary school. From the task-type generators, we define the associated task families. The coordination of the values of the variables VT and Vt allows us to produce families of tasks, on the one hand to motivate and work on the passage from one technique to another by varying their scope and the properties involved, on the other hand to make the techniques evolve by playing on the complexity of the tasks.

The assignment of task-type and task-variable values thus allows for the structuring of task families (Fig. 8) and the description of task paths tailored to students' learning needs.

For first-degree equations without manipulation and of the form $ax + b = cx + d$ $(a - c \neq 0)$ (VT1), having a single solution (VT2), coefficients of the same kind in $\mathbb{Z}$ (VT_C1), and no rewriting (Vt_C2), we define the task families Ft1, Ft2, and Ft3 according to the nature of the solution (Vt_P1): Ft1 with a solution in $\mathbb{Z}$, Ft2 with a solution in $\mathbb{D}$ and Ft3 with a solution in $\mathbb{Q} \setminus \mathbb{D}$. The *target* task family, a common objective of the curriculum concerning all students, is here the Ft3 family.

4.3 Learner Model and Student's Profile

The student's profile is described using the learner model defined in the theoretical part. It situates the learned praxeologies in relation to those aimed at in the institution, based on the modes of justification M1, M2, M3 built on each of the local praxeologies. In the case of elementary algebra, the modes of justification are specified for each of the constituent praxeologies of school algebra: *calculation* (CA), *modelling* (MA), *representation* (TA), *proof* (JA) (Grugeon-Allys, 2016). The student's profile is thus described by a quadruplet (CA_i, MA_j, TA_k, JA_l) with i, j, k and l ∈ {1,2,3} that locates the learned praxeologies relative to a mathematical knowledge, in a given institution, to which is associated an index of success on all the tasks. The selection of task families to be offered to students in the paths according to their learning needs is based on the modes of justification.

4.4 Didactic Path Model

The didactic model of task families characterizes the set of possible families produced by a generator. To select a path for a student, we take into account the student's mode of justification, relative to the praxeology worked on, to bring him/her to work as close as possible to his/her learning needs. From the student's mode of justification, we determine tasks families to work on before the *target task* family to make the student aware of the insufficiency of techniques based on old or non-adapted justifications and to make them evolve towards the targeted techniques (here algebraic techniques). For example, for equations of the first degree, for students in CA3 mode (see Table 3), the task families related to equations $ax + b = c$ and the families Ft1 and Ft2 will be worked on before the *target* family Ft3. It is also the mode of justification that makes it possible to distribute the tasks around the *target* family according to their complexities, and to propose tasks families related to the mobilization of praxeologies outside the regional praxeology. To solve equations with manipulation, it is necessary to bring *develop* or *reduce* types of tasks constitutive of the regional praxeology of algebraic expressions.

The task families are distributed as shown in Table 4 to define the pathways according to the student's grade level and mode of justification.

Table 4 Simplified path model based on justification mode

Justification mode	Ft "Old justification"	Ft "To negotiate the break-up"	Ft "Targeted justification"	Ft "Target, more complex in the RMO worked"	Ft "Target, more complex outside the RMO worked"
Old CA3	10%	30%	60%		
Incomplete CA2		25%	50%	25%	
Adapted CA1			40%	40%	20%

4.5 Computer Representation of Mathematical Knowledge and Activity Through an Ontology

A central issue in the *MindMath* project is to exploit didactic models and to represent in a structured way, both the praxeologies of a mathematical domain and the tasks families, and the elements related to the learner's model. The constructed ontology (Jolivet, 2018; Jolivet, Lesnes-Cuisiniez, & Grugeon-Allys, 2021) serves as a pivot for the description of the manipulated digital objects, communication and information exchanges between the different partners of the *MindMath* project. The ontology reifies the structure of the praxeologies and the associated justifications as well as the relations that link them and are derived from the didactic model of knowledge. The ontology also describes the constitutive elements of the task-type generators that compose the LMO and the associated variables to build the task families and the techniques in relation with the justifications, which are themselves linked by the relations depending on the institutional level. The ontology also integrates the erroneous justifications a priori identified in the didactic study, linked to the associated classes of errors from the learner model, which gives an indication to the system to decide on the types of tasks to make the students work on in priority.

The ontology structure, implemented in WOL (Web Ontology Language), allows to meet the conditions defined by Balacheff to design epistemologically valid ILE. The system leverages the ontology and makes paths recommendations tailored to students' learning needs while considering institutional instructions. The decision-making of the path is carried out by the adaptive AI algorithm, defined by the Domoscio company. The system then calculates the path by assigning a chronology to tasks families or by proposing a task corresponding to a new type of task if necessary to challenge an unsuitable relationship to an object of knowledge. Figure 9 summarizes the *MindMath* structure.

5 Conclusion and Perspectives

In this chapter, we have presented the main results of the *Pépite* and *MindMath* projects aimed at designing and developing diagnostic and regulation tools in online learning environments in order to propose pathways that are tailored to the learning needs of learners (12–16 year-olds) in two mathematical fields: algebra and geometry. The conditions created by Balacheff to represent knowledge in an interactive learning environment by addressing the delicate question of computational transposition largely echo the concerns underpinning this research: *"to create favorable conditions in which the student can construct acceptable knowledge in relation to a learning object, by providing them with relevant feedback"* (Balacheff, 1994, p. 12–13).

Although its strong roots in didactics of mathematics characterize this research, it has nevertheless benefited over a long period (more than twenty years) from a

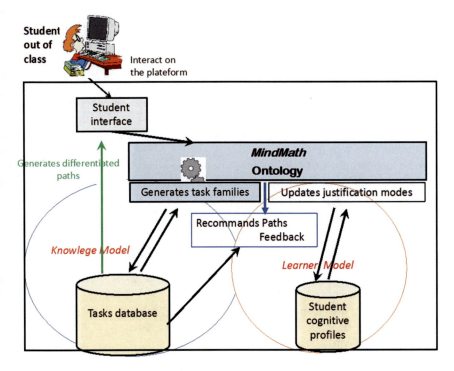

Fig. 9 *MindMath* online learning environment

variety of collaborations involving computer researchers, teachers and companies implicated in the design and implementation of learning environments for teaching and learning. Modelling of the mathematical knowledge involved, as well as that of the learner, is based on the articulation between three complementary approaches (epistemological, institutional and cognitive). Thus, the modelling of mathematical knowledge is based on the Anthropological Theory of Didactics and its evolution with the praxeological model, which is at the heart of this research. The strength of these choices (as demonstrated by the *MindMath* project) is that this approach is reproducible from one mathematical field to another (here, algebra and plane geometry).

Implementation of the diagnosis and regulation carried out in the *Pépite* and *MindMath* environments improved significantly over the course of the projects. In *Pépite*, the automatic diagnosis initiates the teacher's work by providing associated pathways to the groups of students in the class. *MindMath* implements AI techniques and uses an ontology based on the didactic study to carry out a dynamic diagnosis and propose adaptive pathways with individualized feedback. *Pépite* and *MindMath* are intended for use either in class or independently outside the classroom. They are designed for secondary school students with a wide range of profiles, which distinguishes them from the design of other ILE (e.g. Richard et al., 2011; Guin & Lefèvre, 2013) that often focus on average students.[13]

To conclude, we have demonstrated the potential advantages of optimizing learning through the didactics of mathematics. This choice has made it possible to characterize models of knowledge and of learners, essential to the design of a learning environment which analyses and learns information collected over the course of experiments in order to improve proposals for tasks and feedback adapted to the students' learning needs. Analyses of the collection of data obtained during the experiments are ongoing.

In the long term, we wish to continue our collaborations with different teams of researchers (didacticians, computer scientists) and companies, associating teachers, in order to pursue research in the field of ILE in the service of teaching and learning.

Appendix: Algorithm for Calculating Justification Modes for UA

Calculate the **UA** justification modes (modelling and proof) uses two indicators: **MOA** (use of the algebraic tool) and **JA** (type of justification).

MOA is a boolean

a. Expected use of the algebraic tool: MOA = 1

   Success rate (modelling tasks) > 50 % AND Success rate (proof tasks) > 60 %

b. Incorrect or inappropriate use of the algebraic tool: MOA ≠ 1

JA has 5 values: 2, 1, 0, −1, −2.

Validity requirements: at least one coding on the justification type dimension NJ1 + NJ2 + NJ3 + NJ0 ≠ 0 (NJi is the weighted sum of coded responses Ji in the whole test).

a. Algebraic justification, on many tasks (JA = 2)

   Number of tasks performed > Number of tasks not completed AND (NJ1 > NJ2 + NJ3) AND (NJ1 > NJ0)

b. Algebraic justification, on few tasks (JA = 1)

   Number of tasks performed > Number of tasks not completed AND (NJ1 > NJ2 + NJ3) AND (NJ1 <= NJ0)

c. Algebraic justification, on too few tasks (JA = 0)

   Number of tasks performed < Number of tasks not completed AND (NJ1 > NJ2 + NJ3)

---

[13] Following experiments with QED−Tutrix tutor system with 8th graders (13 years olds), the difficulties encountered by the system in assessing the solutions of the most advanced and weak students (Leduc, 2016) could be related to the MIA model calibrated for average students.

d. Incorrect algebraic justification, on many tasks (JA = −1)

   (NJ1≤NJ2 + NJ3) AND (NJ2 < NJ3)

e. Justification based on numerical examples (JA = −2)

   (NJ1≤NJ2 + NJ3) AND (NJ2 ≥ NJ3)

From the values of MOA (0;1) and JA (−2; −1;0;1;2), the following table gives the corresponding ordered justification modes (UA1a > UA1b > UA2 > UA3):

MOA	JA	UA	MOA	JA	UA
1	2	UA1a	0	2	UA1b
1	1	UA1b	0	1	UA1b
1	0	UA1b	0	0	UA2
1	−1	UA1b	0	−1	UA2
1	−2	UA2	0	−2	UA3

This table is used in the algorithm that calculates the ordered justification modes (UA1a > UA1b > UA2 > UA3):

1. UA1a = (MOA = 1) AND (JA = 2)
2. UA1b = (MOA = 0 AND (JA = 2 OU JA = 1)) OR (MOA = 1 AND (JA = 1 OR JA = −1 OR JA = 0))
3. UA2 = (MOA = 0 AND (JA = 0 OR JA = −1)) OR (MOA = 1 AND JA = −2)
4. UA3 = (MOA = 0 AND JA = −2)

References

Balacheff, N. (1994). Didactique et intelligence artificielle. *Recherches en Didactique des Mathématiques 14*(1.2), 9–42.

Bosch, M., & Gascon, J. (2005). La praxéologie comme unité d'analyse des processus didactiques. In A. Mercier & C. Margolinas (Eds.), *Balises pour la didactique: cours de la 12ème école d'été de didactique des mathématiques* (pp. 107–122). Grenoble: La pensée sauvage.

Chaachoua, H. (2018). T4TEL, un cadre de référence didactique pour la conception des EIAH. In J. Pilet & C. Vendeira (Eds.), *Actes du séminaire de didactique des mathématiques 2018* (p. 8–25). IREM de Paris—Université Paris Diderot. https://hal.archives-ouvertes.fr/hal-02421410/document

Chaachoua, H., Bessot, A., Romo, A., & Castela, C. (2019). Developments and functionalities in the praxeological model. In M. Bosch, Y. Chevallard, F. Javier Garcia & J. Monaghan (Éds.), *Working with the anthropological theory of the didactic: A comprehensive casebook* (pp. 41–60). Routledge. https://www.routledge.com/Working-with-the-Anthropological-Theory-of-the-Didactic-in-Mathematics/Bosch-Chevallard-Garcia-Monaghan/p/book/9780367187705

Chenevotot-Quentin, F., Grugeon-Allys, B., Pilet, J., Delozanne, E., & Prévit, D. (2016). The diagnostic assessment Pépite and the question of its transfer at different school levels. In K. Krainer & N. Vondrová (Eds.), Proceedings of the Ninth Congress of the European Society for Research in

Mathematics Education CERME 9, 4–8 February 2015 (pp. 2326–2332). Prague, Czech Republic: Charles University in Prague, Faculty of Education and ERME. http://www.mathematik.uni-dortmund.de/~prediger/ERME/CERME9_Proceedings_2015.pdf

Chevallard, Y. (1985). Le passage de l'arithmétique à l'algèbrique dans l'enseignement des mathématiques au collège—Première partie: L'évolution de la transposition didactique. *Petit X, No, 5*, 51–94.

Chevallard, Y. (1989). Le passage de l'arithmétique à l'algébrique dans l'enseignement des mathématiques au collège—deuxième partie: perspectives curriculaires, la notion de modélisation. *Petit x, 19*, 43–72.

Chevallard, Y. (1992). Concepts fondamentaux de la didactique: perspectives apportées par une approche anthropologique. *Recherches En Didactique Des Mathématiques, 12*(1), 73–112.

Chevallard, Y. (1999). L'analyse des pratiques enseignantes en théorie anthropologique du didactique. *Recherches En Didactique Des Mathématiques, 19*(2), 221–265.

Delozanne, E., Prévit D., Grugeon-Allys, B., & Chenevotot-Quentin, F. (2010). Vers un modèle de diagnostic de compétence. *Revue Technique et Science Informatiques, 29*(8–9), 899–938.

Gantois, J.-Y., & Schneider, M. (2012). Une forme embryonnaire du concept de dérivée induite par un milieu graphico-cinématique dans une praxéologie « modélisation ». *Recherches En Didactique Des Mathématiques, 32*(1), 57–99.

Gruber, T. (2009). Ontology. In L. Liu & M. T. Özsu (Éds.), *Encyclopedia of database systems*. Springer. http://tomgruber.org/writing/ontology-definition-2007.htm

Grugeon, B. (1997). Conception et exploitation d'une structure d'analyse multidimensionnelle en algèbre élémentaire. *Recherches En Didactique Des Mathématiques, 17*(2), 167–210.

Grugeon-Allys, B. (2016). Modéliser le profil diagnostique des élèves dans un domaine mathématique et l'exploiter pour gérer l'hétérogénéité des apprentissages en classe: une approche didactique multidimensionnelle. *E-JIREF, 2*(2), 63–88.

Grugeon-Allys, B., Chenevotot-Quentin, F., Pilet, J., & Prévit, D. (2018). Online automated assessment and student learning: the Pépite project in elementary algebra. In L. Ball, P. Drijvers, S. Ladel, H.-S. Siller, M. Tabach & C. Vale (Eds.), *Uses of technology in primary and secondary mathematics education*. ICME-13 (pp. 245–266). Monographs. Springer, Cham. https://doi.org/10.1007/978-3-319-76575-4

Grugeon-Allys, B., Pilet, J., Chenevotot-Quentin, F., & Delozanne, E. (2012). Diagnostic et parcours différenciés d'enseignement en algèbre élémentaire. In L. Coulange, J.-P. Drouhard, J.-L. Dorier & A. Robert (Eds.), *Recherche en Didactique des Mathématiques, Enseignement de l'algèbre élémentaire* (pp. 137–162). Grenoble: La pensée sauvage.

Guin, N., & Lefèvre, M. (2013). From a Customizable ITS to an Adaptive ITS. In H. C. Lane, K. Yacef, J. Mostow & P. Pavlik (Eds.), *Artificial intelligence in education* (pp. 141–150). Springer. https://doi.org/10.1007/978-3-642-39112-5_15

Jolivet, S. (2018). Modèle de description didactique de ressources d'apprentissage en mathématiques, pour l'indéxation et des services EIAH. Thèse de l'Université Grenoble Alpes. https://tel.archives-ouvertes.fr/tel-02079412

Jolivet, S., Lesnes-Cuisiniez, E., & Grugeon-Allys, B. (2021). Conception d'une plateforme d'apprentissage en ligne en algèbre et en géométrie: prise en compte et apports de modèles didactiques. *Annales de Didactique et de Sciences Cognitives*.

Ketterlin-Geller, L.-R., & Yovanoff, P. (2009). Diagnostic assessment in mathematics to support instructional decision making. *Practical Assessment Research & Education, 14*(16), 1–11.

Kieran, C. (2007). Learning and teaching algebra at the middle school through college levels: Building meaning for symbols and their manipulation. In F. K. Lester (Ed.), *Second handbook of research on mathematics teaching and learning* (pp. 707–762). Greenwich, CT: Information Age Publishing.

Leduc, N. (2016). QED-Tutrix : système tutoriel intelligent pour l'accompagnement des élèves en situation de résolution de problèmes de démonstration en géométrie plane. Thèse de l'École Polytechnique de Montréal. https://publications.polymtl.ca/2450/

Lesnes-Cuisiniez, E., & Grugeon-Allys, B. (2019). Modèle d'exercices et parcours d'apprentissage prenant en compte le raisonnement de l'élève en mathématique au collège. In J. Broisin, E. Sanchez, A. Yessad, F. Chenevotot (Eds.), *Actes de la 9e Conférence sur les EIAH* (pp. 205–210), 4-7 juin 2019. Paris.

Pilet, J. (2015). Réguler l'enseignement en algèbre élémentaire par des parcours d'enseignement différencié. *Recherches En Didactique Des Mathématiques, 35*(3), 273–312.

Pilet, J., Chenevotot, F., Grugeon, B., El-Kechaï, N., & Delozanne, E. (2013). Bridging diagnosis and learning of elementary algebra using technologies. In B. Ubuz, C. Haser & M. A. Mariotti (eds.), Proceedings of the eighth congress of the European research in mathematics education. Antalya, Turquie, du 6 au 10 février 2013 (pp. 2684–2694). Antalya, Turquie: Middle East Technical University of Rzeszow. http://www.erme.tu-dortmund.de/~erme/doc/CERME8/CERME8_2013_Proceedings.pdf

Richard, P. -R., Fortuny, J. -M., Gagnin, M., Leduc, N., Puertas, E., & Tessier-Baillargon, M. (2011). Didactic and theoretical-based perspectives in the experimental development of an intelligent tutorial system for the learning of geometry. *ZDM, 43(3),* 425–439.

Robert, A. (1998). Outils d'analyse des contenus mathématiques à enseigner au lycée et à l'université. *Recherches En Didactique Des Mathématiques, 18*(2), 139–190.

Robert, A. (2010). Une méthodologie pour analyser les activités (possibles) des élèves en classe. In F. Vandebrouck (Ed.), *La classe de mathématiques: activités des élèves et pratiques des enseignants* (pp. 45–57). Toulouse: Octares.

Ruiz-Munzón, N., Bosch, M., & et Gascón, J. (2020). Un modèle épistémologique de référence pour la recherche sur l'algèbre élémentaire. *Nouveaux cahiers de la recherche en éducation, 22*(1), 123–144.

Sirejacob, S., Chenevotot-Quentin, F., & Grugeon-Allys, B. (2018). Pépite online automated assessment and student learning: the domain of equations for grade 8th. In G. Gueudet & T. Dooley (Eds.), *Proceedings of the tenth congress of the european society for research in mathematics education CERME 10, 1–4 February 2017* (pp. 2470–2477). Dublin, Irland: Institute of Education Dublin City University. http://www.mathematik.uni-dortmund.de/ieem/erme_temp/CERME10_Proceedings_final.pdf

Taranto, E., Robutti, O., & Arzarello, F. (2020). Learning within MOOCs for mathematics teacher education. *ZDM, 52,* 1439–1453. https://doi.org/10.1007/s11858-020-01178-2

Vergnaud, G. (1990). La théorie des champs conceptuels. *Recherches en Didactique des Mathématiques, 10*(2.3), 133–170.

Wozniak, F. (2012). Analyse didactique des praxéologies de modélisation mathématique à l'école: une étude de cas. *Education Et Didactique, 6*(2), 65–88.

Combining Pencil/Paper Proofs and Formal Proofs, A Challenge for Artificial Intelligence and Mathematics Education

Julien Narboux and Viviane Durand-Guerrier

1 Introduction

The need for developing research at the interface between mathematics and computer science in education is growing due to the evolution of curriculum, in particular in France but also in many countries.

> According to Churchhouse et al. (1986), the relationship between mathematics and computer science—especially the influence of computer science in mathematics and the role of mathematics in computer science—is an epistemological and didactic issue that transcends school systems and national contexts. The use of computer tools in the teaching of mathematics and informatics, raises questions about the nature of these tools. This can be connected to the particular role played by mathematics in computer science, the proximity of some aspects of both disciplines and the common nature of some of their questions.
>
> (Durand-Guerrier et al., 2019, p. 116)

Among these aspects, proof and logical issues are certainly among the most prominent. The second author has worked for long on the importance and interest of logical analysis of proofs for mathematics education (Durand-Guerrier, 2008). As stressed by Durand-Guerrier and Arsac (2009, p. 148), logical analysis of proof for mathematics education fulfils several functions. The first one, which was the main purpose for logicians since the late nineteenth century, is to check the logical validity i.e. the correctness of the proof. A second function is to understand the proving strategy of the author of the proof. A third one is to contribute to the understanding and appropriation of proofs, as part of the study of the contents of the course in which they appear and as a means to better understand what are mathematical proofs and their

---

J. Narboux (✉)
UMR 7357 ICube CNRS, University of Strasbourg, Strasbourg, France
e-mail: jnarboux@narboux.fr

V. Durand-Guerrier
IMAG, Univ Montpellier, CNRS, Montpellier, France

possible specificities in a given mathematical field on the other hand. In this respect, it contributes to conceptualization. Different tools for realizing such logical analysis have been considered, such as natural deduction or dialogical analysis, which allows intermediate formalization between mathematical proofs and formal proofs. Moving to the design of AI tools for mathematics education such as proof assistant raises new questions.

In this chapter, we will highlight the issues raised by these new questions with two examples. The first one is the proof of the theorem that "the sum of angles of a triangle is two right angles" with a contrastive analysis of the Pythagorean proof as found in Euclid Elements, and a formalization of this proof in the Coq proof assistant.[1] The second one is around the proof of the so-called Varignon's theorem. The second author of the paper uses this theorem as basis for developing proof competencies in mathematics secondary teachers training. We will first motivate the choice of this theorem as food for thoughts on the didactical interest of combining pencil/paper proofs and formal proofs and compare several proofs of this theorem in different mathematical frames: Euclidean geometry, analytic geometry, vector geometry, the area method and their formalization in Coq. We then present the main features of the teacher training session around this theorem that consist in first providing individually at least two proofs of the theorem and to analyse them in line with the questions raised in the introduction, and then explore the "inverse" problem consisting in determining necessary and sufficient conditions for getting a rhombus, a rectangle, a square. We then discuss the interest of introducing in the training session a proof assistant in order to enrich the milieu of the situation.

2 The Sum of Angles of a Triangle is Two Right

2.1 Some Questions and Issues Raised by the Proof by Pythagoras

This is taken from (Durand-Guerrier and Arsac, 2009,) that was presented at the 19th ICMI conference on proof and proving in mathematics education and published in the pre-proceedings of the conference. We recall here the proof attributed to Pythagoreans that the sum of the angles in a triangle is equal to two right angles.

> Given a triangle ABC, let draw DE parallel to BC through A. The alternate angles are equal, on the one hand, the one under DAB to the one under ABC; on the other hand, the one under EAC to the one under ACB. Let add the one under BAC to the two others. The angles DAB, BAC, CAE, that means the ones under DAB, BAE, that means two rights, are hence equal to the three angles of the triangle. Hence, the three angles of the triangle are equal to two rights.

---

[1] A proof assistant is a piece of software which allows the user of the system to state mathematical definition and properties and to prove theorems interactively using a formal language. The proofs are checked mechanically.

Two remarks stand out: 1. A first object is given, a triangle, and nothing is said about hypotheses; 2. A second object is introduced, a parallel line DE to BC through A, that appears as a key for the proof, due to the fact that the whole proof is built on properties of alternate angles. Then, two first questions emerge: Q1. What relationship between data and hypotheses? Do we use a hypothesis in this proof? Q2. What role for the introduction of objects? Could the main ideas of a proof be resumed to the list of objects that have to be introduced? In a middle school's textbook, we can read that it is necessary to take a triangle "absolutely ordinary (scalene)", that means that the proof deals with the general case. This leads to a new question: Q3. How is generality taken into account in Geometry? Is it the same process in Algebra? The proposition is on triangle, so it is natural to introduce a triangle. But the proof relies entirely on the introduction of a second object, a line. Introducing that line can be justified only by the proposition that "one line can be drawn parallel to a given line through any point not on the line". As a triangle is defined by a set of three points not on the same line, we can actually apply this statement. This allows us to answer to Q1: here the data are three points and the hypothesis is that they are not on the same line. So data are objects and hypotheses express relations between these objects. This was hidden in the initial writing of the proof where the necessity of using a hypothesis is masked by the material possibility of doing the construction: drawing a triangle, one determines three points not all on a same line, and then it is actually possible to draw the parallel. The proof could also be completed without relying on the hypothesis that the three points are not on the same line, but by performing a case distinction. This leads to a new question, closely related with Q3: what evidences are used in proofs, particularly in Geometry? And how are we sure to check validity? In fact, there are still other evidences hidden in that proof (Arsac, 1998). The Pythagorean proof above provides an example of such evidence; we are inclined to conjecture that this recourse to evidence is possible because Geometry as a theory has been elaborated in such a manner that those types of evidence, that are expressed by true statements in the drawing register, are logically deducible in the theory (the axiomatic has been built on this purpose). However, it is also clear that some evidences in the drawing register have to be questioned in the theory, this corresponding to the back and forth between objects in an interpretation (here the drawing register) and the theory (here Plane Geometry), and hence between truth and validity (Durand-Guerrier, 2008). Manders has argued that the use of diagrammatic inferences in Euclid is not a lack of rigour as it is restricted to specific statements about the relative position of geometric objects (Manders, 2011). Several authors have proposed formal systems to provide a validity criterion for such inferences or diagrammatic inferences (Avigad et al., 2009; Miller, 2007, 2012; Mumma, 2010; Winterstein, 2004; Winterstein et al., 2000). But, up to our knowledge, Euclid's proofs have never been checked using these formal systems. In the experience of the first author about the mechanical checking of Euclid's proofs of the first book of Euclid's elements (Beeson et al., 2019) and as will be demonstrated by the following formalization of the proof of Pythagoras, it is difficult to justify that diagrammatic inferences are not gaps in the proof because:

1. the diagrammatic inferences are hard to separate from other inferences because statements guarantying the relative position of geometric objects (what Manders calls co-exact attributes) often use as premises exact attributes
2. the diagrammatic inferences sometimes rely on properties which are not even visually evident, they are evident on an instance of the figure, but sometimes the genericity of the validity of the property relies on an exhaustive enumeration of the different possible figures.

On a pragmatic level, it is not possible to prove every "evidence" of the drawing register; hence, to know which "evidences" are (logically) acceptable in a proof is clearly a difficult question that necessitates both mathematical knowledge and logical competencies (in particular to understand what is an axiomatic, and how it is related with interpretation). These questions are at the very core of Tarski's methodology of deductive science (Tarski, 1936) that permits a genuine articulation between form and content, allowing to take into account the powerful methods provided by syntax, without giving up to the advantages of the semantic approach (Sinaceur, 1991). Geometry and figures play a special role in the teaching of proofs: the figure and its declination as interactive experiment using a dynamic geometry system questions the need for a proof for a pupil, the figure is a depiction of the semantics of the statement. Diagrammatic inferences play a crucial role in teaching proof. The teacher claims to do without it, but as we will see this is not the case in practice. Diagrammatic inferences both question the difference between syntax and semantics, and represent an alteration of the didactic contract. Diagrams are pieces of syntax which enjoy some properties of their semantics. For example, a symmetric relation is often depicted by a symmetric symbol.

2.2 The Formal Proof That the Sum of Angles of a Triangle is Two Right Angles

In this section, we describe the formal proof within the Coq proof assistant that the sum of angles of a triangle is two right angles. To describe the formal proof, we need a precise context: an axiomatic setting and some definitions. The proof we describe can be formalized in the context of what Hartshorne (2000) calls an arbitrary Hilbert plane: any model of the first three groups of Hilbert axioms or equivalently Tarski's axioms $A_1 - A_9$ as listed in (Schwabhäuser et al., 1983). This set of axioms describe the results which are valid in both hyperbolic and Euclidean geometry without assuming any continuity axiom. The plane can be non-Archimedean. We also assume in this chapter the postulate of alternate interior angles, stating that if two lines are parallel the alternate interior angles of any secant are congruent. This postulate is equivalent to Euclid 5th postulate (Boutry et al., 2017).

To define the concept of sum of angles of a triangle within a computer, we could define the measure of an angle as a real and use the sum of the reals to define the sum of the angles as the sum of the measures. This is the most common approach in high school. But formally, in a synthetic geometry setting, to define the measure of an angle, the Archimedes postulate is needed, or one need to assume the protractor postulate. In a formal setting, it is interesting as a kind of exercise in reverse mathematics to identify the minimum assumptions needed for the proofs. Therefore, we chose in the library about foundations of geometry in Coq (GeoCoq) to provide a purely geometric definition of the sum of angles which make sense even in a non-Archimedean geometry and without any continuity assumption. More details about the definition of the sum of angles can be found in (Gries et al., 2016).

Instead of proving that the sum of angles is 180°, we prove (as in the proof by Pythagoras above) that it is congruent to a flat angle or equivalently to two right angles.

As noted by modern commentators of Euclid's Elements, the proofs of Euclid lack the justification for the relative position of the points on the figure. Euclid does not even provide the axioms for justifying these kind of reasoning. However, Avigad et al. (2009) claim that these gaps can be filled by some automatic procedure, justifying in some sense the gaps in Euclid's original proofs.

The usual proof that the sum of angles of a triangle is two right, such as the one given by A. Amiot according to French Wikipedia,[2] contains the same kind of gap. It does not provide the proof that the angles are alternate-interior angles, it is stated without proof. In this section, we give a rigorous proof, which is a translation in natural language and simplification of the formal proof which can be found in GeoCoq.[3]

In formal development, we always try to prove the most generic results; that is why in the following, we assume that triangles are not necessarily non-degenerate and for quadrilaterals as well.

To detail the proof, we need a definition of alternate-interior angles. In GeoCoq, we do not have an explicit definition of this concept.[4] But we have a definition to state that two points are on opposite sides of a line. Following Tarski, we say that the points P and Q are on opposite sides of line AB, if there is a point I which lies both on segment PQ and on line AB.

The following property is equivalent to the parallel postulate[5]:

---

[2] The comment in French Wikipedia about Amiot's proof seems to say that the proof is valid only in Euclidean geometry because it uses the construction of parallel line AC trough B. To be precise, the proof does not rely on the uniqueness of this line only on its existence, so this first step of the proof is valid also in hyperbolic geometry (but not in elliptic geometry). The Wikipedia comment fails to notice that essential use of a version of the parallel postulate relies in the use of what we called above the postulate of alternate-interior angles.

[3] http://geocoq.github.io/GeoCoq/html/GeoCoq.Meta_theory.Parallel_postulates.alternate_interior_angles_triangle.html#.

[4] We may add a definition of alternate-interior angles, which would be a shortcut for the predicate TS which states that two points are on opposite sides of a line, but adding more definitions make the formal proofs more cumbersome, that is why we hesitate to introduce a new definition.

[5] Note that the reciprocal is valid in neutral geometry.

Definition 1 *(Alternate interior angles postulate)* If B and D are on opposite sides of line AC and line AB is parallel to line CD then the angles $\angle BAC$ and $\angle DCA$ are congruent.

In Coq's syntax, we have

```
Definition alternate_interior_angles_postulate :=
   forall A B C D, TS A C B D - > Par A B C D - > CongA B A C D C A.
```

TS A C B D means that B and D are on opposite sides of line AC. Par A B C D means that the line AB is parallel (or equal) to line CD. CongA B A C D C A means that the angle BAC is congruent to angle DCA.

To obtain the formal proof, we need two propositions about the relative position of point with regard to a line.

Lemma 1 *If A and C are on opposite side of line PQ, and A and B are on the same side of line PQ then B and C are on opposite side of line PQ*

In Coq's syntax, this lemma (which is present in the ninth chapter of Schwabhäuser et al. (1983) is stated as

```
Lemma 19_8_2 : forall P Q A B C,
   TS P Q A C - > OS P Q A B - > TS P Q B C.
```

OS P Q A B means that A and B are on the same side of line PQ.

We also need the following lemma which is not present in Schwabhäuser et al. (1983):

Lemma 2 *If Y and Z are on the same side of line AX, and X and Z are one opposite sides of line AY then X and Y are on the same side of line AZ.*

```
Lemma os_ts1324__os : forall A X Y Z,
   OS A X Y Z - > TS A Y X Z - > OS A Z X Y.
```

We have now all the properties required to prove the main theorem:

Theorem 1 *Assuming the postulate of alternate interior angles, the sum of angles of any triangle is congruent to the flat angle.*

Proof Let ABC be a triangle, we need to show that the sum of angles is the flat angle. If the points ABC are collinear then the sum of angles is a flat angle.[6] Let l be a parallel to line AC through B[7] (see Fig. 1). Let B_1 be a point on the line l such that B_1 is on the opposite side of A with regard to the line BC. Let B_2 be the symmetric of B_1 through B.[8]

---

[6] We have a separate lemma for this case, we could also assume that we have a proper triangle. In formal development, we always try to prove the most generic results.

[7] Note that we do need "the parallel line", uniqueness is not important here.

[8] We could also use any point B_2 such that B belongs to segment $B_1 B_2$.

Fig. 1 The sum of angles of a triangle

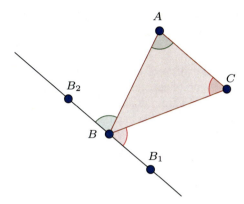

> C and B_2 are on opposite sides of line AB because[9]:
> B_1 and B_2 are on opposite sides of line AB (*) because by construction segment $B_1 B_2$ intersects line AB in B.
> As AC is strictly parallel to line $B_1 B_2$, A and C are on the same side of line $B_1 B_2$.
> By Lemma 2, we have that C and B_1 are on the same side of line AB.
> Hence, using Lemma 1 and fact (*), we can conclude that C and B_2 are on opposite sides of line AB.

By the construction of B_2, $B_1 B B_2$ is a flat angle, hence it suffice to show that the sum of angles is congruent to the angle $B_1 B B_2$.

By the postulate of alternate interior angles, we have that the angle $\angle ABB_2$ is congruent to $\angle CAB$. By construction, the angles $\angle CBB_1$ and $\angle BCA$ are alternate, hence by the postulate of alternate interior angles as the lines BB_1 and CA are parallel, the angle $\angle CBB_1$ is congruent to $\angle BCA$.

We give now a slightly different version of the proof, assuming that angles can be copied on a given side of a line (this is an axiom in Hilbert's foundations of geometry).

Proof Let B_1 be a point on the opposite side of A with regard to the line BC such that the angle $\angle ACB$ is congruent to the angle $\angle CBB_1$. As the alternate angles are congruent, the line AC is parallel to line BB_1. Let B_2 be a point on the line BB_1 such that B belongs to segment $B_1 B_2$. ...

The rest of the proof is the same as in the first version.

---

[9] Maybe there is a simpler proof ? but for sure we need to use the fact that AC is parallel to $B_1 B_2$.

2.3 Some Questions and Issues Raised by the Formalization of the Proof in Coq

The formal proof differs from the proof which is taught in high school, because using a proof assistant, all steps of the proof have to be justified. The proof assistant prevents us from deducing facts from the figure, this reflects the didactic contract between the teacher and the pupil. It is interesting to distinguish in the formal proof, intermediate steps which can be considered as uninteresting details from the steps which can be considered as proper gaps in the informal proof.

We think that the part of the proof which is typeset in a frame can not be considered as an uninteresting detail, it is an important sub-statement, whose justification is not obvious.

From a didactic point of view, there are steps of the proof that should remain *implicit* in a classroom and steps that should be emphasized.

This choice should be made consciously and depending on the context. For example, the transitivity of parallelism can either be explicit or implicit depending on the curriculum/class of the student. The famous example of the fallacious proof that all triangle are isoceles shows that the relative position of the points should not always be taken for granted, but the example of the sum of angles of a triangle is maybe too subtle to be studied rigorously in high school. In the example studied in this section, we believe that in a classroom it should be stated explicitly that the fact the angles are alternate is assumed.

For an integration of this exercise in an AI milieu, the tool-box would display different theorems/construction tools[10]: at least the postulate of alternate-interior angles and the tool to construct parallel lines. Should we have a tool to construct a point on a line on the opposite side of a point? We see here the impact of the AI milieu on the didactic setting.[11]

For an integration in an AI milieu, we would need to automate some steps of the proof which are purely administrative burden, at least the ones which are present in the original Coq code (see appendix) and that we kept implicit in this chapter. It includes: using the fact that the sum of angles is a morphism with regard to congruence of angles ($\alpha \equiv \alpha' \Rightarrow \alpha + \beta = \alpha' + \beta$), the fact that the sum of angles is unique up to congruence, and various permutation properties of the manipulated predicates,...

3 Varignon's Theorem

In this section, we will provide various proofs of Varignon's theorem that can be provided with the knowledge developed in the French secondary curriculum or at the beginning of university, that we will analyse with the question raised in Sect. 2.1.

---

[10] Construction tools correspond to existence theorems.

[11] The proof could also be modified to construct B_1 and B_2 such that B belongs to segment $B_1 B_2$ and then say that at least one of them is on the opposite side of A with regard to the line BC.

Si les côtés AB, BC, CD, DA d'une figure rectiligne de quatre côtés, ſont diviſés chacun en deux parties égales en F, G, H, E, & que les points des diviſions ſoient joints par les lignes droites FE, EH, HG, GF, la figure quadrilaterale $FEHG$ eſt un parallélograme ; car en menant les lignes DB, & AC, comme par l'hypothéſe, $AF = FB$ & $AE = ED$, $AF \cdot FB :: AE \cdot ED$, & ainſi (*Part.* 2.) EF eſt paralléle à DB. De même puiſque, par l'hypothéſe, $BG = GC$, & $DH = HC$; $BG \cdot GC :: DH \cdot HC$, & par conſéquent (*Part.* 2.) GH ſera encore paralléle à la ligne BD. Donc EF & GH ſont paralléles à la même troiſiéme ligne, elles ſont donc auſſi paralléles entre elles.

On peut par la même raiſon prouver que les lignes FG & EH ſont paralléles à la ligne droite AC, & par conſéquent paralléles entre elles. Donc le quadrilatére $EFGH$ eſt un parallélograme.

If the sides AB, BC, CD, DA of a quadrilateral are each divided into equal parts in F, G, H, E, and the division points are joined by lines FE, EH, HG, GF the quadrilateral $FEHG$ is a parallelogram; because by drawing the lines DB and AC, as by hypothesis, $AF = FB$ and $AE = ED$, $AF.FB :: AE.ED$, and hence (Part 2) EF is parallel to DB. Similarly, by hypothesis $BG = GC$ and $DH = HC$; $BG.GC :: DH.HC$ and hence GH is also parallel to line BD. Hence, EF and GH are parallel to the same line and therefore are parallel. Similarly, it can be proved that line FG and EH are both parallel to line AC and therefore are parallel. Hence the quadrilateral $EFGH$ is a parallelogram.

Fig. 2 Original proof of Varignon's theorem and English translation

Varignon's theorem, states that:

Theorem 2 *Let $ABCD$ be a quadrilateral. Let I, J, K and L be the midpoints of AB, BC, CD and AD, then $IJKL$ is a parallelogram.*

3.1 *Logical Analysis of a Classical Proof of Varignon's Theorem*

The usual proof presented in classroom is based on the midpoint theorem as the original proof (see Fig. 2) but this proof suffers from one problem, as we discuss below.

The standard proof is the following:

Proof Consider triangle ABC, by the midpoint theorem, we have that AC is parallel to IJ. Using again the midpoint theorem in triangle ACD, we have that LK is parallel to AC. Hence, by transitivity of parallelism, we have that IJ is parallel to LK. Similarly, we have that IL is parallel to JK. Hence, $IJKL$ is a parallelogram.

A variant of the proof consists in using the characterization of a parallelogram as a quadrilateral with a pair of opposite sides which are parallel, congruent and whose diagonals intersect.

3.2 *Issues and Challenges Raised by the Formalization of the Classical Proof of Varignon's Theorem*

The problem with this proof is at the last step, the theorem which says that if the opposite side of quadrilateral are parallel then it is a parallelogram requires that the parallelism is strict, i.e. the lines do not coincide. But, it could be the case that the points I, J, K and L are on the same line as shown on Fig. 3d. So, in the formal version of *this* proof, we need to add the fact that I, J and K are not collinear. This restriction is not welcome because even in the case where I, J and K are on the same line, then $IJKL$ is a parallelogram in the sense that its diagonals meet in their midpoints, and the opposite side are congruent (in GeoCoq's formalization, we call this figure a flat parallelogram). The formal proof also differs from the informal proof because we assume explicitly that $A \neq C$ and $B \neq D$ to ensure that the sides of $IJKL$ are proper lines.

3.3 Logical Analysis of Alternative Proofs of Varignon's Theorem

Beside the classical proof of Varignon theorem in synthetic geometry, there are other mathematical settings that allow proving this theorem. We provide below some examples.

3.3.1 A Proof Using Vectors

Proofs in this setting rely on the vector characterization of a parallelogram. We assume here just the existence of four points without any hypothesis. It is necessary to introduce vectors that correspond to ordered pair of points. Each ordered pair of points determines a vector, such that for a given parallelogram, there are potentially 12 non-zero vectors. Providing the characterization, as given below, necessitates to identify that only pairs of consecutive points are relevant, and that the two pairs should be in opposite order compared to the initial order of the four points.

Given four points M, N, P, Q, $MNPQ$ is a parallelogram if and only if $\overrightarrow{MN} = \overrightarrow{QP}$ (resp. $\overrightarrow{MQ} = \overrightarrow{NP}$)

In this mathematical setting, the Varignon's theorem can be reformulated as

Given four point A, B, C and D and I, J, K and L the midpoints of the segments AB, BC, CD and AD, $\overrightarrow{IJ} = \overrightarrow{KL}$ (resp. $\overrightarrow{IL} = \overrightarrow{JK}$)

Proof 2

Let A, B, C and D be four points in the plane, and I, J, K and L the midpoints of the segments AB, BC, CD and AD. Prove that $\overrightarrow{IJ} = \overrightarrow{KL}$ (resp. $\overrightarrow{IL} = \overrightarrow{JK}$).
$\overrightarrow{IJ} = \overrightarrow{IB} + \overrightarrow{BJ}$ (Vector addition) (1)
$\overrightarrow{IB} = \frac{1}{2}\overrightarrow{AB}$; $\overrightarrow{BJ} = \frac{1}{2}\overrightarrow{BC}$ (Vector characterization of the midpoint of a segment) (2)
$\overrightarrow{IJ} = \frac{1}{2}\overrightarrow{AB} + \frac{1}{2}\overrightarrow{BC}$ (Substitution) (3)
$\overrightarrow{IJ} = \frac{1}{2}(\overrightarrow{AB} + \overrightarrow{BC})$ (Factorization) (4)

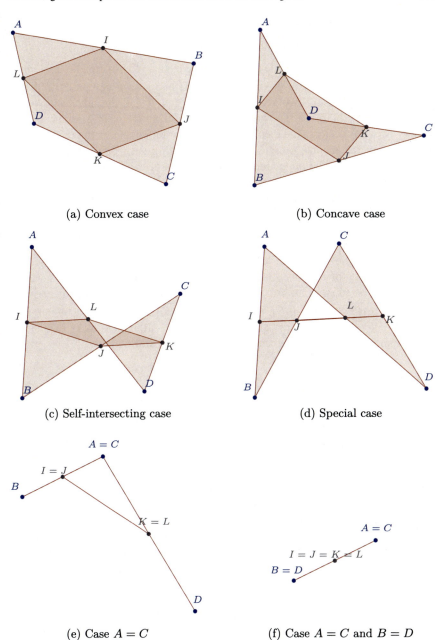

Fig. 3 Varignon's theorem

$\overrightarrow{IJ} = \frac{1}{2}\overrightarrow{AC}$ (Vector addition) (1) (5)
$\overrightarrow{LK} = \overrightarrow{LD} + \overrightarrow{DK} = \frac{1}{2}\overrightarrow{AD} + \frac{1}{2}\overrightarrow{DC} = \frac{1}{2}(\overrightarrow{AD} + \overrightarrow{DC}) = \frac{1}{2}\overrightarrow{AC}$ (6)

From (3) and (6), we conclude that $\overrightarrow{IJ} = \overrightarrow{LK}$ (transitivity of equality relation)

There are several keys in this proof. The first one is to decompose $\overrightarrow{IJ}$ thanks to vector addition, using two vectors, one with end B and the other with origin B. A second key is to use the characterization of a midpoint of a segment as an equality between vectors, and then to perform substitution and factorization, which are more general actions, present in AI. Formalizing such proof will require to explicit the choice to be done along the proof. It is possible to do these transformations without referring to a geometrical drawing, but it seems rather clear that having a drawing of a generic convex quadrilateral, even a freehand drawing is a powerful support for choosing the adequate transformation. Nevertheless, the particular cases do not need to be made explicit, because the vector characterization of a parallelogram, that we have recalled above, does not need any non degeneracy conditions. The conclusion relies on the transitivity of equality. It is noticeable that this property of equality shapes the proof that we provided. Another way of doing would be to continue from step (5) by introducing point D to decompose AC: $\overrightarrow{IJ} = \frac{1}{2}\overrightarrow{AC} = \frac{1}{2}(\overrightarrow{AD} + \overrightarrow{DC}) = \frac{1}{2}(2\overrightarrow{LD} + 2\overrightarrow{DK}) = \frac{1}{2}(2(\overrightarrow{LD} + \overrightarrow{DK})) = \overrightarrow{LK}$.
We could hypothesize that formalizing such proofs will need to make explicit the way of choosing how to implement the successive transformations.

3.3.2 Two Proofs Using Cartesian Coordinates

Proof 3

This proof relies on the characterization of parallelogram as quadrilaterals whose diagonals intersect at their common midpoint, and on the characterization of the midpoint of a segment by the mean of coordinates. The goal is to prove that segments IK and JL have the same midpoint. In this proof, once the four points have been introduced, the first thing to do is to choose three points that will serve as reference for determining the coordinates. In order to make calculations easier, we chose these three points among the four given ones. For example, let us choose A, B and C with coordinates $A : (0, 0)$, $B : (0, 1)$ and $C : (1, 0)$; then D has undetermined coordinates (a, b). Using the fact that the coordinates of the midpoint of a segment are the half of the sum of the coordinates of each point, we get the following:

The coordinates of I, J, K and L are respectively $(0, \frac{1}{2})$, $(\frac{1}{2}, 0)$, $(\frac{a+1}{2}, \frac{b}{2})$ and $(\frac{a}{2}, \frac{b+1}{2})$.

Let O_1 be the midpoint of segment IK. The coordinates of O_1 are $(\frac{a+1}{4}, \frac{b+1}{4})$ (1).
Let O_2 be the midpoint of segment JL. The coordinates of O_2 are $(\frac{a+1}{4}, \frac{b+1}{4})$ (2).
From (1) and (2) we conclude that $O_1 = O_2$ (transitivity of equality) As a consequence $IJKL$ is a parallelogram (characterization of parallelograms using common midpoint of the diagonals).

Remark 1 The choice of three generic points as reference leads to accessible calculations. In the perspective of formalization, it has the advantage of not introducing di-symmetry between the four initial points.

Remark 2 For this proof, once decided to use Cartesian coordinates, the way of doing is rather systematic. In this respect, we may hypothesize that the formalization will fit more the pen/papers proof than it was the case for Proof 2.

Proof 3*

This proof relies on the same characterization by the common midpoint of the diagonals, but using lines equations. The aim is to prove that the two lines IJ and KL intersect at the midpoints of the segments IJ and KL. This necessitates introducing the two lines, their equation, to solve the system of equations and to check that the ordered pair of solutions of the equation is identical to the midpoint of both segments. This method is congruent with the classical way of characterizing a parallelogram as a quadrilateral whose diagonals intersect at their common midpoint, however the calculations are less easy than in proof 3.

Proof 4

This proof relies of the metric characterization of a parallelogram as a quadrilateral with opposite sides having same length. The first step in this case is to choose an origin and two *orthogonal* axes, in order to be able to use the fact that the distance in a Euclidean plane is the square root of the sum of squares of the coordinates differences. In order to make calculations easier, we choose A as origin, and AB with B $(0, 1)$ as first axis. Then, the coordinates of C and D are indeterminate ones. Let them be (c_x, c_y) for C and (d_x, d_y) for D. Then the coordinates of the midpoints are $I : (0, \frac{1}{2})$; $J : (\frac{c_x}{2}, \frac{1+c_y}{2})$; $K : (\frac{c_x+d_x}{2}, \frac{c_y+d_y}{2})$; $L : (\frac{d_x}{2}, \frac{d_y}{2})$

Then using the formulas recalled above:

$IJ = \sqrt{((c_x/2)^2 + (c_y/2)^2}$ (1)
$LK = \sqrt{((c_x/2)^2 + (c_y/2)^2}$ (2)
$JK = \sqrt{((d_x/2)^2 + ((d_y - 1)/2)^2}$ (3)
$LK = \sqrt{((d_x/2)^2 + ((d_y - 1)/2)^2}$ (4)

From (1) and (2) we conclude that $IJ = LK$, and from (3) and (4) we conclude than $JK = KL$. Hence, as the quadrilateral $IJKL$ has his opposite sides with same lengths, we conclude that it is a parallelogram. Both remarks done on proof 3 above holds for this proof.

3.3.3 About Formalization of the Alternative Proofs

The formalization of **Proof 3** highlights again some necessary non degeneracy conditions. Indeed the characterization of parallelograms using midpoints is called `mid_plg` in GeoCoq:

```
mid_plg : A <> C \/ B <> D ->
    Midpoint M A C ->
    Midpoint M B D -> Parallelogram A B C D
```

To use it, we need to prove that either I is different from K or J is different from L. Assuming that A is different from B, C and D, the disjunction can be proved by contradiction. If we had $I = K$ and $J = L$ then both $ACBD$ and $ACDB$ would be parallelograms, which is impossible.

To formalize the analytic proof, there are two solutions: either we consider that the geometric objects are **defined** by the algebraic equations, but then the proof is not about geometry, it is about algebra. It would be possible to prove that the geometric predicates defined by the algebraic equations over their coordinates verify the usual geometric axioms. But this would only prove that $\mathbb{R}^2$ is a model of the axioms. To fully justify the analytic method, it is necessary to prove that the algebraic computations can be performed geometrically, following Descartes (1925). This proof is called the arithmetization of geometry. This is the culminating result of both Hilbert's *Foundations of Geometry* (Hilbert, 1960) and Schwabhäuser, Szmielew and Tarski's book (1983). An analytic proof in geometry, can be seen as a geometric proof, thanks to a meta-theoretical argument: both theories have the same models. Hence, for the formalization of the analytic proofs, starting from a synthetic axiomatic setting, we rely on the Coq formalization of the arithmetization of geometry (Boutry et al., 2019). Technically it means that each arithmetic operation is a shortcut for a geometry construction. Having prove that these operation form a field, then we can forget about the geometry and resort to computations. From a didactic point of view, this raises the question how this link between two different perspective could be presented to present a proof which is both accessible and rigorous. Sometimes, in educational context a bijection between the real line and the geometric line is assumed. In this particular example, and many other geometric statements, we do not need the reals, the formalization shows that the Cartesian plane over a Pythagorean field is sufficient.

In GeoCoq, the change of perspective from synthetic geometry to analytic geometry, can either be performed manually (by using the characterization of midpoint using coordinates) or automatically using a tactic called `convert_to_algebra`. In some sense, the formal proof, as tactic, provides an explanation of the method used to find the proofs. The availability of automation within interactive prover blurs the lines between proofs as explicit objects, and programs generating proofs.

Note that the proving geometric statements using analytic means can lead to some kind of circular proof, if the geometric statement is used in the proof of the arithmetization of geometry. This is the case of the Pythagorean theorem: an analytic proof is straightforward, but the Pythagorean theorem is needed in the proof of the arithmetization of geometry, specifically in the characterization using coordinates of the congruence of segments. Our formal proof of Pythagoras' theorem itself employs the intercept theorem (also known in France as Thales' theorem) so an analytic proof of the intercept theorem would also be somewhat circular.

The analytic proofs by substitution can also either be performed manually by manipulating algebraic equations, or automatically using Gröbner basis algorithm or Gauss elimination. This approach could lead in the future to new presentation of proofs at the frontier between maths and computer science, by making explicit and systematic the heuristic used by the student and turning them in some cases into proper algorithms. Automated theorem proving could play the same role with regards to proof that computer algebra system with regards to computations.

For the formalization of **Proof 4**, the choice of the coordinate axis requires some reasoning. It relies on the invariance of geometric predicates by translation and rotation, so that one can assume simple coordinates for A, B and C, see Sect. 4 of (Genevaux et al., 2011) for an example or (Harrison, 2009). For this proof, once again the non degeneracy conditions complicate the proof. Indeed one should pay attention that for proving that the quadrilateral is a parallelogram using the fact that opposite side have the same length, one needs to prove that the quadrilateral is not crossed, in the sense that the diagonals have a common point, and the figure should be either non flat, or fully flat:

```
Lemma cong_cong_parallelogram:
    forall A B C D P : Tpoint,
    Cong A B C D ->
    Cong B C D A ->
    (~ Col A B C \/ (Col A B C /\ Col A B D)) ->
    B <> D ->
    Col A P C ->
    Col B P D -> Parallelogram A B C D.
```

3.3.4 Alternative Proofs Using Automated Deduction

Area method

In this subsection, we give as example the proof of Varignon's theorem using the area method. The area method is a procedure for a fragment of Euclidean plane geometry (Chou et al., 1994; Janičić, 2012; Zhang et al., 1995). It is based on the concept of signed area of triangles, and can efficiently prove many non-trivial theorems and produces proofs that are often very concise. This method has been taught to student preparing mathematics Olympiad in China. For a quick overview of the method see (Narboux et al., 2018). Using the signed area, a number of geometric predicates can be simply expressed, for instance: Three points are collinear iff the signed area of the triangle is zero:

Col $A B C$ iff $\mathcal{S}_{ABC} = 0$;

Two lines are parallel, if they are at constant distances, and we can characterize it using signed area: $AB \parallel CD$ iff $A \neq B \wedge C \neq D \wedge \mathcal{S}_{ACD} = \mathcal{S}_{BCD}$, etc.

The method consists in eliminating the points one by one using formulas based on the way the point is constructed. In our example, we only have one construction: the midpoint and we use the following lemma about signed areas: if I is the midpoint of AB then for any P and Q: $\mathcal{S}_{PQI} = \frac{\mathcal{S}_{PQA}}{2} + \frac{\mathcal{S}_{PQB}}{2}$.

Then we use the characterization of a parallelogram as a quadrilateral whose opposite sides are parallel. We give here the proof that two sides are parallel. Formally, to prove that it is a parallelogram this piece of proof should be duplicated or an argument of symmetry should be used. We need to show that $IJ \parallel LK$ this is equivalent to $\mathcal{S}_{KIJ} = \mathcal{S}_{LIJ}$.

$$\begin{aligned}
& \mathcal{S}_{KIJ} - \mathcal{S}_{LIJ} \\
= & \frac{\mathcal{S}_{KJB}}{2} + \frac{\mathcal{S}_{KIC}}{2} - \frac{\mathcal{S}_{LJB}}{2} - \frac{\mathcal{S}_{LIC}}{2} & J \text{ Eliminated} \\
= & \frac{\mathcal{S}_{BKA}}{2} + \frac{\mathcal{S}_{BKB}}{2} + \frac{\mathcal{S}_{CKA}}{2} + \frac{\mathcal{S}_{CKB}}{2} - \frac{\mathcal{S}_{BLA}}{2} - \frac{\mathcal{S}_{BLB}}{2} - \frac{\mathcal{S}_{CLA}}{2} - \frac{\mathcal{S}_{CLB}}{2} & I \text{ Eliminated} \\
= & \tfrac{1}{2}(\mathcal{S}_{BKA} + \mathcal{S}_{CKA} + \mathcal{S}_{CKB} - \mathcal{S}_{BLA} - \mathcal{S}_{CLA} - \mathcal{S}_{CLB}) & \text{Simplification} \\
= & \tfrac{1}{2}(\frac{\mathcal{S}_{ABC}}{2} + \frac{\mathcal{S}_{ABD}}{2} + \frac{\mathcal{S}_{ACC}}{2} + \frac{\mathcal{S}_{ACD}}{2} + \frac{\mathcal{S}_{BCC}}{2} & K \text{ Eliminated} \\
& + \frac{\mathcal{S}_{BCD}}{2} - \mathcal{S}_{BLA} - \mathcal{S}_{CLA} - \mathcal{S}_{CLB}) \\
= & \tfrac{1}{2}(\frac{\mathcal{S}_{ABC}}{2} + \frac{\mathcal{S}_{ABD}}{2} + \frac{\mathcal{S}_{ACC}}{2} + \frac{\mathcal{S}_{ACD}}{2} + \frac{\mathcal{S}_{BCC}}{2} + \frac{\mathcal{S}_{BCD}}{2} & L \text{ Eliminated} \\
& - \frac{\mathcal{S}_{ABA}}{2} - \frac{\mathcal{S}_{ABD}}{2} - \frac{\mathcal{S}_{ACA}}{2} - \frac{\mathcal{S}_{ACD}}{2} - \frac{\mathcal{S}_{BCA}}{2} - \frac{\mathcal{S}_{BCD}}{2}) \\
= & \tfrac{1}{4}(\mathcal{S}_{ABC} + \mathcal{S}_{BCA}) & \text{Simplification} \\
= & 0 & \text{Simplification}
\end{aligned}$$

This proof can be obtained automatically.

The same method can be applied to obtain systematically a proof using Cartesian coordinates. As the midpoint relation can be expressed as a linear equality, the goal can be solved by a simple Gauss elimination algorithm (for more involved theorems, one would need to use the method of Wu or Gröbner basis which are also available in proof assistants (Genevauxet al., 2011; Pottier, 2008). The choice of the best characterization of parallelograms for the computational proof (the diagonals intersect in their midpoint) is still to be decided by a human in the current implementation. The AI milieu of GeoCoq is up to our knowledge the unique setting where synthetic and analytic reasoning can be intermixed and validated formally.

3.4 Didactical Implication

The recognition by advanced mathematics students that a given theorem can be stated and proved in a variety of mathematical settings is an important issue. Indeed, as we can experience as university teacher, it is often the case that students think that there is exactly one proof for a given theorem. This is particularly important for prospective secondary mathematics teachers who should be able to recognize the underlying theory of the didactical transposition's choices made in secondary curriculum. This resort of the so-called second discontinuity of Klein that "concerns those who wish to return to school as teachers and the (difficult) transfer of academic knowledge gained at university to relevant knowledge for a teacher" (Winsløw and Grønbæk, 2014). For this purpose, the formalization in Coq raises new issues and opportunities. A first issue has already been mentioned in the first example; it concern the need for considering the question of degeneracy that is often hidden in pencil/paper proofs in Geometry, in particular because it is usual that drawings support our intuition and orient the proofs while "it is clear that some evidences in the drawing register have to be questioned in the theory" (Durand-Guerrier and Arsac, 2009). As shown in Sect. 3.3, this is crucial in the case of Varignon's theorem. A second issue concerns the importance of making explicit the role of axiomatic theories in the proving process, that as stressed by Planchon and Hausberger (2020, p. 162), is not easily recognized by students involved in a prospective mathematics teacher training program. For formalizing geometrical proofs in Coq, one needs to refers explicitly to the assumed axiomatic and open the possibility of discussing it with students, as a contribution to make them aware of the implicit assumptions they have developed during their academic studies. It also opens the possibility of discussing if and why different theories are interpreted by the same models. An important example is the case of synthetic geometry and analytic geometry. This issue is nearly never discussed in undergraduate mathematics studies, while it could offer sound justifications for the omnipresent practice in high school of moving from synthetic geometry to analytical geometry and vice versa. Another example which is not always clearly stated in the undergraduate studies, is the justification of the use of complex numbers in geometry.

In the following section, we present a teacher training session on Varignon's theorem that has been implemented for years by the second author in a pencil/paper modality and we discuss the relevance of introducing formal proof in the milieu of the situation to test the relevance of the link between pencil/paper and formal proofs.

4 A Teacher Training Session on Varignon's Theorem

N.B. This is based on the implementation of this session in a module for future mathematics teachers entitled Didactics and epistemology of mathematics in Master 1 at the University of Montpellier. This module has not been studied in the context of research. This is a perspective to test the relevance of the link between pencil/paper and formal proofs.

4.1 Context, Motivation, Description and a Priori Analysis of the Session

The work around Varignon's theorem that we present below has been implemented during several years in a module of Didactic and Epistemology of Mathematics in the first year of the master degree for prospective mathematics teachers ("Métiers de l'enseignement et de la formation") in France in Montpellier, where 10 h where dedicated to proofs. Motivations of the introduction of a dedicated work on proofs in secondary mathematics teacher training are double-faced. On the one hand, it is necessary to provide prospective teachers with proofs knowledge and skills, that is in general not sufficient to address the professional needs despite the fact that they have practice proofs in their academic studies. On the other hand, as stressed by Durand-Guerrier and Tanguay (2018), we hypothesize that "[...] working with proof is likely to contribute to conceptualization by prompting a work with the mathematical objects at stake, in agreement with the syntax-semantics dialectic in proof and proving (e.g. Weber and Alcock, 2004)." (op. cit, p. 20)

The choice of the Varignon's theorem relies on its potential to deal with both aspects mentioned above. For example, (Durand-Guerrier et al., 2012) considered that Varignon's Theorems : "[provide] a situation relevant to secondary school and teacher education, in which a given definition must be explored in order to identify the entire range of objects that satisfied it" (op. cit. p.380)

4.1.1 Production and Analysis of Varignon's Theorem Proofs

In the first part of the activity, students were invited to provide two proofs of Varignon's theorem in two different mathematical settings. The theorem is given as above: "Given any quadrilateral, the respective midpoints of its sides are the vertices of a parallelogram." Once this done, the students were invited to analyse their proofs from the point of view of objects introduced, properties or relations mobilized, theorems used and methods of reasoning implemented. They were also required to precise in which respect generality was addressed in their proof. In some cases, they worked in pair for the analysis, while in other cases they worked individually for both steps. A main goal of this first part of the activity was to push students to go beyond the classical geometrical proof based on the midpoint theorem, and to engage them in a logical analysis of the proofs in different mathematical settings with a focus on objects, properties, relations on the one hand, theorems and modes of reasoning on the other hand.

4.1.2 Exploring the Inverse Problem

In this part of the activity, students are asked to explore the three inverse problems below: Which necessary and sufficient condition should be satisfied by the initial quadrilateral in order to obtain:

1. a rhombus
2. a rectangle
3. a square

The general idea is here that the property of the sides of the parallelogram comes from the property of the diagonals of the initial quadrilateral. A rhombus is a parallelogram with four isometric sides. As a consequence, in order to get a rhombus, the diagonals of the initial quadrilateral should be isometric. This condition is necessary. It is also sufficient. A rectangle is a parallelogram with perpendicular adjacent sides. As a consequence, in order to get a rectangle, the diagonals of the initial quadrilateral should be perpendicular. This condition is necessary. It is also sufficient. A square is both a rhombus and a rectangle. As a consequence, in order to get a square, it is necessary and sufficient that the diagonal of the initial quadrilateral are isometric and perpendicular. An implicit hypothesis might be that the expected condition be expressed in term of "type of quadrilateral". This would lead to answer with sufficient conditions, such as "starting from a rectangle provides a rhombus", "starting from rhombus provides a rectangle" and "starting from a square provides a square". Questioning the necessity should lead to move to the property of diagonals. Nevertheless, we cannot exclude that the condition be considered as necessary by some students. Indeed, in the French curriculum, characterizing quadrilaterals by the property of their diagonals is considered in general only for parallelograms, with an exception with kites. As Durand-Guerrier (2003) stresses, 60 of 273 students just entering a French university answered that "a quadrilateral with perpendicular diagonals" is a rhombus, some of them adding that the only counterexample they know is the square, but it is a particular rhombus (Durand-Guerrier, 2003, p. 26).

4.2 Account of Naturalist Observation Along Several Years

The classical proof is in general the first one that is produced; the alternative proofs that we have presented above are regularly provided by students, in some cases with redundancy: for example, for the proof using vectors, some students prove the equality of two pairs of vectors instead of one pair. In the classical proof, the midpoint theorem and the inference rule at stake (*modus ponens*) is generally explicit; however other theorems are used without being mentioned. Below is an example of the analysis of the classical proof by a student[12] (E5):

> It is a direct proof starting from a generic quadrilateral and the midpoints of the sides and implying properties until the conclusion of the proof that validate the theorem. The midpoint theorem [not stated] and the definition of a parallelogram are used in the proof. Also the property, implicit here, saying that if two lines are parallel to a same third line, then they are parallel to each other are used here. Of course, the definition of a quadrilateral is implicit. For this proof, we have introduced the name of the points. The generality is taken in account here as we do not give other precision and the initial quadrilateral that remain generic, except the name of the vertices.

Concerning the proofs using vectors or Cartesian coordinates, there are still more implicit assumptions. Indeed, such proofs rely mainly on computation, and in this case, theorems and inferences remain in general implicit. For example, student E6 provided a proof using vectors, and in his analysis he wrote "No theorem is used". Another student (E8) provided a proof using Cartesian coordinates to prove the equality of a pair of vectors. In her analysis, she wrote

> There is no use of theorem, but rather we use the definition of vector equality. The reasoning is an algebraic one in Cartesian geometry.

---

[12] Excerpts from students' productions are translated from French.

While keeping implicit some elements in a proof might not be problematic for experts, we hypothesize that it is important for prospective teachers to identify such implicit steps in their own proof. Indeed, there is research evidence that they might not be shared or recognized by a number of secondary students (e.g. implicit quantification on conditional statements as stressed in Durand-Guerrier (2003). Concerning the second part of the activity with the inverse problem, a majority of students over the years provide the sufficient condition, some of them claiming that it is necessary, and in some cases providing a "proof". Below are two examples from the same corpus as above. Student E2 wrote:

> 4-a) A rhombus is a parallelogram with diagonals intersecting in their midpoint and perpendicular to each other: the initial figure is necessarily a rectangle. b) a rectangle is a parallelogram with diagonals of same length: the initial figure is necessarily a rhombus. c) a square is a parallelogram with diagonals of same length and perpendicular: the initial figure is necessarily a square.

Student E10 proved that if we get a square, the diagonals of the initial quadrilateral have same length and are perpendicular. She concludes as below:

> 4-c) Hence $ABCD$ is a quadrilateral with diagonal of same length and perpendicular, hence it is a square.

She did the same for a) and b). During the session where the answers were collected, only two students among 11 provided the conditions on the diagonals without concluding with a category of parallelogram, which was generally the case in other sessions. An interpretation is that the students reason in the domain of parallelograms, not in the domain of more general quadrilaterals. This might be an effect of the common practice of letting implicit the quantification, with a consequence that the domain of objects at stake remains also implicit, this being reinforced by the fact that the students have encountered mainly the parallelograms and the common particular ones (rhombus, rectangle, square).

4.3 Evolution of the Teacher Training Session by Introducing a Proof Assistant

The brief account we give of the students work is in line with the naturalistic observation done along years with this activity. We hypothesize that introducing in the activity the possibility of checking the proof with a proof assistant might help students to identify more precisely the logical structure on the one hand (e.g. recognizing the use of theorems in proofs using vectors or Cartesian coordinates), and discussing the generality in a more accurate way. We remark that using a proof assistant founded on a logical framework which allows separation of reasoning from computation such as type theory of deduction modulo (Dowek, 2014) or simply a proof language allowing the description of automatic procedure, could give sense to the assertion: "there is no use of theorem". The fully automatic proof of Varignon's theorem using the area method which can be found in the appendix is an example of such a 'theorem less' proof. The contrastive analysis of the pencil/paper proofs with the formal ones open paths for designing an adaptation of this teacher training session.

We propose the following scenario: after a familiarization with a proof assistant such as Coq, we ask for two different proofs of Varignon's theorem and to choose one and to formalize it within Coq using the GeoCoq library. Then we can discuss what the formalization brings in the analysis of the proof. Our hypothesis is that formalization will produce a finer and more rigorous analysis of the proof, making clear reference to the underlying properties or axioms which are necessary for the change of mathematical settings. Then, we propose to solve the inverse problem on pencil/paper and later check the proof using the proof assistant. It will be interesting to study the impact of this alternation between the two modalities on the clarification of the concepts involved in the proofs and on the concept of proof per se. We must acknowledge that having an audience acquainted to a proof assistant is a demanding prerequisite of the proposed scenario.

5 Conclusions

In this chapter, we have presented an exploratory study aiming to discuss the relationships between pencil/paper proofs and formal ones in a didactic perspective. We first discuss the classical proof that the sum of the angles of a triangle is two right angles, putting in evidence the points raised by the formalization in a proof assistant, and discussing which of the implicit steps would be relevant to clarify in an educational setting, in particular in teacher training.

Our second example, the theorem of Varignon, has been chosen for its potentiality a priori to feed the discussion on the interaction between pencil/paper proofs and formal ones. This has been evidenced for the proof in synthetic geometry, and for the alternative proofs, the question raised being far beyond the specific example of the Varignon theorem, for example, for what concerns the back and forth between synthetic and analytic geometry, which is taken for granted in secondary curriculum in France, and certainly in other educational systems.

We hypothesize that the teacher training briefly presented in Sect. 3 will be improved by the introduction of a proof assistant in the milieu, in order 1/ to enrich the experience of prospective teachers for what concerns proofs, a professional competence that need to be developed as it is evidenced in international literature ; 2/ to improve their knowledge on the relationships between Synthetic Geometry and Analytic Geometry, thanks to the clue questions raised by formalization.

Appendix: Verbatim of the Formal Proofs

We first give the formal proof of the fact that the sum of angles of a triangle is congruent to the flat angle.

```
Section alternate_interior_angles_postulate_triangle.

Context '{T2D:Tarski_2D}.

Lemma alternate_interior__triangle :
 alternate_interior_angles_postulate -> triangle_postulate.
Proof.
  intros aia A B C D E F HTrisuma.
  elim(Col_dec A B C).
    intro; apply (col_trisuma__bet A B C); auto.
  intro HNCol.
  destruct HTrisuma as [D1 [E1 [F1 []]]].
  destruct(ex_conga_ts B C A C B A) as [B1 [HConga HTS]]; Col.
  assert (HPar : Par A C B B1)
     by (apply par_left_comm, par_symmetry, 112_21_b; Side; CongA).
  apply (par_not_col_strict _ _ _ _ B) in HPar; Col.
  assert(HNCol1 : ~ Col C B B1) by (apply (par_not_col A C); Col).
  assert(HNCol2 : ~ Col A B B1) by (apply (par_not_col A C); Col).
  assert(HB2 := segment_construction B1 B B1 B).
  destruct HB2 as [B2 [HBet HCong]].
  assert_diffs.

  assert(HTS1 : TS B A B1 B2).
  { repeat split; Col.
    intro; apply HNCol2; ColR.
    exists B; Col.
  }
  assert(HTS2 : TS B A C B2).
  { apply (19_8_2 _ _ B1); auto.
    apply os_ts1324__os; Side.
  }
  apply (bet_conga_bet B1 B B2); auto.
  apply (suma2__conga D1 E1 F1 C A B); auto.
  assert(CongA A B B2 C A B).
  { apply conga_left_comm, aia; Side.
    apply par_symmetry, (par_col_par _ _ _ B1); Col; Par.
  }
  apply (conga3_suma__suma B1 B A A B B2 B1 B B2); try (apply conga_refl); auto.
    exists B2; repeat (split; CongA); apply 19_9; auto.
  apply (suma2__conga A B C B C A); auto.
  apply (conga3_suma__suma A B C C B B1 A B B1); CongA.
    exists B1; repeat (split; CongA); apply 19_9; Side.
Qed.

End alternate_interior_angles_postulate_triangle.
```

Proof of Varignon's theorem using the midpoint theorem:

```
Lemma varignon :
 forall A B C D I J K L,
  A<>C -> B<>D -> ~ Col I J K ->
  Midpoint I A B -> Midpoint J B C ->
  Midpoint K C D -> Midpoint L A D ->
  Parallelogram I J K L.
```

```
Proof.
intros.
assert_diffs.
assert (Par I L B D)
(* Applying the midpoint theorem in the triangle BDA. *)
  by perm_apply (triangle_mid_par B D A L I).
assert (Par J K B D)
(* Applying the midpoint theorem in the triangle BDC. *)
  by perm_apply (triangle_mid_par B D C K J).
assert (Par I L J K)
(* Transitivity of parallelism *)
  by (apply par_trans with B D;finish).
assert (Par I J A C)
(* Applying the midpoint theorem in the triangle ACB. *)
  by perm_apply (triangle_mid_par A C B J I).
assert (Par L K A C)
(* Applying the midpoint theorem in the triangle ACD. *)
  by perm_apply (triangle_mid_par A C D K L).
assert (Par I J K L)
(* Transitivity of parallelism *)
  by (apply par_trans with A C;finish).
apply par_2_plg;finish.
(* If in the opposite side of quadrilateral are parallel and
   two opposite side are distinct then it is a parallelogram. *)
Qed.
```

Alternative proof using characterization of parallelogram using midpoints and coordinates:

```
Lemma varignon : forall A B C D I J K L,
  A<>B -> A<>C -> A<>D ->
  Midpoint I A B -> Midpoint J B C ->
  Midpoint K C D -> Midpoint L A D ->
  Parallelogram I J K L.
Proof.
intros A B C D I J K L HAB HAC HAD HI HJ HK HL.
destruct (midpoint_existence I K) as [O HO].
assert (I<>K \/ J<>L).
 {
 destruct (eq_dec_points I K).
 subst;right.
 intro.
 treat_equalities.
   assert (Parallelogram A C B D).
     apply mid_plg with O;auto.
   assert (Parallelogram A C D B).
     apply mid_plg with J;finish.
 apply (plg_not_comm_1 A C B D);auto.
 auto.
 }
assert (Midpoint O J L).
{
 revert HI HJ HK HL HO.
 convert_to_algebra.
 decompose_coordinates;intros;spliter.
 split;
 nsatz;prove_discr_for_powers_of_2.
}
apply mid_plg with O;assumption.
Qed.
```

A completely automatic proof using the area method:

```
Theorem varignon:
 forall A B C D I J K L,
 is_midpoint I A B ->
 is_midpoint J B C ->
 is_midpoint K C D ->
 is_midpoint L D A ->
 parallel I J K L /\ parallel J K I L.
Proof.
area_method.
Qed.
```

A detailed proof script using the area method, the tactics highlights the key idea of eliminating points one by one from the goal, but the actual computation is implicit:

```
Theorem varignon:
 forall A B C D I J K L,
 is_midpoint I A B ->
 is_midpoint J B C ->
 is_midpoint K C D ->
 is_midpoint L D A ->
 parallel I J K L /\ parallel J K I L.
Proof.
geoInit.
eliminate I.
eliminate J.
eliminate K.
eliminate L.
Runiformize_signed_areas.
field_and_conclude.
eliminate I.
eliminate J.
eliminate K.
eliminate L.
Runiformize_signed_areas.
field_and_conclude.
Qed.
```

References

Arsac, G. (1998). *L'axiomatique de Hilbert et l'enseignement de la géométrie au collège et au lycée.* Lyon: Aléas.

Avigad, J., Dean, E., & Mumma, J. (2009). A formal system for Euclid's elements. *The Review of Symbolic Logic, 2,* 700–768.

Beeson, M., Narboux, J., & Wiedijk, F. (2019). Proof-checking Euclid. *Annals of Mathematics and Artificial Intelligence, 85*(2–4):213–257. Publisher: Springer.

Boutry, P., Gries, C., Narboux, J., & Schreck, P. (2017). Parallel postulates and continuity axioms: A mechanized study in intuitionistic logic using Coq. *Journal of Automated Reasoning* 68.

Boutry, P., Braun, G., & Narboux, J. (2019). Formalization of the arithmetization of Euclidean plane geometry and applications. *Journal of Symbolic Computation, 98,* 149–168.

Chou, S. -C., Gao, X. -S., & Zhang, J. -Z. (1994). *Machine proofs in geometry.* Singapore: World Scientific.

Churchhouse, R. F., Cornu, B., Howson, A. G., Kahane, J. -P., van Lint, J. H., Pluvinage, F., Ralston, A., & Yamaguti, M.,(Eds.) (1986). *The influence of computers and informatics on mathematics and its teaching: Proceedings from a symposium held in Strasbourg, France in March 1985 and sponsored by the international commission on mathematical instruction*. Cambridge University Press, 1 edition.

Descartes, R. (1925). *La géométrie*. Chicago: Open Court.

Dowek, G. (2014). *Deduction modulo theory*. Wien, Austria: In All about proofs. Proofs for all.

Durand-Guerrier, V., & Arsac, G. (2009). Analyze of mathematical proofs. Some questions and first answers. In F. -L. Lin, F. -J. Hsieh, G. Hanna & M. D. Villiers (Eds.), *ICMI study 19 conference: Proof and proving in mathematics education* (Vol. I, pp. 148–153). Taipei, Taiwan. The Department of Mathematics, National Taiwan Normal University, Taipei, Taiwan.

Durand-Guerrier, V., & Tanguay, D. (2018). Working on proof as contribution to conceptualisation—The case of R-completeness. In A. J. Stylianides & G. Harel (Eds.), *Advances in mathematics education research on proof and proving. An international perspective*, ICME-13 Monographs (pp. 19–34). Springer International Publishing.

Durand-Guerrier, V., Boero, P., Douek, N., Epp, S. S., & Tanguay, D. (2012). Examining the role of logic in teaching proof. In G. Hanna & M. D. Villiers (Eds.), *Proof and proving in mathematics education*, number 15 in New ICMI Study Series (pp. 369–389). Springer Netherlands.

Durand-Guerrier, V., Meyer, A., & Modeste, S. (2019). Didactical issues at the interface of mathematics and computer science. In *Proof technology in mathematics research and teaching*. Springer.

Durand-Guerrier, V. (2003). Which notion of implication is the right one? From logical considerations to a didactic perspective. *Educational Studies in Mathematics, 53*(1), 5–34.

Durand-Guerrier, V. (2008). Truth versus validity in mathematical proof. *ZDM, 40*(3), 373–384.

Genevaux, J. -D., Narboux, J., & Schreck, P. (2011). Formalization of Wu's simple method in Coq. In J. -P. Jouannaud & Z. Shao (Eds.), *CPP 2011 First International Conference on Certified Programs and Proofs*, volume 7086 of *Lecture Notes in Computer Science* (pp. 71–86). Kenting, Taiwan: Springer.

Gries, C., Boutry, P., & Narboux, J. (2016). Somme des angles d'un triangle et unicité de la parallèle: une preuve d'équivalence formalisée en Coq. In *Les vingt-septièmes Journées Francophones des Langages Applicatifs (JFLA 2016)*, Actes des Vingt-septièmes Journées Francophones des Langages Applicatifs (JFLA 2016), page 15, Saint Malo, France. Jade Algave and Julien Signoles.

Harrison, J. (2009). Without loss of generality. In *TPHOLs*, pp. 43–59.

Hartshorne, R. (2000). *Geometry: Euclid and beyond*. Undergraduate texts in mathematics. Springer.

Hilbert, D. (1960). *Foundations of Geometry (Grundlagen der Geometrie)*. Open Court, La Salle, Illinois. Second English edition, translated from the tenth German edition by Leo Unger. Original publication date, 1899.

Janičić, P., Narboux, J., & Quaresma, P. (2012). The area method: A recapitulation. *Journal of Automated Reasoning,48*(4), 489–532.

Manders, K. (2011). *The Euclidean diagram (1995)*. In The Philosophy of Mathematical Practice: Paolo Mancosu, Oxford University Press edition.

Miller, N. (2007). *Euclid and his twentieth century rivals: Diagrams in the logic of Euclidean geometry*. Studies in the theory and applications of diagrams. CSLI Publications, Stanford, Calif. OCLC: ocm71947628.

Miller, N. (2012). On the inconsistency of Mumma's Eu. *Notre Dame Journal of Formal Logic, 53*(1), 27–52.

Mumma, J. (2010). Proofs, pictures, and Euclid. *Synthese, 175*(2), 255–287.

Narboux, J., Janicic, P., & Fleuriot, J. (2018). Computer-assisted theorem proving in synthetic geometry. In M. Sitharam, A. S. John & J. Sidman (Eds.), *Handbook of geometric constraint systems principles*. Discrete Mathematics and Its Applications: Chapman and Hall/CRC.

Planchon, G., & Hausberger, T. (2020). Un problème de CAPES comme premier pas vers une implémentation du plan B de Klein pour l'intégrale. In *INDRUM 2020*, Cyberspace (virtually from Bizerte), Tunisia. Université de Carthage, Université de Montpellier.

Pottier, L. (2008). Connecting Gröbner bases programs with Coq to do proofs in algebra, geometry and arithmetics. In G. Sutcliffe, P. Rudnicki, R. Schmidt, B. Konev & S. Schulz (Eds.), *Knowledge exchange: Automated provers and proof assistants, volume 418 of CEUR Workshop Proceedings, Doha.* CEUR-WS.org: Qatar.

Schwabhäuser, W., Szmielew, W., & Tarski, A. (1983). *Metamathematische Methoden in der Geometrie.* Berlin: Springer.

Sinaceur, H. (1991). *Corps et modèles: essai sur l'histoire de l'algèbre réelle.* Vrin, Paris: Mathesis. J.

Tarski, A. (1936). *Introduction to logic and to the methodology of deductive sciences.* New York: Dover Publications.

Weber, K., & Alcock, L. (2004). Semantic and syntactic proof productions. *Educational Studies in Mathematics, 56*(3), 209–234.

Winsløw, C., & Grønbæk, N. (2014). Klein's double discontinuity revisited: What use is university mathematics to high school calculus? arXiv:1307.0157 [math]. arXiv: 1307.0157.

Winterstein, D. (2004). Dr. Doodle: A diagrammatic theorem prover. In *Proceedings of IJCAR 2004.*

Winterstein, D., Bundy, A., & Jamnik, M. (2000). A proposal for automating diagrammatic reasoning in continuous domains. In *Diagrams,* pp. 286–299.

Zhang, J.-Z., Chou, S.-C., & Gao, X.-S. (1995). Automated production of traditional proofs for theorems in Euclidean geometry. *Ann. Math. Artif. Intell., 13*(1–2), 109–138.

Interaction Between Subject and DGE by Solving Geometric Problems

Jiří Blažek and Pavel Pech

1 Introduction

Teaching mathematical proof is complex, interdisciplinary topic, also involving psychology. It is acknowledged that DGE (Dynamic Geometry Environment) can enhance the teaching of mathematics in several aspects, Mariotti (2006):

1. DGE can provide to a student a concrete model and meaning of abstract mathematical notions. Thus, it enhances understanding of complex mathematical considerations and prevents the peril of merely formal understanding of mathematical justification (as algorithm or a game with symbols).
2. During a solution of a geometrical problem, DGE can offer significant help. Thanks to accurate sketches, the possibility of dynamical changes of a construction and other advanced tools of the software, students can grasp the key facts considerably facilitating the solution.

Although it would seem that the help of DGE in solving problems is significant, the situation is not so clear. The fact is that the help of DGE for an average student is not always sufficient; see Mariotti (2006), Blažek and Pech (2019), Oflaz et al. (2016). Today numerous researches point out that the help of software must be accompanied with a suitable choice of learning materials (which include appropriate problems for solution, Fahlgren and Brunström (2014)) and assistance of the teacher. Answers to the questions: "What does interaction of a student with DGE look like?" and "What are the typical obstacles preventing getting to the solution?" could be useful not only for identification of problems suitable for solution with the help of DGE, but they

J. Blažek · P. Pech (✉)
Faculty of Education, University of South Bohemia, České Budějovice, Czech Republic
e-mail: pech@pf.jcu.cz

J. Blažek
e-mail: blazej02@pf.jcu.cz

will also help us to understand why some students succeed in solving a problem whereas others fail despite the potential help of the software.

The purpose of this paper is to answer the first of the above stated questions. We will focus on the interaction between a subject and software. The account of the interaction is based on introspection: the authors (of this article) looked into students' solving procedures of two geometrical problems, and during this process, they recorded the students' steps and motivation for each action carried out in DGE, in this case in GeoGebra.

The structure of the article is as follows: we start by formulating the theoretical background and, on its basis, describe our theoretical model. The purpose of the model is to classify motivation for a subject's action while solving a geometrical problem with the help of DGE. Then, we apply the model to two particular cases, in which the help of DGE in the solution is crucial. In conclusion, we state (open) questions related to the impact of the model on the selection of problems suitable for being solved with the support of DGE.

2 Theoretical Background

It is possible to divide the process of problem solving into the phase of making conjectures and proving, Arzarello et al. (2002). The first activity consists of making sense of the problem and developing an enquiry in which students or mathematicians formulate conjectures. The latter activity of proving consists of developing a chain of logical consequences. The phase of making conjectures is not less important or demanding than the phase of proving. As Villiers (2004) points out, "proof is not necessarily a prerequisite for conviction—to the contrary, conviction is probably far more frequently a prerequisite for the finding of a proof." Likewise, the distinguished Hungarian mathematician Polya (1954), p. 83–84, wrote:

> ... having verified the theorem in several particular cases, we gathered strong inductive evidence for it. The inductive phase overcame our initial suspicion and gave us a strong confidence in the theorem. Without such confidence we would have scarcely found the courage to undertake the proof which did not look at all a routine job. When you have satisfied yourself that the theorem is true, you start proving it.

The right conjecture facilitates the solution or proof of a problem significantly. If the subject is not able to produce a relevant conjecture, they do not know what to focus on and they get stuck in the solving process. The case when the subject tries to prove a conjecture that is not valid is even worse (in practice of the authors a well-known situation). In that case, the solver often spends a lot of time until they derive a contradiction or give up any further attempts to prove it. The production of relevant conjectures is a crucial part of mathematics work and if a subject has an instrument facilitating it, it is a considerable advantage.

The instrument, which we have in mind, is DGE software. The hallmarks of DGE are the accurate construction and the possibility of dynamical changes in the construction. Modern DGE software (like GeoGebra) have many efficient tools for grasping key information leading to the solution of a problem. How can the help of DGE be described?

A general and straightforward description of how to work with DGE is given in Guven (2008):

I. Experimental Results (*"In which I explore the problem empirically using Cabri Geometry"*).
II. Towards a Proof (*"In which I use Cabri Geometry to make observations that will lead to a deductive proof"*).
III. Proof (*"In which I prove some results about the locus deductively, using all the experimental results acquired from Cabri Geometry"*).
IV. Some generalizations (*"In which I extend the proof using additional observations supported by Cabri Geometry"*).

On the basis of this description, it can be deduced that to solve a problem successfully, the student must be able to

1. discover relevant facts and hypotheses related to the solution (with or without the help of software),
2. arrange the facts to a deductive chain (within the frame of a mathematical theory).

In order to accomplish these two steps, the student must fulfil the following prerequisites:

The primary prerequisite is the knowledge of the relevant mathematical theory. As Mariotti (2006) states:

> *It is not possible to grasp the sense of a mathematical proof without linking it to the other two elements: a statement and overall a theory.* Likewise Polya (2014) states that it is difficult to get a good idea when we know little about the topic of the problem and it is impossible if we know nothing.

Another factor is the subject's experience and cognitive abilities—creativity, talent and deductive reasoning. These factors are from the domain of psychology and we will not pursue this topic further in this paper.

The above-mentioned factors are important in both the phases of making conjectures and proving. Let us now focus on the process of making conjectures with the help of DGE. The key question is as follows: in what way, is it possible to use software in order to grasp facts related to the solution of the problem?

We start with general procedures that can be useful, see Fahlgren and Brunström (2014) and Baccaglini-Frank and Mariotti (2010):

1. We modify the construction for a particular case in order to expose the hidden relationships.
2. We find out (approximately or precisely) the set of certain points in which the construction has a given property.
3. We modify the structure dynamically and observe those relations between the objects remaining invariant.

An expert uses these general procedures automatically and subconsciously, but for beginners, they are not a matter of course. In the following section, these procedures are organized into a more general classification based on the motivation of the solver.

However, these procedures have one common property—they require the subject's ingenuity and strategy. Even in the case of a mere observation of a dynamically changing construction, the knowledge of theory and visual thinking plays a crucial role, Magajna (2017). Synergy comes only if the tools of the software interact with the student's knowledge and a strategy. Finding experimental facts is by no means a routine process.

In order to understand the interplay between the subject's cognitive processes and tools of the software better, Arzarello et al. (2002) describe two opposing processes that they refers to as ascending and descending:

> These computer-supported practices can be framed within a cognitive evolution back and forth from perceptions to abstract ideas; in fact, there are two main cognitive typologies, which can be differently faded according to the concrete situation:
> Ascending processes, from drawings to theory, in order to explore freely a situation, looking for regularities, invariants, etc.
> Descending processes from theory to drawings, in order to validate or refute conjectures, to check properties, etc.

Arzarello et al. (2002) focus on the dragging tool. On this tool, they illustrate two processes. In their examples, students solve problems of elementary (upper secondary school) geometry. The crucial steps of their solution were as follows:

I. The students began to explore the problem in Cabri. They used the so-called "wandering dragging", which means moving the basic points randomly, without a plan. They only hoped to spot some regularity. This phase falls under the "ascending process".

II. Having discovered a geometrical property of the construction, they tried to find out what conditions were necessary for the property to be valid. They moved the basic point in such a way that the construction retained this property. Arzarello denotes the strategy as "Dummy locus dragging". In most cases, the basic points follow a curve invisible to the subject. If the subject is able to carry out the operation correctly, the curve becomes visible. In more recent literature, this strategy is referred to as "Maintaining dragging model", Baccaglini-Frank and Mariotti (2010). As before, the phase belongs to the "ascending process".

III. In this phase, the students formulated a conjecture of what the locus gained in the preceding phase was. In other words, they tried to identify the locus with a known object. After identification, they used the strategy that Arzarello refers to "linked dragging"—based on their conjecture, they constructed the locus (the tacit assumption of Arzarello is that the locus is not too complicated) and investigated experimentally whether any arbitrary point of the constructed locus has the desired property. This phase represents the "descending process"—the students formulated a conjecture and in order to verify it used the Cabri tools.

IV. Arzarello concludes that it is useful in many cases to carry out a modality of "dragging test". The purpose of the modality is to find out whether the conjecture is valid "for all cases", or whether it is only a random coincidence or optical illusion. For example, we have a special case of the construction for which it only seems that the conjecture is valid, but in reality, the conjecture is false or is valid only for this special case.

The article of Arzarello et al. (2002) constitutes the first pillar we come out of. The second pillar is the "Toulmin model", Pedemonte (2002, 2007). The Toulmin model was proposed in order to compare the structure of argumentation leading to the formulation of a conjecture on the one hand and the structure of mathematical proof of the conjecture on the other hand. The model is important for our classification because it takes into account not only the solver's conjectures, but also their motivation leading to the formulation of these conjectures. The motivation is referred to as "warrant" in the model. As Magajna (2017) states: ... *dynamic geometry software is equipped with exceptionally nice and powerful methods (e.g. dragging) for obtaining warrants for the hypothesized claims.*

2.1 The Toulmin Model

For our purposes, it is sufficient to describe the "simplified Toulmin model", Pedemonte (2007). The solution of the geometric problem can be seen as a sequence of several steps from premises through transient claims to the final statement. In order to be able to take the transient statement into account, a certain justification indicating its correctness is needed. This justification is called "Warrant" in the Toulmin model.

The reasoning leading to a Claim can be divided into three categories:

(1) Deductive warrant. The claim is deduced from premises.
(2) Heuristic warrant. It is not easy to define a heuristic method. According to Puig (1996): "Heuristic are the means used in the resolution process that are independent of the content and do not necessarily lead to the solution of the problem." In other words, heuristic strategies are general procedures and can be used when looking for lost car keys or in the process of solving a mathematical problem. According to Polya (2014), a heuristic assertion also can be seen as such an assertion for which there are logical indications which are not sufficient in themselves. Let us illustrate this with the following example: Two triangles have one angle identical → these triangles are (could be) similar.
(3) Warrant based on an empirically observed fact. There are usually no logical indications for such a claim. It is based only on visual perception or on results derived from other empirical facts.

If it turns out that the hypothesized claim plays an important role in the proof, a formal argumentation must be provided. This step is called "backing".

The Toulmin model scheme is shown in Fig. 1.

In short: Students can discover a claim relevant to the solution deductively, by heuristic arguments or in an empirical way, e.g. thanks to an accurate sketch. It is obvious that the deductive justification is valid only in the final proof of the claim.

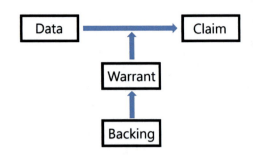

Fig. 1 Simplified Toulmin model

The situation is further complicated by the fact that it is not easy (or even possible) to make a sharp distinction between heuristic and deductive arguments for a claim. As Polya states (Polya, 2014), the solver has in many cases logical arguments that are not sufficient. The arguments can refer in part to the mathematical theory, but in part, they can be only random or plausible. They often border on specific deductive arguments and general heuristics. Nevertheless, the production of such arguments is often necessary in order to achieve the solution.

If the subject finds the claim in an empirical way, consequences can be inferred from it. This procedure is called abduction:

> Abduction is an inference that allows the construction of a claim starting from an observed fact, Peirce (1960).

Abduction is based on observation and hence it is not possible to use it in a deductive proof. Let us illustrate inferences on an example, Peirce (1960), p. 372: A bag is full of white beans. Consider the sentences: (A) these beans are white; (B) the beans of that bag are white; (C) these beans are from that bag.

Deduction is a concatenation of the logical form: B and C, hence A; abduction is: A and B, hence C (Peirce called hypothesis "abduction"). Induction is: A and C, hence B.

In both classical and DGE-supported approaches, the production of hypothetical claims is mainly driven by abduction. However, while in the former case (paper and pencil environment), abductions are produced mostly because of the ingenuity of the subjects, in the latter case, the tools of DGE (like dragging or dynamic transformation of a construction) mediate them, Arzarello et al. (2002). In this case, the importance of visual reasoning and perception increases and it can facilitate finding the solution considerably.

The Toulmin model takes into consideration "warrant", which precedes the formulation of a claim. Arzarello describes the interaction of a solver with software as a two-way dialogue, in which the solver's knowledge and strategies are intertwined with tools of the software and synergy occurs. Based on these two conceptions, we will create a theoretical classification of ways in the environment of dynamic geometry that lead to empirical facts relevant to the solution of a problem and its justification.

3 The Model: Classification of Experimental Facts Attained with the Help of DGE

In the above-mentioned Toulmin model, we stated three kinds of "warrants" that lead the solver to take into account a claim: (1) deductive warrant, (2) heuristic warrant and (3) warrant based on empirical data. While in the "environment" of paper and pencil, the solver must rely on their ingenuity, the support of DGE tools facilitates the production of right conjectures considerably. Grasping empirical facts with DGE is not a matter of course, the solver's knowledge and logic must be involved in the process. In this section, the ways that lead the solver to relevant empirical facts are classified.

Before introducing this classification, we must emphasize that a conjecture is not always the outcome of argumentation. Such conjecture is referred to as a "fact" by Pedemonte (2007). Taking this into account, she classifies the argumentation into two categories:

> The argumentation named constructive argumentation, contributes to the construction of a conjecture, thus it precedes the statement;
>
> On the other hand, the argumentation named structurant argumentation, justifies a conjecture, previously constructed as a 'fact', and so it comes afterwards.

It is possible to say that in the case of "constructive argumentation", the subject knows the indications why the conjecture could be true. To speak in the words of the Toulmin model, the subject knows deductive or heuristic warrants for the conjecture.

The main question is as follows: What argumentation or motivation precedes discovering a conjecture (relevant empirical fact or claim in the Toulmin model) if a subject uses the means of DGE?

There are two ways of finding empirical facts in DGE: by visual perception or with the help of an "experiment". By the term "experiment", we mean an activity that goes beyond mere visual perception and in which the solver's ingenuity and strategy are involved. For example, the strategy "dummy locus dragging" mentioned in the previous section (moving the basic point in a way that the construction keeps the given property) is an example of "experiment" and not "visual perception".

Let us describe this classification in detail:

- **Discovering a fact by visual perception**
  The student changes the construction (more or less randomly) and observes the consequences. They look for invariants or patterns of the construction.
- **Discovering a fact due to an experiment**
  The solver's experience and knowledge are involved in this way of grasping facts because the experiment must be prepared. However, what is the motivation (or warrant) for the proposition of the solver's experiment? We distinguish two cases:

  – **Constructive motivation:** The subject knows that the experiment is related to the solution of a problem. They pursue a specific idea related to the solution. In many cases, the subject is aware of (logical) arguments supporting the importance of the experiment.

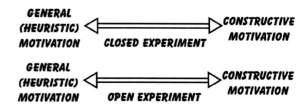

Fig. 2 Two poles of motivation for executing an experiment

- **General (heuristic) motivation:** In this case, the subject is not sure whether their idea is related to the solution. Although they could have some arguments for the validity of a conjecture, the most common is the case when they are simply making guesses or the reason for conduction an experiment is quite general and heuristic.

Another division is possible with respect to the form of the experiment. In our classification, we distinguish between the following two cases:

- **Closed experiment (CE):** The subject has a specific idea that is verified using the software. In other words, the subject knows in advance what they look for. The subject formulates a "closed question" and gets an answer in the true/false form. It would not be entirely correct to confine the role of software in this category to mere verification. When a person solves a problem, many ideas have the form of "hazy hypotheses". Most of them turn out to be incorrect. The software tools help to decide which of these ideas can be correct and are worth proving. Especially in open problems (problems without a clear goal), the student does not know what to look for and their motivation is guided by guessing numerous random conjectures.
- **Open experiment (OE):** The subject conducts an experiment with an unexpected output (it means that the output consists of new information). We say that the student formulates an "open question". They get an answer that is valuable and new for them.

It is obvious that there is no sharp border between "constructive" and "general" motivation. In many situations, the subject just hopes that the output of an experiment could be useful but is not entirely sure. This is the reason why we do not divide motivation into two separate categories and rather consider it is a continuous dimension, Fig. 2. The more arguments the subject has for the relevance of an experiment to the solution, the more their motivation is "constructive". And on the contrary, the fewer arguments for the relevance of the experiment the subject has, the more their motivation is "general" or "heuristic".

4 Illustrations of the Model

The functionality of the model is demonstrated in two examples. The examples have the form of subjective records of the solving process. Although the problems are more difficult, they are still within the frame of upper secondary grammar school mathematics. The first problem, generally unknown, was published in Blažek (2020). The second problem is common in textbooks focussing on Euclidean geometry.

4.1 The First Example

An ellipse can be defined in two (upper secondary) ways:

- **The foci definition of an ellipse**
  An ellipse ϱ is the locus of point P such that the sum of distances of point P to two given points F_1, F_2 (the foci) is a given constant $2a$, Fig. 3.
  This definition is equivalent to the definition using the director circle: Let k be a circle centred at F_2 with radius $2a$ and F_1 a point inside k. Then the locus of the point P such that the distances of P to the circle k and to F_1 are equal is an ellipse, Fig. 4. We leave the justification to the reader.
- **The focus-directrix definition of an ellipse**
  Ellipse τ is the locus of the point P such that the ratio ε of distances of P to the directrix line and to focus F_1 is constant and is equal to $\varepsilon = \frac{|F_1 F_2|}{2a} < 1$, Fig. 5.

We will prove the following problem synthetically (i.e. without the help of analytic expressions):

Problem 1: The locus τ, defined by the directrix line, is identical with the locus ϱ, defined by the director circle. Prove that these two definitions are equivalent.

Fig. 3 Foci definition of an ellipse

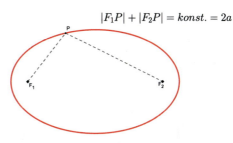

$|F_1 P| + |F_2 P| = konst. = 2a$

Fig. 4 Definition using the director circle

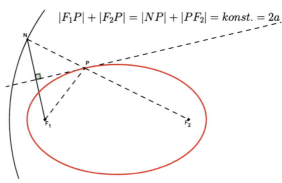

$|F_1 P| + |F_2 P| = |NP| + |PF_2| = konst. = 2a$

Fig. 5 Definition using the directrix line

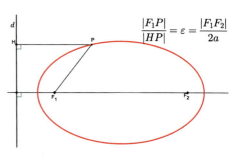

$$\frac{|F_1P|}{|HP|} = \varepsilon = \frac{|F_1F_2|}{2a}$$

Fig. 6 The assignment of the problem

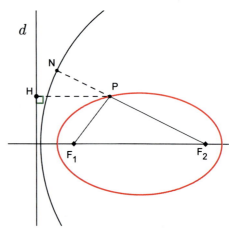

Exploration of the problem

The solver started by construction of the assignment of the problem (they constructed an ellipse, its directrix line and the director circle). Then the solver selected an arbitrary point P on the ellipse and constructed the point N on the director circle and the point H on the directrix line, Fig. 6.

After a moment of thinking, the solver came to the question:

Question 1: What is the "relation" between the points H and N?
The question is indefinite, it is not clear how to reply to the question in a mathematical way. It is a "heuristic question" because the solver did not have any concrete arguments for its relation to the solution. The solver conducted an "open experiment" (without knowing what the output would be) and their motivation was "general" (they were aware of no arguments supporting the relation between the experiment and the solution of the problem).

Experiment (open, motivation "general"): Let us construct line HN and turn on its "trace" using the GeoGebra command `Trace On`. Let us move by a point P along the ellipse. What shall we observe, Fig. 7?

The first claim: The line HN passes through a fixed point E lying on the line F_1F_2.

Fig. 7 The first experiment, discovering a fixed point

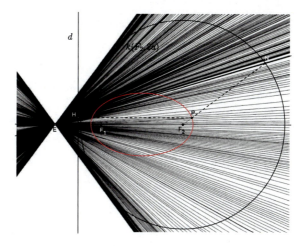

The construction of the point E using line HN is not elegant from the mathematical point of view (it lacks symmetry). This opens a new question:

Question 2: Is it possible to construct the point E directly, without constructing the line HN?

The relation of this question to the solution is not clear (because the relation between the "First claim" and the solution is not clear). Hence, as before, the solver had no conscious intention related to the solution and their motivation was "general" or "heuristic". They carried out two "closed experiments". In the first one, they determined the intersection of the circle k with the line F_1F_2. Let us denote it X (the point is not depicted in Fig. 7). Then they constructed the symmetric image of the point F_1 with respect to the point X. Let us denote it Y. They found out that $Y \neq E$; therefore, the construction was not correct. What could be tested next? The inversion came to the solver' mind and they found out the following:

Experiment (closed, motivation "general"): The subject constructed the image of point F_1 in inversion with respect to the circle k. The image coincided with the point E.

The second claim: The point E is the image of the point F_1 in inversion with respect to the circle k.

Note: The inverse image E of a point F_1 with respect to the circle with centre F_2 and radius R is a point satisfying the relation $|EF_2||F_1F_2| = R^2$ and lying on the ray F_2F_1.

Figure 8 shows the current situation.

Looking at the picture motivated the following question:

Question 3: Does the midpoint of segment F_1E lie on the directrix line d?

The motivation for this question (or conjecture) is purely visual. As it turned out, the conjecture was correct: By applying the command Midpoint or Centre on the segment F_1E, it was verified that the midpoint lies on the line d.

Fig. 8 Situation after discovering the "Second Claim"

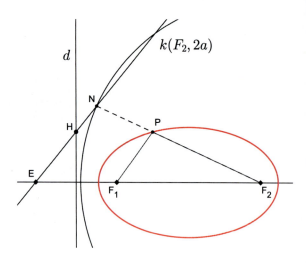

The third claim: The midpoint S of the segment $F_1 E$ lies on the line d. In other words, the line d is a perpendicular bisector of the segment $F_1 E$.

Finally, the solver asked one more question, although they did not know if this question would be related to the solution of the problem.

Question 4: Are any four points of the construction concyclic?

Experiment (closed, motivation "general"): The solver tested whether any four points of the construction lie on a circle. He found out that the points F_1, H, N, P of the construction lie on a circle.

The fourth claim: The quadrilateral $F_1 H N P$ is cyclic.

At this moment, the phase of exploration ended and the phase of justification began. All the facts discovered empirically are summarized in Fig. 9. Before stating the justification, let us analyse the help of GeoGebra in the exploratory phase.

As mentioned earlier, it is possible to compare the interaction between the subject and software to a dialogue. The subject formulates a question and the software answers. The question must be formulated in the language of the software; therefore, not every question is allowed. Let us look back at the gained facts (denoted as claims):

- The first claim was preceded by a "random question". Its answer was new for the solver.
- The second question concerned the construction of the point obtained from the first claim. The second attempt of the construction, based on the solver's experience and guess, was successful.
- The third claim sprouted from visual perception.
- The fourth claim was motivated by the question of whether any four points of the construction are concyclic.

Let us emphasize that the solvers were not sure whether the gained facts were related to the solution of the problem—it was clarified by the following phase of justification.

Fig. 9 Summary of all facts discovered empirically

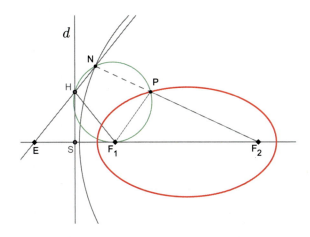

So they could not be sure if they were on the right track. Similarly, the formulations of questions, which motivated the experiments, rested on the solvers' mathematical experience. So, what was the crucial help of the software based on? The software answered the questions experimentally. If a subject did not have software available for the first question, they would have to consider two possibilities: (1) the line HN has a fixed point or (2) the line HN does not have a fixed point. Due to the software tools, they determined the correct answer immediately. Similarly, in the second question, a subject without the help of software would have to consider both presented constructions as possible, and moreover, they would have to take into account that neither of them is correct. The fourth question had many possible answers: among seven points, there are 35 foursomes and to consider all of them is virtually impossible. After excluding the foursomes with three collinear points, one gets 14 foursomes, which is still too many. As was emphasized by Polya (2014), it is common in mathematical work that conviction about validity of a claim precedes its proof. In our case, the software indicated what hypotheses could be valid. It illuminated the path, which, as it turned out, led to the solution of the problem.

Solution of Problem 1: First we select a suitable strategy. Coming out from the definition of ellipse by the director circle, we construct the line d according to experimental facts as follows:

1. We construct director circle k centred at F_2 with radius $R = 2a$,
2. On the ray $F_1 F_2$, we construct point E such that $|EF_2||F_1 F_2| = R^2$,
3. We construct a perpendicular bisector d of segment EF_1.

Now we prove that the line d is a directrix line. In order to do this,

4. We construct an arbitrary point P on ellipse and its assigned point N on the director circle,
5. We construct the intersection H of lines NE and d.

Now, it is necessary to prove that

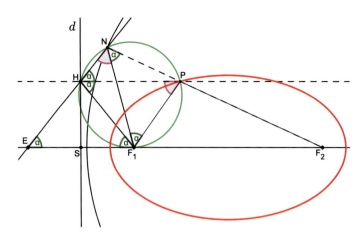

Fig. 10 Proof of the equivalence

(i) The length of segment HP is equal to the distance of the point P to the line d (in other words, lines HP and d are perpendicular).
(ii) The ratio of segment lengths fulfils $\frac{|F_1P|}{|HP|} = \varepsilon = \frac{|F_1F_2|}{2a}$.

All subsequent considerations are related to Fig. 10.

According to the construction of the point E, the equality $|EF_2||F_1F_2| = R^2$ or $|EF_2|/R = R/|F_1F_2|$ holds. Due to this equation and a common angle at the vertex F_2, the triangles F_2F_1N and F_2NE are similar: $\triangle F_2F_1N \sim \triangle F_2NE$. Triangles NF_1P and HEF_1 are isosceles (the point H lies on the perpendicular bisector d), and hence, the following equalities are valid:

$$\alpha = \angle NF_1P = \angle F_1NF_2 = \angle NEF_2 = \angle HEF_1 = \angle HF_1E.$$

From this, it follows $\angle F_1HN = 2\alpha$ (exterior angle theorem) and $\angle F_1PN = 180° - 2\alpha$, and hence, the quadrilateral $HNPF_1$ is cyclic. Applying the angle inscribed theorem, one gets $\alpha = \angle NF_1P = \angle NHP = \angle NEF_1$, hence the line HP is parallel to the line EF_1, which is perpendicular to the line d. The proposition (i) is proved.

The proof of the second proposition immediately follows from the similarity of triangles ENF_1 and HPF_1. Their similarity results from $\angle HPF_1 = \angle HNF_1 = \angle ENF_1$ and $\alpha = \angle F_1NP = \angle F_1HP = \angle F_1EN$. Now, since $\triangle F_2F_1N \sim \triangle F_2NE$, one gets

$$\frac{|F_1N|}{|EN|} = \frac{|F_1F_2|}{|NF_2|} = \frac{|F_1F_2|}{2a} = \varepsilon.$$

From similarity $\triangle ENF_1 \sim \triangle HPF_1$, it follows that $\varepsilon = \frac{|F_1N|}{|EN|} = \frac{|F_1P|}{|HP|}$, which is (ii).

As a final remark, we note that the proof above deals with one-way implication from one definition to the other, not with their equivalence. It is not difficult to prove the converse by contradiction.

4.2 The Second Example

The problem is the following:

Problem 2: Consider an acute-angled triangle ABC. Determine how to construct a point P in the plane of the triangle such that the feet of perpendiculars D, E, F dropped from the point P to the sides a, b, c respectively, are vertices of an equilateral triangle.

A triangle whose vertices are feet of perpendiculars dropped from a point P on the sides of the given triangle is called a "pedal triangle", Fig. 11.

Exploration of the problem

With the support of DGE, a natural question arises: can we find the point P by the means of this software?

Modern DGE software like GeoGebra have tools that fall under the class of CAS programs (Computer Algebra System). Among them, the command LocusEquation is able to determine the equation of an unknown locus based on given conditions. (The command Locus is not so powerful since it only displays the locus without its equation).

The command LocusEquation cannot be applied directly as the locus of point P depends on the lengths of the three sides. To express the equality of three sides, one needs two Boolean conditions (for example, $a == b$, $b == c$), but the command allows only one condition. It is not a disadvantage in this case as the position of points tells us nothing about the construction. A better strategy is needed. More details about various possibilities of the command are described in Hohenwarter et al. (2019) and Kovács et al. (2020).

Experiment (open, motivation "constructive"): What is the locus of point P if we simplify the condition of the equilateral pedal triangle and demand only an isosce-

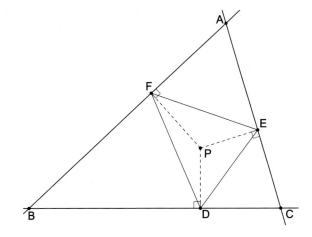

Fig. 11 Pedal triangle of an arbitrary point P

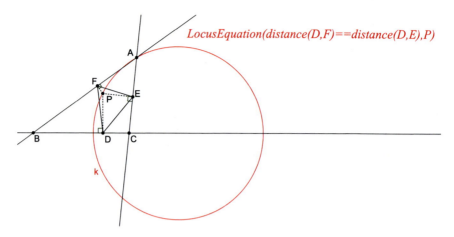

Fig. 12 Locus of point P fulfilling the condition $|DF| = |DE|$

les pedal triangle? Is it possible to determine such defined locus by the command `LocusEquation`?

The experiment above was "open", because the solver did not know what the output of the experiment would look like. At the same time, the experiment was "constructive" because the solver pursued a specific idea related to the solution of the problem. After the experiment, the solver came with the following claim:

The first claim: Locus of the point P such that for its pedal triangle $|DF| = |DE|$ holds is a circle k passing through vertex A, Fig. 12.

The immediate question is: How to construct the circle? We will start with a more modest question: Where is the centre of the circle? Looking at the picture, there is a strong suggestion: The centre of the circle lies on the line BC. Application of the command `Midpoint` or `Centre` confirms this hypothesis.

The second claim: The centre of the circle k lies on the line BC.

Despite the help, the solver was still unable to identify the circle. They tried to identify the intersection of the circle with the line segment BC. It was a closed and constructive experiment because its relation to the solution was obvious.

Experiment (closed, motivation "constructive"): The solver tried to identify the intersection of the circle k with the segment BC. The foot of the altitude from A to the side a was excluded immediately. Then the solver tried the midpoint of the segment BC. Finally, they constructed the bisector of the angle BAC and it passed through the intersection. Hence,

The third claim: k is an Apollonius circle, i.e. the locus of the point P such that $|PB|/|PC| = |BA|/|CA|$.

Fig. 13 Pedal triangle DEF of the point P is equilateral

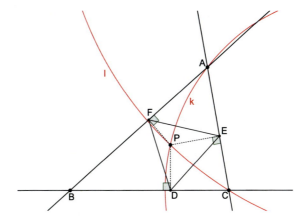

Then the construction of the desired point P was obvious. For feet of perpendiculars of a point P of the Apollonius circle k (which passes through the vertex A) is $|DF| = |DE|$. For point P of another Apollonius circle l (passing through the vertex C) is $|DF| = |FE|$. Hence, the intersections of these two circles are the solutions to the problem, Fig. 13.

It remains to show why for point P of the circle k $|DF| = |DE|$.

Proof:
Let us start with a trivial statement.

Lemma: Let a chord of length d subtend an angle α at a point of circle. Then the length d is given by the relation $2R \sin \alpha = d$, where $2R$ is diameter of the circle.

The quadrilateral $DPFB$ is cyclic (the sum of the opposite angles is 180°), and the segment BP is the diameter of its circumcircle. Applying the Lemma, one gets

$$|BP| \sin \beta = |FD|,$$

where β is the angle at the vertex B. Similarly,

$$|CP| \sin \gamma = |DE|.$$

If we put these equalities into ratio, one obtains

$$\frac{|FD|}{|DE|} = \frac{|BP| \sin \beta}{|CP| \sin \gamma} = \frac{|BA| \sin \beta}{|CA| \sin \gamma} = \frac{v_a}{v_a} = 1,$$

where v_a is the length of altitude on the side a.

Note: In literature, the intersections of the Apollonius circles are referred to as "isodynamic points".

5 Conclusions

Arzarello et al. (2002) noted that it was not easy to separate cognitive actions of a subject from the tools of software and that their intertwining would occur. Pedemonte (2007) distinguished between two cases: the subject knows the arguments contributing to the construction of a conjecture (thus, they precede the statement), and the subject works with a conjecture for whose validity they have no support. The model of interaction of a subject and software is based on uniting these two theses.

Using examples, we showed that some actions that the solver carries out with the software have a genuine justification—the subject knows the purpose why to do the actions and pursues a specific idea related to the solution. In other words, they know why the output of the experiment is important. We refer to this case as "Constructive motivation".

A situation can occur when the subject conducts an experiment the reason for which is haphazard, indeterminate or heuristic. The subject does not know if and how the facts gained in the experiment are related to the solution. This case is referred to as "General" or "Heuristic" motivation.

In both examples, there was a case when an experimental fact was revealed by pure visual perception. But while in the case of the ellipse (third claim: the midpoint of the segment $F_1 E$ lies on the line d), the experimental fact solely followed from visual perception, in the second case (second claim: the centre of the circle k lies on the line BC), the visual perception was guided by the intention to identify the circle. It is an illustration of the following statement by Magajna (2017):

> Observing is not just gazing at a geometric configuration. It means also looking for specific objects or relations in accordance to the observers solving strategy, conceptual understanding, knowledge base and experience.

Thus, it is possible to conclude that facts gained by visual perception can be divided into two categories in exactly the same way as "experimental facts". In the category of "constructive motivation", the subject pursues a specific intention related to the solution, and in the category of "general motivation", the subject is just "gazing" at the construction and looking for patterns or invariants.

Summarizing the motivations, in the first case, the motivation for DGE experiments was "hazy", based on individual experience, heuristics and guessing. In the second case, the motivations were purely constructive, pursuing specific intentions.

The drawback of the paper is that the model is tested on an experienced solver. It does not discuss the role of the model in the process of looking for a solution by a less experienced student (prospective teacher). The authors conducted one experiment with students with the aim of applying the model in real situations. However, due to a small sample (four students participating in the experiment) and the scope of the article, this experiment is not presented in detail. Based on outcomes of the experiment, it is possible to conclude that successful students discover relevant facts much more likely, thanks to constructive motivation such as abduction and deduction. On the other hand, conducting experiments by students who are not able to solve the problem is random. One possible reason for their failure is that they do not consider the randomly discovered facts as relevant (they do not know the logical arguments connecting the facts to the solution of a problem). Also, most of the facts that students discover are discovered by constructive argumentation, not by simple "guesses". The authors will explore this topic further in more reliable experiments.

We must highlight the increasing importance of merging DGE tools (based on accurate graphics and numerical computation) and CAS tools (based on symbolic computation). The synergy of these two approaches provides a very powerful instrument, see Kovács et al. (2020) (One of the leading programs in this direction is GeoGebra.). For example, article Hohenwarter et al. (2019) speculates that work with such tools could affect students' learning: These tools may suppress the need for logical validation and support work with machine, (cooperative reasoning), based on slightly different abilities than classical geometric reasoning. With respect to our model, if a solver uses a command like `LocusEquation`, it can indicate that they know that the output of GeoGebra will be important, but does not know what such output looks like ("constructive" and "open" experiment). Such command gives the solver an answer but without (logical) arguments indicating its correctness. It is up to the solver to justify this answer. However, as we stated in the preceding paragraph, discovering relevant facts is mostly dependent on the student's constructive motivation (like abduction and deduction), i.e. on their mathematical reasoning and knowledge.

Some relatively new (and not so known) programs try to facilitate or even avoid the difficult process of finding relevant experimental facts by the function of "automated observation". The principle of this function is that the program randomly changes the construction (by moving independent points of the construction) and looks for invariants that were preserved throughout the motion. A promising development of Java Geometry Expert (JGEX) JGEX (2021) equipped with this (and many other) functions was stopped in 2011, and nowadays, it is even impossible to download the program. The program OK Geometry by Z. Magajna is still being updated and improved. The principles of the program are described in Magajna (2017). The program has, besides automatic observation, many other functions, for example, it enables to produce special files for training deductive proofs. The impact of the function "Automated observation" on the student's ability to deduce proofs, as far as the authors know, is not charted. The possible disadvantage of this function could be either that the number of the "observed" facts is considerably greater than it is optimal and the solver must decide which facts to exclude and which to take into account, or, on the contrary, the crucial facts are missing. Based on our personal experience, we conclude the more the subject knows about geometry, the more they appreciate and exploit the (advanced) tools of OK Geometry. After all, this is generally valid for other DGE software.

References

Arzarello, F., Olivero, F., Paola, D., & Robutti, O. (2002). A cognitive analysis of dragging practices in Cabri environments. *ZentralBlatt für Didaktik der Mathemetik, 34*, 66–72.

Baccaglini-Frank, A., & Mariotti, M. A. (2010). Generating conjectures in dynamic geometry: The maintaining dragging model. *International Journal of Computers for Mathematical Learning, 15*, 225–253.

Blažek, J. (2020). Důkaz ekvivalence dvou definic elipsy. *South Bohemia Mathematical Letters, 27*, 1–9.

Blažek, J., & Pech, P. (2019). Synthetic proof with the support of dynamic geometry. *International Journal for Technology in Mathematics Education, 26*, 107–112.

Duval, R. (1991). Structure du raisonnement déductif et apprentissage de la démonstration. *Educational Studies in Mathematics, 22*(3), 233–263.

Fahlgren, M., & Brunström, M. (2014). A model for task design with focus on exploration, explanation, and generalization in a dynamic geometry environment. *Technology, Knowledge and Learning, 19*, 287–315.

Guven, B. (2008). Using dynamic geometry software to gain insight into a proof. *International Journal of Computers for Mathematical Learning, 13*, 251–262.

Hohenwarter, M., Kovács, Z., & Recio, T. (2019). Using GeoGebra Automated Reasoning Tools to explore geometric statements and conjectures. In Hanna, G., de Villiers, M., & Reid, D. (Eds.), Proof Technology in Mathematics Research and Teaching, Series: Mathematics Education in the Digital Era, Vol. 14, pp. 215–236.

JGEX https://en.wikipedia.org/wiki/GeometryExpert.

Kovács, Z., Recio, T. Richard, P. R., Van Vaerenbergh S., & Vélez, M. P. (2020). Towards an ecosystem for computer-supported geometric reasoning. *International Journal of Mathematical Education in Science and Technology*. https://doi.org/10.1080/0020739X.2020.1837400

Leung, A. (2011). An epistemic model of task design in dynamic geometry environment. *ZentralBlatt für Didaktik der Mathematik, 43*, 325–336.

Magajna, Z. (2017). Automated observation of dynamic constructions. *International Journal for Technology in Mathematics Education, 24*, 115–120.

Mariotti, M. (2006). Proof and proving in mathematics education. In *Handbook of research on the psychology of mathematics education* (pp. 173–204).

Oflaz, G., Bulut, N., & Akcakin, V. (2016). Pre-service classroom teachers' proof schemes in geometry: A case study of three pre-service teachers. *Eurasian Journal of Educational Research, 63*, 133–152.

Pedemonte, B. (2002). *Etude didactique et cognitive des rapports de l'argumentation et de la démonstration, co-tutelle*. Grenoble: Universitè di Genova and Université Joseph Fourier.

Pedemonte, B. (2007). How can the relationship between argumentation and proof be analyzed? *Educational Studies in Mathematics, 66*, 23–41.

Peirce, C. S. (1960). *Collected Papers, II*. Elements of Logic: Harvard, University Press.

Polya, G. (1954). *Mathematics and plausible reasoning. Induction and analogy in mathematics*, Vol. 1. Princeton: Princeton University Press.

Polya, G. (2014). *How to solve it?* Princeton University Press.

Puig, L. (1996). *Elementos de resolucion de problemas*. Spain, Comares: Granada.

Villiers M. (2004). Proof in Dynamic Geometry: More than Verification. http://math.unipa.it/~grim/21_project/21_charlotte_deVilliersPaperEdit.pdf

Creative Use of Dynamic Mathematical Environment in Mathematics Teacher Training

Roman Hašek

1 Introduction

The chapter does not deal with the concept of artificial intelligence (AI) as it is usually understood, i.e. as a comprehensive agent, as it is dealt with, for example, in Russell et al. (2010) or du Sautoy (2019). It relates to AI indirectly, through the particular properties of currently available free-of-charge dynamic geometry software (DGS), namely GeoGebra (2020) and OK Geometry (2020), that undoubtedly move the capabilities of this software into the realm of AI. With full awareness of the relevant difference in the conceptual behaviour of this software and of the fact that the practices we present are not the only solution, the possibility of a purely automatic solution using GeoGebra is becoming more and more real, Botana et al. (2020), we specifically discuss the use of the software's ability to independently assess an individual geometric sketch, applying the specific method of automated observation of a dynamic construction and the principles of the automatic proving and deriving of geometric theorems. Based on this assessment, the software provides a user with specific feedback on her or his approach to the solution of a particular task. Such features definitely link this software to the world of AI. We believe in the usefulness of the connection of the worlds of DGS and AI, and we are convinced that the wide fulfilment of their potential in educational practice is very real. An educational environment with AI in the background and equipped with the features of reading dynamic constructions and automated reasoning could be the right means to balance the teacher's supervisory role in the classroom with the individualisation of learning in terms of the goals formulated by Bloom in his well-known treatise on mastery learning, Bloom (1968, 1971), Levin (2017).

R. Hašek (✉)
University of South Bohemia, Faculty of Education, Jeronýmova 10,
371 15 České Budějovice, Czech Republic
e-mail: hasek@pf.jcu.cz

From the above perspective, using examples from geometry based on historical problems, we will show specific educational procedures utilising the environment of the mentioned software providing users with feedback that leads them in their independent creative work, the result of which can be both a dynamic geometric model of the respective phenomenon and a real object, e.g. the physical model printed on a 3D printer. All are based on practical findings from the preparation of prospective mathematics teachers of lower and upper secondary schools. The use of the software to interpret historical geometrical subject matter from the perspective of up to date mathematics, to create a dynamic model of the respective phenomenon and also to serve as a basis to create its physical model has proven to be a functioning component of mathematics teacher education, Dennis (2000), Clark (2012), Furinghetti (2007), Hašek et al. (2017).

2 Historical Context

The story underlying the chapter is based on two contributions submitted by Josef Rudolf Vaňaus (1839–1910), a Czech grammar school mathematics teacher and one of the leading personalities of both the professional and social life of his time, to the *Journal for the Cultivation of Mathematics and Physics*; a paper *Trisektorie* (*Trisectrix* in English) on the use of an oblique strophoid to trisect an angle, Vaňaus (1881), published in 1881, and an assignment of a geometry task for the journal's problem corner, Vaňaus (1902), the solution of which was based on the trisection of an angle, Ostermann and Wanner (2012), published in 1902.

Josef Rudolf Vaňaus was born in 1839. In 1862, he graduated from the Faculty of Arts of Charles University in Prague, and then, for more than thirty years, he worked as a grammar school teacher. He died in 1910, after fourteen years of retirement. All his life J. R. Vaňaus was very active in promoting mathematics and its teaching. Starting as a young university student, in 1862, he became one of four founders of the Union of Czech Mathematicians and Physicists and he continued doing research and publishing papers on findings in mathematics and its teaching in relevant Czech journals. He paid significant attention to supporting students with mathematical talent at the secondary school level through assigning them problems in the problem corner of the Czech *Journal for the Cultivation of Mathematics and Physics* (with an original Czech title *Časopis pro pěstování matematiky a fysiky*), Folta and Šišma (2003).

A beneficial and a creative way of using GeoGebra and OK Geometry to solve the problem assigned by Vaňaus and to model and analyse his original method of trisecting an angle with contemporary students of mathematics teaching is treated in this chapter.

3 Problem 36

The following problem authored by J. R. Vaňaus was, as Problem No. 36, set in the problem corner of the third issue of the *Journal for the Cultivation of Mathematics and Physics* Vaňaus (1902), published in Czech. The target group of the problem assignment was students of an upper secondary school, ages 15–18.

> Given a line segment AB. Circular arcs, both with the radius $|AB|$, are drawn around points A and B, passing through points B and A, respectively, and intersecting at point C. The task is to set points M and N at arcs AC and BC, respectively, so that the line segment MN is parallel to AB and the angle $\angle MAN$ is equal to a given acute angle; see Fig. 1.

Solution: First, we add a few more elements to our sketch, see Fig. 2, an angle $\beta = \angle BAN$, the knowledge of which would immediately lead to point N, and segment MB, the diagonal of trapezoid $ABNM$ and at the same time the leg of the isosceles triangle $\triangle\, MAB$, which has two sides $|AB|$ and $|MB|$ of equal length due to the fact that both given arcs have the same radii.

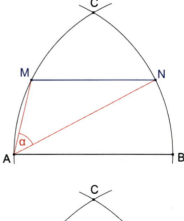

Fig. 1 Problem 36: Determine the line segment MN; $MN \parallel AB$, for a given angle α

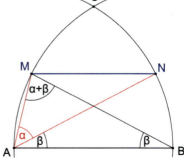

Fig. 2 Problem 36: Elements β and MB added for solution, $\triangle\, MAB$ is an isosceles triangle

Following Fig. 2, we express β in terms of α as follows. Since it applies to the interior angles of $\triangle MAB$ that $2\alpha + 3\beta = 180°$, angle β can be written as

$$\beta = 60° - 2\frac{\alpha}{3}. \tag{1}$$

Therefore, obtaining angle β, which leads to solving the problem, is subject to the trisection of an angle, namely the given angle α, a task that is not solvable by using just a straightedge and compass.

Trisecting an angle together with squaring the circle and doubling the cube are three classical problems of Greek mathematics that were proved to be impossible constructions using just a straightedge and compass. The impossibility of this so-called Euclidean construction of a trisection of an angle was proved by French mathematician Pierre Laurent Wantzel in 1873. For more information, see Impossible constructions (2020), Ostermann and Wanner (2012).

Three solutions to this problem, leading to the trisection of an angle, all in a similar manner to the one above, sent by students of upper secondary schools, were published in the last issue of the journal volume Vaňaus (1902). All three authors were aware that the solution is not constructable using only a straightedge and compass. One of them offered to complete the solution analytically, converting it into the problem of the intersection of conic sections, namely the circle and hyperbola, as recorded in Vaňaus (1902). Obviously, the upper secondary school students at that time were familiar with the non-Euclidean techniques of the trisection of the angle, among others using curves called 'trisectrix', namely, for example, the trisectrix of Maclaurin, Trisectrix of Maclaurin (2020), see Fig. 3, named after Scottish mathematician Colin Maclaurin

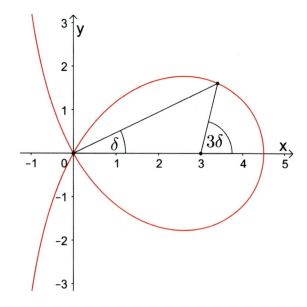

Fig. 3 Trisectrix of Maclaurin given by the Cartesian equation $2x(x^2 + y^2) = a(3x^2 - y^2)$, for $a = 3$. Processed in GeoGebra

(1698–1746). A list of curves that can be used as a trisectrix, i.e. as an additional tool with compass and ruler to trisect an arbitrary angle, can be found at Wikipedia, Trisectrix (2020). An angle can also be trisected using other non-Euclidean methods; see Angle trisection (2020) and Ostermann and Wanner (2012).

4 Problem 36 from the Contemporary Perspective

The possible use of the current free software in the field of Problem 36 will be presented in this section through the two software mentioned in the introduction, GeoGebra and OK Geometry. Our ambition is simply to share our specific findings and experience with the reader. We do not aim to provide a comparative evaluation of the software used and do not claim that this is the only possible way to solve the discussed problems with the currently available software.

4.1 GeoGebra

We assigned Problem No. 36, as a problem for volunteers, to students of the first year of the study of mathematics teaching at lower secondary school. Like the authors of the solutions published in the journal, in 1902, most of the current solvers arrived at a solution corresponding to (1). Yet there was a difference. Contemporary students are not as familiar with the non-Euclidean ways of trisecting an angle as their peers from the early twentieth century. On the contrary, they are well acquainted with the available mathematical software, which was clearly reflected in their approach to the solution of the problem. Once they found relation (1), they used GeoGebra to construct a solution based on the numerically calculated trisection of a given angle. GeoGebra, thus, served primarily as an environment for creating a dynamic model of a numerically calculated solution to a given problem, one such model being shown in Fig. 4.

Another approach to the use of GeoGebra to solving the problem, which we could identify among the students' solutions, went to the essence of a dynamic geometry system. Its author employed the dynamic features of GeoGebra to try to find a solution by manipulating the construction; see Fig. 5. She created a movable transversal MN between given arcs, $M \in \widehat{CA}$, $N \in \widehat{BC}$, visible from A at a given angle α. Moving M the midpoint S of MN draws a curve, the locus of point S. The intersection of this curve with the axis of symmetry of the line segment AB determines the position of MN we are looking for to solve the problem. This dynamic investigation of the nature of the locus curve gives rise to the question of which curve it is. Can we determine its equation? Yes, using GeoGebra CAS, or any other suitable computer algebra system, it is possible, without any special knowledge of differential geometry of curves.

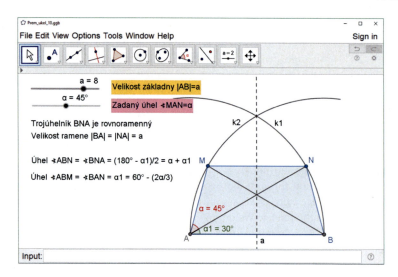

Fig. 4 Problem 36: Student's solution, a dynamic model created in GeoGebra

Fig. 5 Problem 36: Dynamic investigation of the solution. Processed in GeoGebra

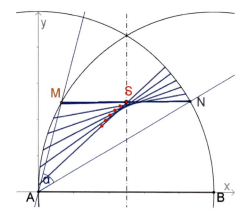

First, we place the given construction into the coordinate system appropriately so that $A[0, 0]$, $B[r, 0]$, $M[m_1, m_2]$, $N[n_1, n_2]$, $k = \tan \beta$ and $a = \tan \alpha$, see Fig. 6, and express its configuration by symbolic equations $e_1, e_2, \ldots, e_6$ as follows: From the right triangle with hypotenuse AM and an internal angle $\alpha + \beta$, we get the first equation $e_1 : \tan(\alpha + \beta) = \dfrac{m_2}{m_1}$, which can be written as

$$e_1 : (a + k)m_1 - (1 - ak)m_2 = 0, \qquad (2)$$

where $k = \tan \beta$ and $a = \tan \alpha$. Analogously, from a right triangle with the hypotenuse AN and an internal angle β, we get the second equation

Fig. 6 Problem 36: Placement in the coordinate system

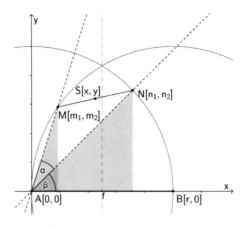

$$e_2 : kn_1 - n_2 = 0. \tag{3}$$

The other two equations e_3 and e_4 are based on the fact that $S[x, y]$ is the midpoint of the line segment MN:

$$e_3 : m_1 + n_1 - 2x = 0, \tag{4}$$
$$e_4 : m_2 + n_2 - 2y = 0. \tag{5}$$

The last two equations reflect the fact that M and N are the points at circles $k(B; r)$ and $l(A; r)$, respectively, where $r = |AB|$:

$$e_5 : m_1^2 - 2m_1 r + m_2^2 = 0, \tag{6}$$
$$e_6 : n_1^2 + n_2^2 - r^2 = 0. \tag{7}$$

The notation of these equations and their further processing in the *CAS* environment are shown in Fig. 7. For equations, see lines 1 to 6. To get the general algebraic polynomial representation of the locus curve, we use the Eliminate command, based on the method of the Groebner bases, Kovács (2017), Hašek (2019). Part of its result can be seen on line 7 of the *CAS* view. The complete resulting algebraic equation of the locus curve is as follows:

$$\begin{aligned}
& 16x^6 a^2 - 32x^5 r a^2 + 8x^4 r^2 a^2 + 16x^3 r^3 a^2 - 7x^2 r^4 a^2 - 2xr^5 a^2 + r^6 a^2 \\
& + 48x^4 y^2 a^2 - 64x^3 r y^2 a^2 + 16xr^3 y^2 a^2 - 3r^4 y^2 a^2 + 48x^2 y^4 a^2 \\
& - 32xry^4 a^2 - 8r^2 y^4 a^2 + 16y^6 a^2 + 8x^2 r^3 ya - 2r^5 ya + 8r^3 y^3 a \\
& + 16x^6 - 32x^5 r + 8x^4 r^2 + 8x^3 r^3 - 3x^2 r^4 + 48x^4 y^2 - 64x^3 ry^2 \\
& + 8xr^3 y^2 + r^4 y^2 + 48x^2 y^4 - 32xry^4 - 8r^2 y^4 + 16y^6 = 0.
\end{aligned} \tag{8}$$

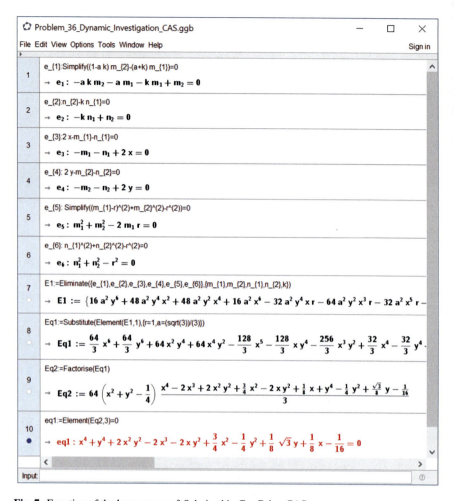

Fig. 7 Equation of the locus curve of S derived in GeoGebra CAS

Equation (8), thus, completes the phase of the symbolic solution of the problem. Now, in order to plot a particular curve given by (8) in the *Graphics View* of GeoGebra, we have to substitute some specific values of its parameters into this equation, i.e. the radius r of arcs and the angle α, precisely, the tangent of the angle α as the value of a; see line 8 in Fig. 7.

Let us use the values $r = 1$ and $a = \tan\left(\frac{\pi}{6}\right) = \frac{\sqrt{3}}{3}$. Factorising the resulting polynomial using the `Factorise` command, we get two curves, a circle drawn by S when M coincides with A and the curve that interests us, which appears to be the limacon, Limacon (2020), plotted in the Graphics view of GeoGebra; see Fig. 8.

The solutions mentioned so far have always been based on an idea. But what if we have no idea? Can any software assist us in such a way that it gives us an impetus to

Fig. 8 Problem 36: Limacon, the locus of midpoints of MN for $\alpha = \frac{\pi}{6}$. Processed in GeoGebra

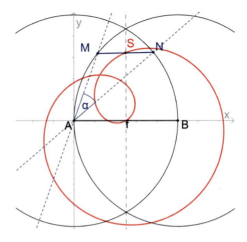

start a solution? And, moreover, can such software reveal to us a hitherto unknown solution? In the next section, we will show how OK Geometry can serve us in this way.

4.2 OK Geometry

OK Geometry is a dynamic geometry software that has a unique ability to analyse a dynamic geometric construction, either created directly in it or imported from another DGS, and to provide a list of properties of this construction, not proved, but determined with a high degree of probability. OK Geometry was conceived by Zlatan Magajna and is available free of charge from OK Geometry (2020), where an interested user can find all the necessary information on its use and functionality.

By drawing just a sketch of the assignment of Problem 36 and letting OK Geometry analyse it, a user receives a number of properties that with high probability pay for this geometric construction. Focussing only on those of them that relate to the stated task, she or he almost certainly obtains a base for developing some ideas on the problem's solution, some of which may be not entirely obvious even to an experienced solver. A small portion of the result of such an analysis is shown in Fig. 9. The software indicates that the sizes of angles $\angle CAM$ and $\angle NAM$, where $|\angle NAM|$ is the given acute angle, see α in Fig. 1, are in the ratio 1 : 3. If we manage to prove this hypothesis, we can design a new way of constructing the segment MN according to Problem 36 assignment, of course, again based on the trisection of a given angle. Having the line AC determined with the fixed points A and C, we simply find the point M, so that $|\angle CAM| = \frac{1}{3}\alpha$. So, is the relationship between angles $\angle CAM$ and $\angle NAM$, stated by OK Geometry, true? If it is not obvious from Fig. 9, we can 'ask' the software for more information on the relations of involved angles. Among

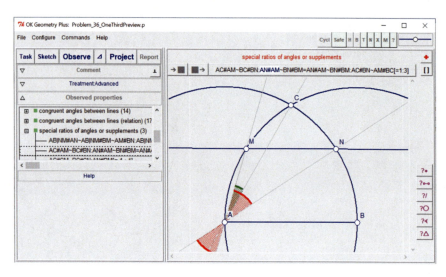

Fig. 9 Problem 36: Analysis using OK Geometry; $|\angle NAM| = 3|\angle CAM|$

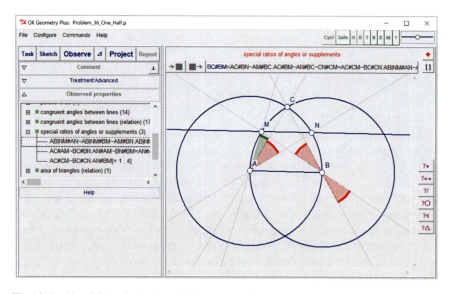

Fig. 10 Problem 36: Analysis using OK Geometry; $|\angle NAC| = 2|\angle CAM|$

them, we then find the key to the proof, 1 : 2 ratio of measures of angles $\angle CAM$ and $\angle NAC$, where $\angle NAC \cup \angle CAM = \angle NAM$; see Fig. 10. Both relationships between angles are, therefore, equivalent, with the latter being a clear consequence of the relationship between the inscribed angle $\angle CAM$ and the central angle $\angle CBM$, where $|\angle CBM| = 2|\angle CAM|$. Consequently, due to $|\angle NAC| = |\angle CBM|$, the relation $|\angle NAC| = 2|\angle CAM|$ pays. Therefore, the former, $|\angle NAM| = 3|\angle CAM|$,

5 Vaňaus' Trisectrix

In Sect. 4, the part dealing with the 'dynamic geometry' approach to solving Problem No. 36, we used the dynamic geometry and computer algebra features of GeoGebra to create a dynamic model of the respective geometric construction and to derive an equation of the corresponding locus curve.

Here, in the section devoted to the first of two Vaňaus' publications covered in this text, a paper *Trisektorie* from 1881, Vaňaus (1881), we will apply this approach again and complete it with the creation of a dynamic geometric model of Vaňaus' trisector, a mechanical linkage implementing his method of trisection.

Let's move back to 1902 to complete the story of solving Problem No. 36. In his comment to the solutions of the three students, published in Vaňaus (1902), Vaňaus recommended his 1881 paper in which he introduced a method of doing a trisection using the cubic curve shown in Fig. 11. This cubic curve, currently known as the *oblique strophoid*, Gibson (1998), Lockwood (2007), Strophoid (2020), is presented by Vaňaus as follows.

> The locus of points M for B moving along the line l, a secant to the circle c, so that $|MD| = |DB|$, where D is the intersection of the line OB with c.

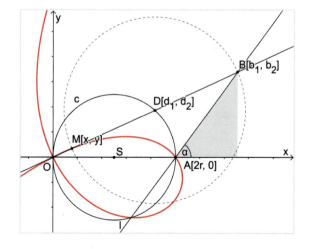

Fig. 11 Vaňaus' trisectrix, the *oblique strophoid*, for $r = 1$ and $a = 3$. Processed in GeoGebra

He derives the Cartesian equation of this curve

$$a(y^2(2r+x) - x^2(2r-x)) = y(y^2 + x^2 - 4rx), \tag{9}$$

where r is the radius of the circle c and a is the slope of the line l (i.e. a is the tangent of the angle of incline α of the line l), and describes a simple way of using it to trisect an angle, namely the angle u in Fig. 15. To learn which position of the curve with respect to the Cartesian coordinate system corresponds to (9), see Fig. 11.

Analysing this curve as a locus curve with students, GeoGebra allows us to apply different approaches to the derivation of its algebraic equation. On the one hand, we let the software do it automatically, from the perspective of a user in the hidden 'black box' mode, and simply ask it to derive the equation based on the geometric construction created in the 'Graphics' view. To do so, we apply the LocusEquation command, which utilises the algorithms of the automated theorem proving to compute an equation of the locus of a given point; see Fig. 12. On the other hand, we can do it manually, using the 'CAS' of GeoGebra, with its functions and tools, as the environment to control the process. In Fig. 13, the use of the Eliminate command to derive the locus equation in a manner analogous to the derivation of the limacon equation in Sect. 4 is shown.

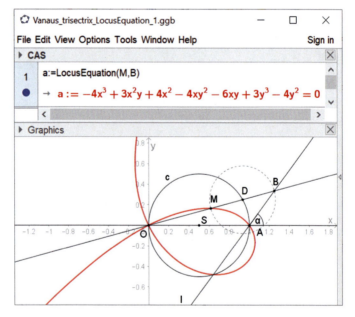

Fig. 12 Derivation of the algebraic equation of Vaňaus' trisectrix using the LocusEquation command of GeoGebra

Creative Use of Dynamic Mathematical Environment in Mathematics Teacher Training

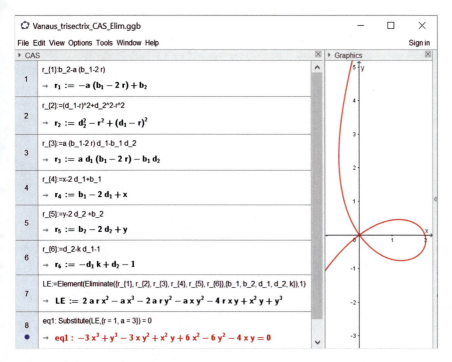

Fig. 13 Derivation of the algebraic equation of Vaňaus' trisectrix using the `Eliminate` command of GeoGebra

To get acquainted with the meaning of the equations $r_1, r_2, \ldots r_6$ that we used to symbolic derivation of the locus curve equation, see Fig. 11. They are derived as follows: From the right triangle with hypotenuse AB and an internal angle α, we get the first equation $r_1 : \tan(\alpha) = \dfrac{b_2}{b_1 - 2r}$, which can be written as

$$r_1 : (b_1 - 2r)a - b_2 = 0, \tag{10}$$

where $a = \tan \alpha$. The fact that the point $D[d_1, d_2]$ lies on a circle c with centre $S[r, 0]$ and radius r led to the second equation

$$r_2 : (d_1 - r)^2 + d_2^2 - r^2 = 0. \tag{11}$$

The third equation reflects the condition of collinearity of points O, D and B, i.e. $\dfrac{d_2}{d_1} = \dfrac{b_2}{b_1}$, where the expression $a(b_1 - 2r)$ is substituted for b_2:

$$r_3 : ad_1(b_1 - 2r) - b_1 d_2 = 0. \tag{12}$$

The other two equations r_4 and r_5 are based on the fact that $D[d_1, d_2]$ is the centre of the line segment MB, with the coordinates $M[x, y]$, $B[b_1, b_2]$ of its endpoints:

$$r_4 : b_1 - 2d_1 + x = 0, \tag{13}$$
$$r_5 : b_2 - 2d_2 + y = 0. \tag{14}$$

The last equation represents a non-degenerate condition preventing the case $D[0, 0]$, which would lead to the curve's asymptote instead of the curve itself:

$$r_6 : -d_1 k + d_2 - 1 = 0, \tag{15}$$

where k is the real parameter.

As already mentioned, Vaňaus identified this curve as a trisectrix, i.e. the curve that can be used, together with compass and ruler, to trisect an arbitrary angle. Specifically, he uses the property of equidistance among the points of the given circle, its secant and the curve, respectively. For a detailed illustration, see Fig. 14. Let us remember that B is the mover in Vaňaus' definition of trisectrix; moving B the point M draws the curve. Then, decisive for the trisection is a configuration of B and consequently M, where M lies on a circle centred at O passing through A. Only in this position, it holds that $|\angle HOM| = \frac{1}{3}|\angle HOA|$. To prove it, we will deal with Fig. 15, where the trisectrix is rotated, in comparison with the position used so far, so that the ray OH of the angle $\angle HOA$, the trisection of which is the subject of our interest, is horizontal. Let us focus on the triangle $\triangle OBA$ and its exterior angle $\angle MBA$, the measure of which is $|\angle MBA| = \alpha + 2x$. Due to the exterior angle theorem, which states that an exterior angle of a triangle is equal to the sum of the opposite interior angles,

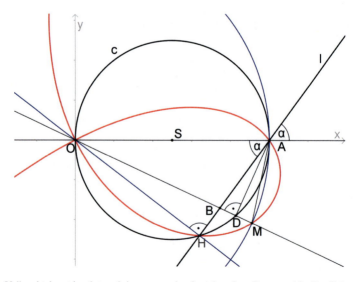

Fig. 14 Vaňaus' trisectrix; determining properties for trisection. Processed in GeoGebra

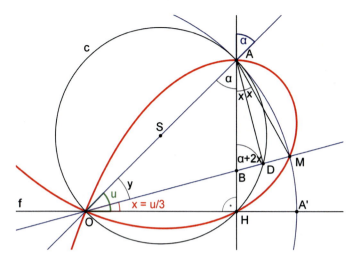

Fig. 15 Using the 'Vaňaus curve' to trisect an angle u

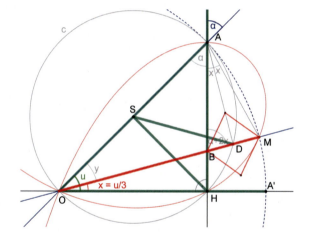

Fig. 16 Proposal of a possible design of Vaňaus' mechanism

we know that $\alpha + 2x = \alpha + y$. After simplification, we get $y = 2x$. Consequently, because $u = x + y$, it follows that $x = \frac{1}{3}u$.

In conclusion to his paper, Vaňaus (1881), Vaňaus mentions that he designed a quite simple mechanism to implement his method of trisection which he had made with a satisfactory result by a skilled mechanic. However, no such mechanism was found in his estate, so nobody knows what it looked like. A possible design of this mechanism is indicated in Fig. 16 using Vaňaus' trisection method in the background. Its dynamic model created in GeoGebra is available at Hašek (2020). The author is not entirely satisfied with this proposal as he believes that there could be a simpler mechanical implementation of Vaňaus' method. Of course, there are a number of other mechanisms for angle trisection; see, e.g. Soluzione (2021).

Fig. 17 Conchoid of Nicomedes; the locus of points D, the midpoints of MB, when M moves along the circle centred at O and passing through A

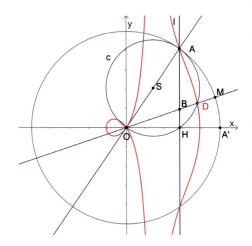

The use of DGS brought us another interesting revelation in this story of a method of trisection. When searching for the way of the mechanical trisection of an angle using Vaňaus' strophoid, a question appeared; what is the locus of points D, the midpoints of the segment BM, when M moves along the circle centred at O and passing through A; see Fig. 12. The resulting curve, shown in Fig. 17, is the conchoid of Nicomedes, Lawrence (2014), the Cartesian equation of which is $(x^2 + y^2)(x - a)^2 = b^2 x^2$, where $a = \frac{1}{2}|OH|$ and $b = \frac{1}{2}|OA|$. Thus, applying the GeoGebra Tools, we have found a close relationship of Vaňaus' trisectrix to this conchoid, a curve which is well known for its use in trisecting an angle Lockwood (2007).

6 Conclusion

Through a real story from the history of the study and teaching of mathematics concerning the trisection of an angle and related problems, we have shown the properties of contemporary dynamic geometric software, such as the ability to immediately respond to user's demands, the ability to provide individual feedback tailored to the user's needs, equipment with the environment supporting creative approach to solve the given problem and to find new ways of doing it, the possibility of sharing ideas and approaches, among others. This all predetermines this software for its use in contemporary mathematics teaching. Its potential to be implemented into an educational environment controlled by artificial intelligence is obvious and undoubtedly calls for detailed research.

Theoretical bases of AI implementation in mathematics education are stated in Balacheff (1993). We are convinced that now all the necessary components of this implementation, whether of a technical, software or didactic nature, are sufficiently mature to assemble them all together for the application of AI to assist a teacher in supporting and streamlining the school education of pupils according to their individual needs and to focus on their skills, knowledge and demands.

References

Angle trisection (2020). In: *Wikipedia.*https://en.wikipedia.org/wiki/Angle_trisection. Accessed 16 Oct 2020.

Balacheff, N. (1993). Artificial intelligence and mathematics education: Expectations and questions. In *14th Biennal of the Australian Association of Mathematics Teachers* (VOl. 2020, pp. 1–24). Perth, Australia.

Bloom, B.S. (1971). Mastery learning. In: J. H. Block (Ed.), *Mastery learning: Theory and practice* (pp. 47–63). New York: Holt, Rinehart & Winston.

Bloom, B. S. (1968). *Learning for mastery. Evaluation Comment, 1*(2), 1–12.

Botana, F., Kovács, Z., Recio, T. (2020). A mechanical geometer. *Mathematics in Computer Science.* https://doi.org/10.1007/s11786-020-00497-7. Accessed 25 Mar 2021.

Botana, F., Hohenwarter, M., Janičić, P., Kovács, Z., Petrović, I., Recio, T., & Weitzhofer, S. (2015). Automated theorem proving in GeoGebra: Current achievements. *Journal of Automated Reasoning, 55*(1), 39–59.

Clark, K. M. (2012). History of mathematics: Illuminating understanding of school mathematics concepts for prospective mathematics teachers. *Educational Studies in Mathematics, 81*(1), 67–84.

Dennis, D. (2000). The role of historical studies in mathematics and science educational research. In Kelly and Lesh (Eds.), *Research design in mathematics and science education.* Mahwah, NJ: Lawrence Erlbaum.

du Sautoy, M. (2019). *The creativity code: How AI is learning to write, paint and think* (1st ed.). 4th Estate.

Folta, J., Šišma, P. (2003) Josef Rudolf Vaňaus. In: *Významní matematici v českých zemích.* (In Czech) https://web.math.muni.cz/biografie/josef_rudolf_vanaus.html. Accessed 16 Oct 2020.

Furinghetti, F. (2007). Teacher education through the history of mathematics. *Educational Studies in mathematics, 66*(2), 131–143.

GeoGebra, *free mathematics software for learning and teaching.* http://www.geogebra.org . Accessed 16 Oct 2020.

Gibson, C. G. (1998). *Elementary geometry of algebraic curves: An undergraduate introduction.* Cambridge: Cambridge University Press.

Hašek, R. (2020). *Possible Vaňaus' trisector.* GeoGebra online material. https://www.geogebra.org/m/jdafysqv. Accessed 16 Oct 2020.

Hašek, R. (2019). Dynamic geometry software supplemented with a computer algebra system as a proving tool. *Mathematics In Computer Science, 13*(1–2), 95–104.

Hašek, R., Zahradník, J., & Kovács, Z. (2017). Contemporary interpretation of a historical locus problem with the use of computer algebra. In I. S. Kotsireas & E. Martínez-Moro (Eds.), *Springer proceedings in mathematics & statistics: Applications of computer algebra.* Springer.

Impossible constructions. In: *Wikipedia.*https://en.wikipedia.org/wiki/Straightedge_and_compass_construction#Impossible_constructions. Accessed 16 Oct 2020.

Kovács, Z. (2017). Automated reasoning tools in GeoGebra: A new approach for experiments in planar geometry. *South Bohemia Mathematical Letters, 25*(1), 48–65.

Lawrence, J. D. (2014). *A catalog of special plane curves.* New York: Dover Publications Inc.

Levin, D. (2017). How Artificial Intelligence Can Help Us Solve the 33-Year-Old "Two-Sigma Problem". In: *Huffpost (09/20/2017).*www.huffpost.com. Accessed 16 Oct 2020.

Limacon. In: *Wikipedia.*https://en.wikipedia.org/wiki/Lima%C3%A7on. Accessed 16 Oct 2020.

Lockwood, E. H. (2007). *Book of curves (Reprint)*. Cambridge: Cambridge University Press.
OK Geometry. https://www.ok-geometry.com . Accessed 16 Oct 2020.
Ostermann, A., & Wanner, G. (2012). *Geometry by its history* (1st ed.). Berlin: Springer.
Russell, S. J., Norvig, P., & Davis, E. (2010). *Artificial intelligence: a modern approach* (3rd ed.). Pearson.
Soluzione a problemi. In: *Associazione Macchine Matematiche*.http://www.macchinematematiche.org. Accessed 26 Mar 2021.
Strophoid. In: *Wikipedia*.https://en.wikipedia.org/wiki/Strophoid. Accessed 16 Oct 2020.
Trisectrix of Maclaurin. In: *Wikipedia*.https://en.wikipedia.org/wiki/Trisectrix_of_Maclaurin. Accessed 16 Oct 2020.
Trisectrix. In: *Wikipedia*.https://en.wikipedia.org/wiki/Trisectrix. Accessed 16 Oct 2020.
Vaňaus, J.R. (1881). Trisektorie. *Časopis pro pěstování matematiky a fyziky* (Vol. 10, No. 3, pp. 153–159). JČMF, Praha.
Vaňaus, J.R. (1902). Úloha 36. *Časopis pro pěstování matematiky a fyziky* (Vol. 31, No. 3, p. 262). JČMF, Praha. https://dml.cz/handle/10338.dmlcz/122611. Accessed 16 Oct 2020.
Vaňaus, J.R. (1902). Úlohy. Řešení úloh. *Časopis pro pěstování matematiky a fyziky* (Vol. 31, No. 5, p. 471–474). JČMF, Praha. https://dml.cz/handle/10338.dmlcz/122172. Accessed 16 Oct 2020.

Experimental Study of Isoptics of a Plane Curve Using Dynamical Coloring

Thierry Dana-Picard and Zoltán Kovács

1 Tradition Versus Experimental Mathematics

For a long time, the traditional way to convey mathematics was structured according to the definition-theorem-proof-example scheme. A few decades ago, digital tools appeared: calculators working numerically, then graphical features (also numerically based) were added. With the developments of algorithms for symbolic computations, things changed profoundly. Computer Algebra Systems (CAS) are now important systems involving symbolic and numerical features, together with efficient graphical abilities. The affordances of symbolic computations had a great influence on the proof of new theoretical results. Software has been developed focusing on specific mathematical domains, such as Macaulay for Algebraic Geometry,[1] GAP[2] for Group Theory, and Felix[3] for commutative and non-commutative rings and modules. Sometimes, specific programs have been written to address a very focused question; for example, see Dana-Picard and Schaps (1993).

The graphical features of CAS had already some dynamical aspects, but dynamical graphics (especially when driven by the user with the mouse, but not only) is the specific strength of other packages which appeared later: we mean Dynamical Geometry Systems (DGS). All these can be either freely downloadable (GeoGebra,

---

[1] https://faculty.math.illinois.edu/Macaulay2/.
[2] https://www.gap-system.org/.
[3] https://www.swmath.org/software/1048.

T. Dana-Picard (✉)
Jerusalem College of Technology, Havaad Haleumi 21 St., 9116011 Jerusalem, Israel
e-mail: ndp@jct.ac.il

Z. Kovács
The Private University College of Education of the Diocese of Linz, Salesianumweg 3, 4020 Linz, Austria
e-mail: zoltan@geogebra.org

© The Author(s), under exclusive license to Springer Nature Switzerland AG 2022
P. R. Richard et al. (eds.), *Mathematics Education in the Age of Artificial Intelligence*, Mathematics Education in the Digital Era 17,
https://doi.org/10.1007/978-3-030-86909-0_11

Desmos, etc.) or commercial packages (Cabri, Sketchpad, etc.). The same is true for CAS. Of course, this dynamics added reinforced visualization and its influence on mathematical thinking, and enabled not only to prove but also to first explore geometrical situations, then to conjecture, and finally, to prove results.

Recently, CAS has been included in DGS. For example, the DGS *GeoGebra* includes the CAS *Giac* (Kovács & Parisse, 2015). This inclusion makes the dialog between a CAS and a DGS more efficient, and more developments are foreseen in that direction. Examples of the importance of this dialogue between the two kinds of packages, especially for the study of plane curves (envelopes, isoptics, etc.) can be found in the literature. For instance, experiments in Cieślak et al. (1991) use a CAS only. Later works such as Dana-Picard (2019), Dana-Picard et al. (2020), and Dana-Picard and Kovács (2018) rely on networking between the different kinds of technology.

Networking between technologies calls for automation. First experiments on connecting different systems always require a human researcher. After considering typical case studies, the human controller may be substituted by a mechanical process. A mechanical algorithm can be, however, very complex and may require techniques from artificial intelligence (AI) like machine learning to select from a portfolio of the available approaches. In fact, AI subsystems may need periodic human inspections to verify if the underlying processes are still up-to-date and their operation is still optimal, or some improvements should be performed by human intervention. In our paper, we focus on the human's first experiments and the possible intervention that may promote a better networking on new inputs.

Technology provided environments to change the approach from the traditional one, evoked above, to a more experimental approach. This is the approach we present in this chapter for the study of isoptics of plane curves. Actually, the usage of a CAS enabled to describe (algebraic presentation, using polynomials and algorithms from the theory of Gröbner basis; see Adams and Loustaunau (1994); Cox et al. (1992)) and visualize the isoptics of a given plane curve (Dana-Picard et al., 2011, 2014). But the visualization was provided for one isoptic at a time.

Some time ago, we began to develop a different approach (Dana-Picard & Kovács, 2019). Here, we present another development: a new approach, based on the Newton root-finding algorithm, with the following main features:

- for a given curve, several isoptics are determined and displayed at the same time;
- different colors are used to enhance the difference between the isoptic angles, in particular making a clear distinction between acute and obtuse angles;
- the data can be changed dynamically in order to explore the influence of the changes in various parameters.

We describe activities built on joint usage of a CAS and a DGS. Sometimes, a CAS is implemented within a DGS, such as Giac in GeoGebra (Kovács & Parisse, 2015). These technologies are always under further development, and may be available on various platforms, offering specific features. In any case, we have here rich environments supporting the 4C's of Education in 21st century (Chiruguru, 2021; OECD, 2021): Critical Thinking, Creativity, Communication, and Collaboration.

The dynamical colorings that we offer foster critical thinking, to analyze the output and explore the changes when changing the parameters. These changes promote creativity. And more than that, we will see that Collaboration and Communication are made possible at 3 different levels:

- between humans, during the activities;
- between kinds of software, within each activity;
- between man and machine at every step of the work.

2 Isoptic Curves of Plane Curves—A Short Historical Survey

A plane curve $\mathscr{C}$ and an angle θ are given. A θ-isoptic of $\mathscr{C}$, denoted by Opt($\mathscr{C}, \theta$), if it exists, is the geometric locus of points in the plane through which passes a pair of tangents to $\mathscr{C}$ making an angle equal to θ. The angle θ is called isoptic angle. In other words, Opt($\mathscr{C}, \theta$) is the set of points in the plane from which the curve $\mathscr{C}$ isn viewed under the fixed angle θ. A $\frac{\pi}{2}$-isoptic of the curve $\mathscr{C}$ is called its *orthoptic* curve (or shortly its orthoptic).

The orthoptics of conics are well known (see Fig. 1):

- The orthoptic of a parabola is called its directrix;
- The orthoptic of an ellipse is its *director circle*. If a and b denote half-axes of the ellipse, then the director circle has the same center as the ellipse and its radius is equal to $\sqrt{a^2 + b^2}$;
- The orthoptic of a hyperbola, if it exists, is a circle whose center is the center of the hyperbola. Its existence depends on the angle between the asymptotes, and also on the parts of the plane where the hyperbola is situated.

Isoptic curves have been studied for a long time, since the end of the nineteenth century, with (Taylor, 1884). We refer also to Wunderlich's works (Wunderlich, 1937, 1971a; 1971b), to Yates's book (Yates, 1952) and to the works by Cieślak, Mozgawa, and their team (Cieślak et al., 1991; Miernowski & Mozgawa, 1997). Their approach

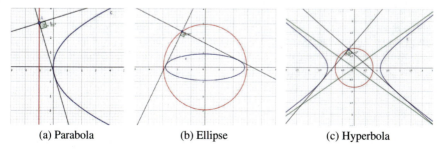

(a) Parabola   (b) Ellipse   (c) Hyperbola

Fig. 1 Orthoptics of conics

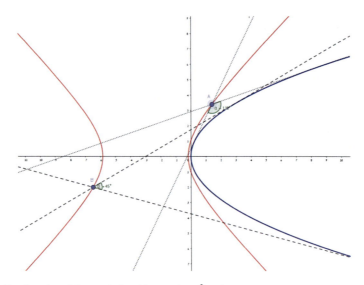

Fig. 2 Two isoptics of the parabola with equation $y^2 = 4x$

uses support functions, enabling the study of isoptics for smooth closed curves, and deriving a criterion for convexity of the isoptic. Sałkowski studied isoptics of open rosettes, i.e. of some non-closed curves, relating to some periodicity (Szałkowski, 2005, 2007, 2014).

More recently, isoptics of conics have been studied for any angle, using a Computer Algebra System (CAS). The following results have been obtained (or re-discovered, that time in a technology-rich environment). For $\theta \neq \frac{\pi}{2}$, the following holds:

- The θ-isoptic of a parabola is an arc of hyperbola; see Dana-Picard et al. (2011). Figure 2 shows the parabola whose equation is $y^2 = 4x$ and the hyperbola whose equation is $x^2 + 6x - y^2 + 1 = 0$. The right (resp. left) arc of the hyperbola is the $\frac{3\pi}{4}$-isoptic (resp. $\frac{\pi}{4}$-isoptic) of the parabola. This picture has been obtained with GeoGebra; the angles in the picture are measured in degrees, as measures in radians are not given by the software as fractions of π but in decimal format.[4]
- The θ-isoptic of an ellipse is a Spiric[5] of Perseus (a generalization of a Cassini oval), i.e. a specific quartic; see Dana-Picard et al. (2011). Actually, as the algebraic computations involve squaring of both sides of an equation, the θ-isoptic and the $(\pi - \theta)$-isoptic are obtained simultaneously. Moreover, plotting the obtained equation of degree 4 yields both components, not enabling an algebraic distinction between them. Therefore, the name *bisoptic* has been coined there. The bisoptic

---

[4] A GeoGebra applet for an interactive study of the isoptics of a parabola is available at https://www.geogebra.org/m/RSmsrFCp.

[5] Spiric curves appear in various fields. Before Kepler laws have established, there was an attempt to describe the orbits of planets with Cassini ovals. More recently, a gravitation theory based on spiric curves has been proposed in Nieto and Beltrán (2015).

curve of an ellipse is a quartic with two components, one of them being in the interior of the other. The internal one is an isoptic for an obtuse angle, and the exterior one is the isoptic for its acute complement. It is well known that a spiric curve is the intersection of a torus with a plane parallel to the axis of the torus. In the case of bisoptics, the torus is self-intersecting (i.e. it is obtained by revolving a circle around a line which intersects it). Figure 3 shows isoptics of an ellipse whose equation is $x^2 + 4y^2 = 1$, for $\theta = \frac{3\pi}{4}$ (dotted tangents) and $\theta = \frac{\pi}{4}$ (dashed tangents).

An interactive GeoGebra applet[6] is available at https://www.geogebra.org/m/xBeqhrtD. It shows the dependence of the spiric curve on the angle and the eccentricity of the ellipse. Moreover, it shows that in the pair torus-plane, both components change according to the angle: the toric intersection is not defined by a fixed torus and moving the plane only. A screen snapshot is displayed in Fig. 4.

- Using the same algebraic approach as for ellipses, it appears that bisoptics of hyperbolas are also spiric curves; see Dana-Picard et al. (2014). That time, the asymptotes decompose the plane into four zones; let us call them N-W-S-E (the meaning is obvious). For an angle θ, the θ-isoptic is the union of arcs N-S of the external component and E-W of the internal one, the union of the four remaining arcs being the $(\pi - \theta)$-isoptic. Figure 5 shows a hyperbola whose equation is $x^2 - 4y^2 = 1$. The bisoptic curve is a subset of the spiric curve whose equation is $4x^4 + 4y^4 + 8x^2y^2 - 5x^2 - 10y^2 + 1.25 = 0$, showing the following properties:

  - The $63.43°$-isoptic is N-S on the external loop and E-W on the internal one.
  - The $116.57°$-isoptic is E-W on the external loop and N-S on the internal one.
  - The points of intersection of the spiric with the asymptotes of the hyperbola do not belong to either isoptic, as through these points passes only one tangent to the hyperbola.

Figure 6 illustrates the intersection of a self-intersecting torus with a plane parallel to the torus axis.

Note that the usage of a Dynamical Geometry System (DGS), here the mouse-driven **Move** command of GeoGebra, enables to convince the user that the right answer has been found: the chosen point of the isoptic can be moved interactively. By that way, a good visualization of two phenomena is obtained:

(i) Moving the point along the loop shows that between the asymptotes, the angle is constant;
(ii) The point of intersection of the loop with an asymptote is a jump point for the isoptic angle.

These results have been obtained using two technologies:

a. A CAS has been used to perform the algebraic computations and to derive a general equation for isoptics. Entering the equation of the given curve and of

---

[6] Thanks to George Ghantous, from the Center for Educational Technology, in Tel Aviv, for his help in developing the applet.

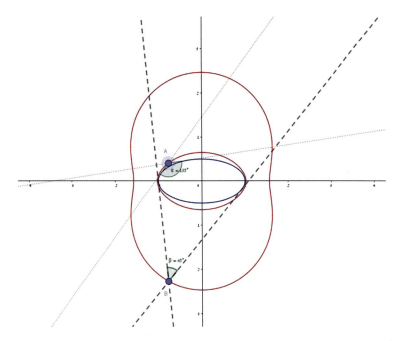

Fig. 3 A bisoptic of an ellipse

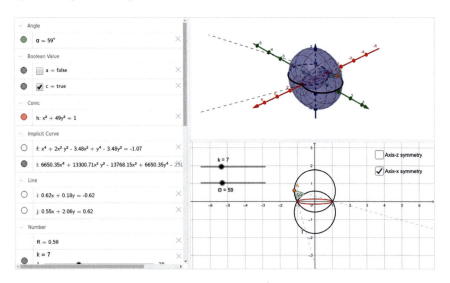

Fig. 4 Dynamical visualization of isoptics as toric intersections

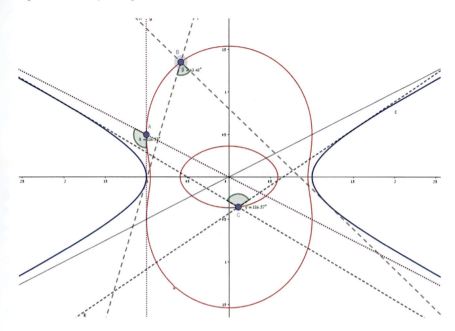

Fig. 5 A bisoptic of a hyperbola

the computed isoptic, a single picture can be obtained. By substitution of several values of the angle, several pictures can be obtained, sometimes plotted together.
b. A DGS enables to introduces two degrees of dynamics:

- A point can be marked on the isoptic and then moved along the isoptic, showing the conservation of the angle between the tangents.
- A general equation for the isoptic can be entered, we mean an equation containing the isoptic angle (actually the tangent of the angle) as a parameter. Then the slider bar enables to change the angle and to have an insight into the modifications of the shape of the isoptic.

Further experiments provided a possibility to visualize more than one isoptic at a time. For this, a new software tool has been used (Montag & Richter-Gebert, 2018), and work has been based on the principles described in Losada (2014), Losada et al. (2010). In the next section, we present applets built with CindyJS, which enable to visualize at one glance several isoptics for different curves: conics and open quartics. This exploration way enables to study isoptics even without computing "exact" equations.

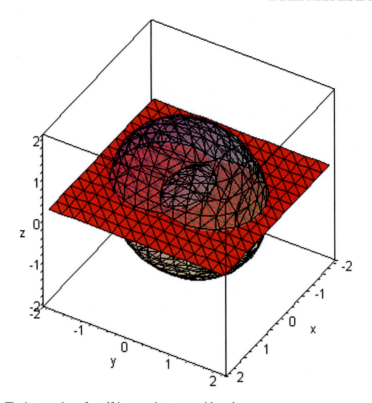

Fig. 6 The intersection of a self-intersecting torus with a plane

3 Experiments with Dynamical Coloring

We present here an experimental approach to the discovery of the various areas in the plane, according to the possible angles between possible tangents. The work is based on a dynamical coloring of the plane using CindyJS Montag and Richter-Gebert (2018).

3.1 First Experiments

We begin our investigation by letting $F(x, y) = 0$ be the equation of a convex curve. Through an external point $P(x_P, y_P)$ pass two tangents to the curve; we denote by $A(x_A, y_A)$ and $B(x_B, y_B)$ the points of contact of the curve with the two tangents, denoted, respectively, by t_A and t_B. See Fig. 7.

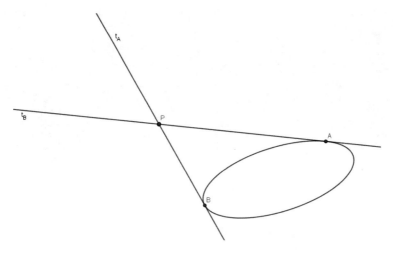

Fig. 7 The two points of contact of the tangents to a convex curve through an external point

The equation of a tangent at the point $P(x_P, y_P)$ is as follows:

$$t_P : F'_x(x, y) \cdot (x - x_P) + F'_y(x, y) \cdot (y - y_P) = 0. \tag{1}$$

This can be used to express the coordinates A and B as functions of the coordinates of P without using heavy computer algebra, that is, only by *derivation, substitutions,* and *numerical equation solving* in one variable, if the following properties hold:

1. F is a polynomial in the variables x and y.
2. F can be written in explicit form, that is, for example, as $y = f(x)$.

For instance, when considering the example $F(x, y) = x^2 + 2 - y$, the formula

$$x_{A,B} = \frac{2x_P \pm \sqrt{4x_P^2 - 4y_P + 8}}{2} = x_P \pm \sqrt{x_P^2 - y_P + 2}$$

can be derived and, from this, we obtain immediately $y_A = x_A^2 + 2$ and $y_B = y_B^2 + 2$.

At this stage, computing the angle $\angle ACB$ is a simple numerical operation, which can be performed for each point P in the plane, or in a bounding box that fits the user's screen.

Two possible outputs are shown in Fig. 8, one for the parabola described above, and one for a hyperbola with the input $F(x, y) = x^2 - xy + 1$. In these figures, acute angles are displayed in blue and obtuse angles in red. Right angles will be obtained when the color is black, and this corresponds to the directrix of the parabola, and the director circle for the hyperbola (if it exists, as we explained previously), respectively. In fact, we use a technique similar to the technique described in Losada (2014), Losada et al. (2010).

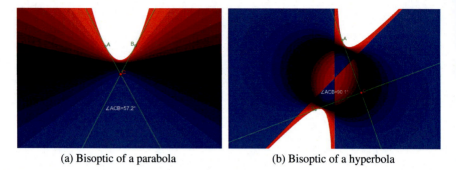

(a) Bisoptic of a parabola   (b) Bisoptic of a hyperbola

Fig. 8 Simultaneous and dynamical plotting of isoptics of conics

3.2 Automated Way of Obtaining Dynamical Colored Plots

Our approach could be generalized by embedding a Computer Algebra System in CindyJS. However, in the examples presented at the beginning of this section, we focused on keeping the computations as quick as possible, to provide the users with immediate feedback from the computer's side. On the other hand, for a generally given input formula, some symbolic steps cannot be avoided, even if they can be performed by a simple CAS as well. CindyJS' tutorial suite suggests the usage of Davide Della Casa's *Algebrite* (see http://algebrite.org), but for substitutions we may need some more complicated systems.

At this point, we refer to our GitHub project (Dana-Picard & Kovács, 2019) that performs the process above but in a general way, by using Bernard Parisse's Giac (see https://www-fourier.ujf-grenoble.fr/~parisse/giac.html) on the top of the HTML web page. In the file `isoptics.html`, Giac preprocesses the input formula by solving an equation in one variable, differentiating the formula, and simplifying the result—finally after substitution, a symbolic solution of another equation is provided. This final formula will be then inserted into the CindyJS function space, and it will be evaluated for a high number of occurrences by the CindyJS engine. (Giac is also used to convert the intermediate formulas to LaTeX and finally MathJax is invoked to display them in a well-readable way.)

It is important to mention that Giac provides a formula that is equivalent to the well-known solution of a general quadratic equation, and this is evaluated by the web browser in an efficient way. (Actually, CindyJS and its underlying WebGL system perform the computations in a *very* efficient method.)

Figure 9 shows the workflow of the algorithm.

We highlight that this process can be automatized only for parabolas and hyperbolas, but not for ellipses. Luckily, some more careful computations will help us extend our method to obtain useful outputs for ellipses as well, if the ellipse is defined by its canonical equation $F(x, y) = 0$, where $F(x, y) = ax^2 + by^2 - 1$. In particular, for the case of an ellipse, we need to classify the points P of the plane; different

Fig. 9 The workflow of the visualization algorithm

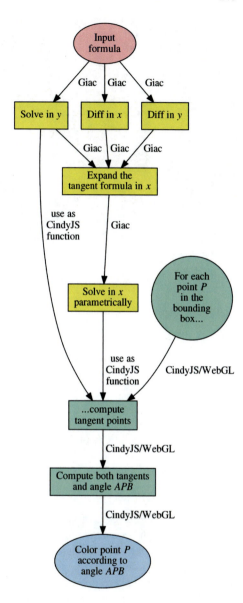

positions may require different signs in the exact definition of y_A and y_B (see Fig. 10 for a typical run of the script ellipse.html). Roughly speaking, we compare the x-coordinate of point P with $-1/\sqrt{a}$ and $1/\sqrt{a}$, and check the y-coordinate of it as well. Then a piecewise operation is used.

We illustrate this idea by computing the isoptics for certain quartic inputs. Instead of using the symbolic way of solving a quartic equation via Ferrari's method

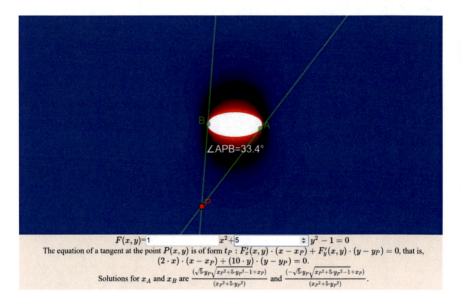

Fig. 10 Simultaneous and dynamical plotting of isoptics of an ellipse

(Cardano, 1545), we use Newton's root finder algorithm (Newton, 1669). For example, by assuming $F(x) = x^4 + x + y$, the formula (1) is of the form

$$(4x^3 + 1) \cdot (x - x_P) + (-1) \cdot (y - y_P) = 0 \tag{2}$$

which leads to find solutions of an equation of degree 4. Now the Newton iteration

$$X_{n+1} = X_n - \frac{3X_n^4 - 4X_n^3 x_P - x_P + y_P}{12X_n^3 - 12X_n^2 \cdot x_P}$$

can be used to find the roots of (2) if the initial value X_0 is well chosen.

Unfortunately, no convergence can be guaranteed in general, as it is well known (Kovács, 2011). Our experiments resulted in choosing X_0 from the interval $I = \left[-\frac{4}{3}, \frac{16}{3}\right]$ by using a deterministic pseudo-random number generator (Vickers, 1980). In this way, we are able to color the plane in the viewport of Fig. 11 for almost all pixels. For each X_0, at most 20 different iterations are checked—after that another pseudo-random value will be chosen for X_0. After 300 iterations altogether, the loop for one pixel ends with an unsuccessful result—otherwise the computation delivers both roots of (2).

The algorithm described above illustrates what challenges are to be solved by an AI-based network. Different inputs can have different degrees—conic inputs can be processed by a purely symbolic algorithm, but higher degrees (in our case, 4) should trigger a numerical computation. In the symbolic case, a sophisticated recipe will cover all cases properly, while in the numerical case the pre-configured interval I

Experimental Study of Isoptics of a Plane Curve ...

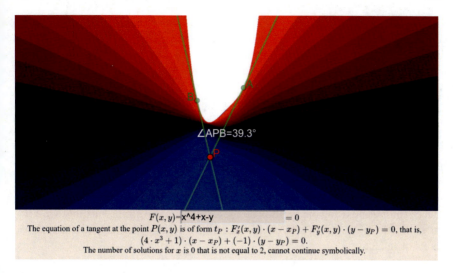

Fig. 11 Simultaneous and dynamical plotting of isoptics of a quartic

should perform well enough for *many* inputs. In fact, the numerical case has not yet been tested with a large scale of inputs, therefore, an update may be required after a higher amount of tests, provided by end users, is available. After such results, the ad-hoc algorithm of setting the interval I (and eventually, configuring the pseudo-random number generator) may be improved to ensure even better results. Here, we highlight that optimal selection of I (and other refinements) can be supported by machine learning methods as well.

4 Final Remarks and Directions for Future Work

4.1 What Has Been Earned with the New Approach

The isoptics of conic sections have been already described in Dana-Picard et al. (2011, 2014). There the algebraic work, performed either by hand or using a CAS, was a prerequisite to visualization of the isoptics. This visualization has then been obtained in one of the following ways:

(i) Using the plotting commands of the CAS.
(ii) Copying the equations into the DGS, which are automatically translated into plots in the graphical window. This copy has to be performed by hand (copy-paste) as generally the two kinds of technologies have no possibility to have a dialog. This is analyzed in an ongoing work (see Dana-Picard & Kovács, 2021).

The output results in the given conic and one bisoptic. Using GeoGebra's **Move** command, it is possible to verify that the obtained curve is really a bisoptic. The difference of behavior between isoptics of ellipses and of hyperbolas is made clear by these experiments. We emphasize that only one bisoptic can be plotted by that way.

With our new approach, we visualize at the same time a certain number of isoptics, making their general shape obvious. The difference between the isoptics for either acute or obtuse angle is emphasized by the color changes.

A purely algebraic approach is possible from a theoretical point of view: for example, if there exist points of inflection on the isoptic Opt($\mathscr{C}, \theta$), then these are points of intersection of $\mathscr{C}$ with its Hessian curve (see Bix, 1998, Theorem 12.1). A CAS may help to compute the solution of the needed system of polynomial equations, but understanding and using the solution on the display may be unilluminating. Working with a DGS together with a CAS may contribute to an experimental discovery of points of inflection.

We mentioned in Sect. 2 that an isoptic of a parabola (for a non-right angle) is an arc of hyperbola. With the packages in use there, both CAS and DGS, the user was unable to make the distinction between both without an explicit mention of the angle (given by the DGS, not by the CAS). Actually, the algebraic computations which have to be performed before plotting involved squaring both sides of a non-polynomial equation in order to get rid of a square root. This is the algebraic reason why both the θ-isoptic and the $(\pi - \theta)$-isoptic are obtained at the same time, hence the name *bisoptic*; see Dana-Picard et al. (2011, 2014). No algebraic test can distinguish between the components. With the new technology, things are different.

1. The coloring in Fig. 8a makes this distinction.
2. In Fig. 8b, the coloring makes the alternate N-S and E-W obvious.

Moreover, in all situations, the point A can be moved, showing the conservation (or the non-conservation) of the isoptic angle.

4.2 What We Are Still Missing

Our exploration of isoptics involves both pure mathematical work and technological developments. This is not a surprise, as we are in line with Artigue's claim in Artigue (2002) that the new technological knowledge is an integral part of the newly acquired mathematical knowledge. As the study of isoptics is a live topic, it is not a surprise that numerous open issues exist.

4.2.1 Developments of the Technology

Monaghan and Trouche write in Monaghan et al. (2016, p. 8) that "the artefact shapes the way the user is acting; in the reverse sense, the user shapes ... the artefact that s/he appropriates". If in the beginning, we worked with existing software and applied it, despite the lack of communication between the kinds of available packages, in the present paper we showed how the user can develop new technological tools for the study of isoptics. This too is part of the construction of knowledge. Nevertheless, with a DGS, we still miss the option of moving the point from which the curve is seen, while the isoptic angle remains constant, which could be an important addition to the set of commands for automated discovery using a DGS. A trial to bypass this problem has been described in Dana-Picard and Kovács (2018).

4.2.2 Open Mathematical Questions

As mentioned in Sect. 2, the isoptics of a parabola are arcs of hyperbolas. In that case, the convexity issue is easy to handle. Considering the two components of a spiric curve obtained as a bisoptic of an ellipse, one seems to be convex and the other one not, but until now, we do not have a characterization for the values of the isoptic angle for which an isoptic is convex here. Conditions for convexity of the isoptics are derived in Miernowski and Mozgawa (1997). These conditions are not easy to check, even with a CAS. Using a DGS, we can derive conjectures, but until now, no proof is known. Even for well-described isoptics, such as isoptics of closed Fermat curves (i.e. Fermat curves of even degree; Dana-Picard et al. (2020)), questions related to convexity are still open. Of course, this question is connected to the determination of points of inflection mentioned previously.

Another open question is as follows. If the given curve $\mathscr{C}$ is smooth and convex, then through any external point P passes exactly a pair of tangents to $\mathscr{C}$, as shown in Fig. 7. Moreover, it has isoptics for any angle. Otherwise, it may have isoptics for specific angles only. Non-convexity of the curve $\mathscr{C}$ may create totally different situations. This can be seen experimentally using a DGS. To find boundaries for the interval of existence of isoptics is still an open question; computation may be really hard, but (numerical-visual) experiments help.

Let us consider two examples:

Example 1 Denote the Fermat curve of degree 3 by $\mathscr{F}_3$ and take a point A in the plane. Depending on the position of the point A, there may be one tangent, or 2 tangents, or 4 tangents, or there may be no tangent at all to $\mathscr{F}_3$ through A. See Fig. 12.

Example 2 Isoptic curves of an astroid have been studied in Dana-Picard (2019). This curve is given parametrically by $(x, y) = (\cos^3 t, \sin^3 t)$, $t \in [0, 2\pi]$. It can be given by an implicit equation $x^{2/3} + y^{2/3} = 1$ if we allow fractional powers of negative numbers, which is not allowed by all CAS. If not, we have to use absolute values, which makes the algebraic treatment by a CAS more complicated, for sure non-polynomial. An astroid is non-convex and non-smooth; it has 4 cusps. As every closed non-self-intersecting curve (generally, we call this a loop), it divides the plane into three disjoint regions. The following claims have been obtained experimentally, using the DGS.

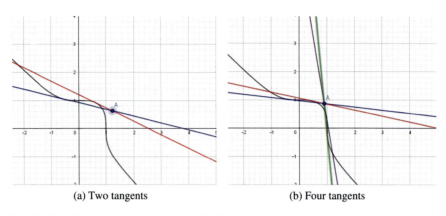

(a) Two tangents (b) Four tangents

Fig. 12 Variable number of tangent to a cubic Fermat curve

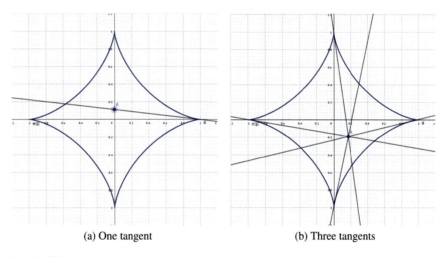

(a) One tangent (b) Three tangents

Fig. 13 Different numbers of tangents: true or software flaw?

(i) Exterior: through an external point, there may pass one or two tangents, the points of contact not being on the "closest" arc of the astroid.
(ii) The curve itself: through every point on $\mathscr{C}$ passes a unique tangent.
(iii) Interior: through an internal point, there may pass either one tangent, or 2 tangents or 3 tangents; see Fig. 13.

The claims in these examples have to be proven with algebraic computations, with respect to the constraints both of the hardware and of the software; see Trouche (2000); Dana-Picard (2007).

Figure 12 shows a Fermat curve of odd degree. This curve is open. Exploration of points through which passes a pair of tangents making a given angle reveals that the plane can be divided at least into three zones:

- The set of points through which a single tangent passes. This set contains at least the curve itself.
- The set of points through which no tangent passes.
- The set of points through which more than one tangent passes. If more than two tangents pass through the given point, we have to choose a suitable pair (if it exists), as what happened with an astroid (Dana-Picard, 2019).

Here too, the determination of the regions in the plane for whose points no tangents to the curve exists is an important question. Isoptics may exist for some angles, and not exist for other angles, as seen in Dana-Picard and Kovács (2018). These open curves have been called open rosettes by Szałkowski; he explores the isoptics of open rosettes in Szałkowski (2005, 2007, 2014). A special case is described in Dana-Picard and Kovács (2018). For the case shown in Fig. 12, this question is still open.

4.2.3 Transition Toward 3D

For decades, the exploration of isoptics has been made for plane curves. The transition toward surfaces in the 3D space seemed nonsense, because of the problem to have a spatial analogue for the isoptic angles. Such a transition from 2D to 3D is always a non-trivial issue. It has been evoked in Dana-Picard and Zehavi (2019) from a neighboring domain, namely envelopes of parametric families of curves and surfaces. Recently, studies of isoptics of surfaces in the 3D space have been performed, and we refer to Csima and Szirmai (2014, 2016); Nagy et al. (2018).

The technological part of the transition has numerous sides: general visualization of surfaces and of their singularities (see Zeitoun and Dana-Picard (2019)). New tools for a fruitful dialog between DGS and CAS, studied in Roanes-Lozano et al. (2003) and Dana-Picard and Kovács (2021), are needed for 3D.

4.3 Conclusions

In a standard curriculum, the study of plane curves is often limited to lines and conics, with some extension toward other curves, either algebraic or not, as appearing in the study of functions. The tools which are developed as "Applications of Calculus to Geometry" enable to study a large set of plane curves. These curves may be the basis of developments in STEAM education, of course for the M—Mathematics and also strongly for the A—Arts. This is not specific to plane curves, and other mathematical fields may be involved, such as Linear Algebra (rotations, reflections, etc.). We wish to emphasize that, among the 4 C's mentioned in Sect. 2, Creativity has here a strong benefit when dealing with Mathematics and Arts.

Important mathematical databases are devoted to plane curves, either totally[7] or parts of them.[8] Searching databases is nowadays an integral part of educative activities. The students may enjoy the animations presented in these databases, and may reproduce them with their own technological tools. After all, the technological knowledge acquired during these activities is an integral part of their new mathematical knowledge; see Artigue (2002).

In this chapter, we presented new kinds of activities. Not only dynamical constructions are proposed but also coloring is a central issue. It enables to distinguished different cases in the problem, giving a concrete appearance to abstract notions.

Finally, we wish to emphasize that visualization in 3D is often a difficult issue for students. After having worked out many examples in 2D, the students are accustomed to using technology (more than one kind of software). The transition toward 3D is not less non-trivial, but it is made easier by the use of technology. The 4 C's mentioned in Sect. 2 are very important here. Each student reinforces his/her Critical Thinking, analyzing new situations with the abstract presentation together with the concrete realization using software. Then Collaboration and Communication help students to analyze and to try to explain to each other what they see. These activities foster Creativity in an evident way.

Acknowledgements The authors wish to thank Aaron Montag for fruitful discussions and help with various technological issues. The second author was partially supported by FEDER/Ministerio de Ciencia, Innovación y Universidades—Agencia Estatal de Investigación/MTM2017-88796-P (Symbolic Computation: new challenges in Algebra and Geometry together with its applications).

References

Adams, W., & Loustaunau, P. (1994). *An introduction to Gröbner bases, graduate studies in mathematics 3*. Providence: American Mathematical Society, RI.

Artigue, M. (2002). Learning mathematics in a CAS environment: The genesis of a reflection about instrumentation and the dialectics between technical and conceptual work. *International Journal of Computers for Mathematical Learning, 7*(3), 245–274.

Bix, R. (1998). *Conics and cubics: A concrete introduction to algebraic curves*. New York: Springer.

Cardano, G. (1545). *Artis Magnæ, Sive de Regulis Algebraicis Liber Unus*.

Chiruguru, S. (2021). *The Essential Skills of 21st Century Classroom (4Cs)*. https://www.researchgate.net/publication/340066140_The_Essential_Skills_of_21st_Century_Classroom_4Cs. https://doi.org/10.13140/RG.2.2.36190.59201.

Cieślak, W., Miernowski, A., & Mozgawa, W. (1991). Isoptics of a closed strictly convex curve. In *Global differential geometry and global analysis*. Lecture Notes in Mathematics (Vol. 1481, pp. 28–35).

Cox, D., Little, J., & O'Shea, D. (1992). *Ideals, varieties, and algorithms: An introduction to computational algebraic geometry and commutative algebra*. Undergraduate Texts in Mathematics. Springer Verlag.

---

[7] For example, http://www.mathcurve.com.

[8] Such as the MacTutor site for the history of mathematics https://mathshistory.st-andrews.ac.uk/Curves/.

Csima, G., & Szirmai, J. (2014). Isoptic curves of conic sections in constant curvature geometries. *Mathematical Communications, 19*(2), 277–290.
Csima, G., & Szirmai, J. (2016). Isoptic surfaces of polyhedra. *Computer Aided Geometric Design, 47,* 55–60. https://doi.org/10.1016/j.cagd.2016.03.001.
Dana-Picard, T. (2007). Motivating constraints of a pedagogy embedded computer algebra system. *International Journal of Science and Mathematics Education, 5*(2), 217–235.
Dana-Picard, T. (2019). Automated study of isoptic curves of an astroid. *Journal of Symbolic Computation, 97,* 56–68.
Dana-Picard, T., & Kovács, Z. (2018). Automated determination of isoptics with dynamic geometry. In F. Rabe, W. Farmer, G. Passmore & A. Youssef (Eds.), *Intelligent computer mathematics.* Lecture Notes in Artificial Intelligence (a subseries of Lecture Notes in Computer Science) (Vol. 11006, pp. 60–75). Springer.
Dana-Picard, T., & Kovács, Z. (2019). *Experiments on isoptics by dynamic coloring*, a GitHub Project. https://github.com/kovzol/isoptics.
Dana-Picard, T., & Kovács, Z. (2021). Networking of technologies: A dialog between CAS and DGS. *Journal of Mathematics and Technology, 14* (1).
Dana-Picard, T., Mann, G., & Zehavi, N. (2011). From conic intersections to toric intersections: The case of the isoptic curves of an ellipse. *The Montana Mathematical Enthusiast, 9*(1), 59–76.
Dana-Picard, T., Naiman, A., Mozgawa, W., & Cieślak, V. (2020). Exploring the isoptics of Fermat curves in the affine plane using DGS and CAS. *Mathematics and Computer Science, 14,* 45–67.
Dana-Picard, T., & Schaps, M. (1993). *A computer assisted project: Classification of algebras, computational algebraic geometry and commutative algebra (Cortona, 1991).* Symposia Mathematica (Vol. XXXIV, pp. 71–83). Cambridge: Cambridge University Press.
Dana-Picard, T., & Zehavi, N. (2019). Automated study of envelopes: The transition from 2D to 3D. *The Electronic Journal of Mathematics, 13*(2), 121–135.
Dana-Picard, T., Zehavi, N., & Mann, G. (2014). Bisoptic curves of hyperbolas. *International Journal of Mathematical Education in Science and Technology, 45*(5), 762–781.
Kajetanowicz, P. (2014). *Hypocycloids.* https://www.geogebra.org/material/show/id/wkB3Ydqe.
Kovács, Z. (2011). Understanding convergence and stability of the Newton–Raphson method. In R. Vajda & J. Karsai (Eds.), *Interesting mathematical problems in sciences and everyday life.* University of Novi Sad: University of Szeged.
Kovács, Z., & Parisse, B. (2015). Giac and GeoGebra-improved Gröbner basis computations. In *Computer algebra and polynomials.* Lecture Notes in Computer Science (Vol. 8942, pp. 126–138).
Losada, R. (2014). El color dinámico de GeoGebra. *La Gaceta de la RSME, 17*(3), 525–547. http://gaceta.rsme.es/abrir.php?id=1220.
Losada, R., Valcarce, J. L., & Recio, T. (2010). *On the automatic discovery of Steiner–Lehmus generalizations.* http://geogebra.es/pub/adg2010def1.pdf.
Miernowski, A., & Mozgawa, W. (1997). On some geometric condition for convexity of isoptics. *The Rendiconti del Seminario Matematico Università e Politecnico di Torino, 55*(2), 93–98.
Monaghan, J., Trouche, L., & Borwein, J. (2016). *Tools and mathematics: Instruments for learning.* Switzerland: Springer International Publishing.
Montag, A., & Richter-Gebert, J. (2018). *Bringing together dynamic geometry software and the graphics processing unit.* https://arxiv.org/abs/1808.04579.
Nagy, F., Kunkli, R., & Hoffmann, M. (2018). New algorithm to find isoptic surfaces of polyhedral meshes. *Computer Aided Geometric Design.* https://doi.org/10.1016/j.cagd.2018.04.001.
Newton, I. (1669). *De analysi per aequationes numero terminorum infinitas.*
Nieto, J. A., & Beltrán, L. A. (2015). Towards an alternative gravitational theory. *Modern Physics Letters A, 30*(13). https://doi.org/10.1142/S0217732315500686. https://www.researchgate.net/publication/277231197_Towards_an_alternative_gravitational_theory.
OECD. (2021). *Future of Education and Skills 2030, Conceptual learning framework.* https://www.oecd.org/education/2030-project/teaching-and-learning/learning/skills/Skills_for_2030.pdf.

Roanes-Lozano, E., Roanes-Macías, E., & Vilar-Mena, M. (2003). A bridge between dynamic geometry and computer algebra. *Mathematical and Computer Modelling, 37,* 1005–1028.

Szałkowski, D. (2005). *Isoptics of open rosettes.* Annales Universitatis Maria Curie – Sklodowska Lublin Polonia Vol. LIX Section A, 119–128.

Szałkowski, D. (2007). *Isoptics of open rosettes II.* (English) Zbl 1150.53002, Analele Stiintifice ale Universitatii Al I Cuza din Iasi, *New Series Mathematics, 53*(1) 167–176.

Szałkowski, D. (2014). *Singular points of isoptics of open rosettes.* (English) Zbl 1299.53004, Analele Stiintifice ale Universitatii Al I Cuza din Iasi, *New Series Mathematics, 60*(1) 85–98.

Taylor, C. (1884). Note of a theory of orthoptic and isoptic Loci. *Proceedings of the Royal Society London, 37,* 138–141.

Trouche, L. (2000). La parabole du gaucher et de la casserole á bec verseur: étude des processus d'apprentissages dans un environnement de calculatrices symboliques. *Educational Studies in Mathematics, 41,* 239–264.

Vickers, S. (1980). *ZX81 BASIC programming.* Cambridge: Sinclair Research Limited.

Wunderlich, W. (1937). Die isoptischen Kurven der Zykloiden. *Zeitung für angewandte Mathematische Mechanik, 17,* 56.

Wunderlich, W. (1971a). Kurven mit isoptischem Kreis. *Aequationes Mathematicae, 6,* 71–81.

Wunderlich, W. (1971b). Kurven mit isoptischer Ellipse. *Monatschrifte Mathematik, 75,* 346–362.

Yates, R. C. (1952). *Isoptic curves: A handbook on curves and their properties* (pp. 138–140). Ann Arbor, MI: J. W. Edwards.

Zeitoun, D., & Dana-Picard, T. (2019). On the usage of different coordinate systems for 3D plots of functions of two real variables. *Mathematics in Computer Science, 13,* 311–327.

Teaching Programming for Mathematical Scientists

Jack Betteridge, Eunice Y. S. Chan, Robert M. Corless, James H. Davenport, and James Grant

1 Background

This volume is part of the fast-growing literature in a relatively new field—being only about 30 years old—namely Artificial Intelligence for Education (AIEd). The survey (Luckin et al., 2016) gives in lay language a concise overview of the field and advocates for its ambitious goals. For a well-researched discussion of an opposing view and of the limitations of Artificial Intelligence (AI), see Broussard (2018). This chapter is concerned with AI in mathematics education in two senses: first, symbolic computation tools were themselves amongst the earliest and most successful pieces of AI to arise out of the original MIT labs already in the sixties,[1] and have had a significant impact on mathematics education. This impact is still changing the field of mathematics education, especially as the tools evolve (Kovács et al., 2017).

---

[1] For example, (Slagle, 1963), which took a "Good Old-Fashioned Artificial Intelligence (GOFAI)" approach, and concluded "The solution of a symbolic integration problem by a commercially available computer is far cheaper and faster than by man". Of course, this was from the era when people still believed in GOFAI. We are grappling with different problems today, using much more powerful tools. Yet some important things can be learned by looking at the effects of the simpler and older tools. The riposte to Slagle (1963) was the development of Computer Algebra (Davenport, 2018) as a separate discipline.

J. Betteridge · James H. Davenport · J. Grant
The University of Bath, Bath, England
e-mail: jdb55@bath.ac.uk

J. Grant
e-mail: rjg20@bath.ac.uk

E. Y. S. Chan · R. M. Corless (✉)
Western University, London, Canada
e-mail: rcorless@uwo.ca

E. Y. S. Chan
e-mail: echan295@uwo.ca

© The Author(s), under exclusive license to Springer Nature Switzerland AG 2022
P. R. Richard et al. (eds.), *Mathematics Education in the Age of Artificial Intelligence*, Mathematics Education in the Digital Era 17,
https://doi.org/10.1007/978-3-030-86909-0_12

Second, and we believe more important, the existence of these tools, and similarly the existence of other AI tools, has profoundly changed the affordances of mathematics and therefore *should change the content of mathematics courses, not just their presentation methods* (Corless, 2004a). That last paper introduced the phrase "Computer-Mediated Thinking", by which was meant an amplification of human capability by use of a computer. The idea itself of course is not new, and later we will give a quotation from the 1970s expressing the same thought. In Hegedus et al. (2017), we find this idea beautifully articulated and set in the evolving sequence of human methods for mediating their thinking: symbolic marks on bone, through language, to symbols on paper, to where we are today. One of our theses—surely a fact obvious to all—is that such tool use is not innate; instead, people need to be given opportunities to learn how best to use these tools. This chapter discusses how to help people to learn how to use computational tools in mathematics, and why this is a good thing.

This chapter reflects our experiences in changing mathematical course syllabi to reflect these new tools and affordances, and may serve as a reference point in discussing future curricular changes (and what should not change) owing to the ever-shifting technological ground on which we work. Our methodology is to consider mathematics education and computer programming together. Our thesis is that the effect of computational tools, including AI, is greater and more beneficial if students are taught how to use the tools effectively and even how to create their own.

The connection between mathematics and computer programming is widely recognized and profound. Indeed, most members of the general public will (if they think about it) simply assume that all mathematicians can, and do, program computers. When mathematics is used instrumentally in science, as opposed to purely for aesthetic mathematical goals, this is in fact nearly universally true. This is because computational tools are ubiquitous in the mathematical sciences. Such tools are nowadays becoming increasingly accepted in pure mathematics, as well: see, for example, Borwein and Devlin (2009). Modern tools even approach that most central aspect of pure mathematics, the notion of mathematical proof.[2] See Richard et al. (2019) and its many references for a careful and nuanced discussion of the notion of proof in a modern technologically assisted environment, and the implications for mathematical education.

One lesson for educators is that we *must* teach students in the mathematical sciences how to use computational tools responsibly. We earlier said that the public assumes that all mathematicians can program; with a similar justification, many students assume that they themselves can, too. But programming computers *well* (or even just *using* them well) is a different story. The interesting thing for this chapter is that learning to use computers well is a very effective way to learn mathematics well: by teaching programming, we can teach people to be better mathematicians

---

[2] Much interest in computation and proof for pure mathematics was generated by the very successful *polymath project*. Because computation has always been perceived as instrumentally important, a corresponding but much larger scale project on the Applied Mathematics side might be the Intergovernmental Panel on Climate Change.

and mathematical scientists (Papert, 1993). This thesis is well-supported by modern research: see, for instance, (Monaghan et al., 2016) or, even for earlier levels of education, the project Learning Math through Coding (Tepylo & Floyd, 2016).

To be specific, learning "iteration" helps to understand dynamical systems; learning "recursion" helps to understand mathematical induction; learning "numerical analysis" helps to understand real (and complex) analysis, including approximation theory; learning "computational linear algebra", numerical or symbolic, helps to understand linear algebra; learning "algorithms for numerical quadrature" helps to understand calculus (both integral and differential); and learning to *write a program that works, and can be shown to work*, helps to understand how to construct a mathematical proof. We have seen this happen in our students, and felt it happen in ourselves.

Of course, this works the other way, too. For a scientist to work with Machine Learning, they need to know *linear algebra*, *optimization*, and *statistics*, perhaps even above knowing calculus. Indeed, Gilbert Strang has been saying for years that our current curriculum has "Too Much Calculus" (Strang, 2001). One of the instrumental reasons that we teach mathematics is so that scientists can be effective and productive, and this is deeply connected in the modern world with programming (Wilson et al., 2014).

We used the word "must", above: we *must* teach students how to . . . . . Why "must"? For what reason? We contend that this is the *ethical* thing to do, in order to prepare our students as best we can to be functioning and thinking adults. This is more than just preparation for a job, even as a scientist: we are aiming at *eudaemonia* here (Flanagan, 2009). This observation has significant implications for the current revolution in AI-assisted teaching. We will return to this observation after discussing our experiences.

Our experience includes teaching programming to mathematical scientists and engineers through several eras of "new technology", as they have flowed and ebbed. Our early teaching experience includes the use of computer-algebra-capable calculators to teach engineering mathematics[3]; calculators were a good solution at the time because we could not then count on every student having their own computer (and smartphones were yet a distant technological gleam). Some of the lessons we learned then are still valid, however: in particular, we learned that we must teach students that they are *responsible* for the results obtained by computation, that they *ought to know* when the results were reliable and when not, and that they should *understand the limits of computation* (chiefly, understand both complexity of computation and numerical stability of floating-point computation; today, we might add that generic algebraic results are not always valid under specialization, as in Camargos Couto et al. (2020)). We claim that these lessons are invariant under shifts in technology,

---

[3] We had intended to give the reference (Rosati et al., 1992) for this; however, that journal seems to have disappeared and we can find no trace of it on the Web, which is a kind of testimony to ephemerality. Some of the lessons of that article were specific to the calculator, which was *too advanced* for its era and would be disallowed in schools today. We shall not much discuss the current discouragingly restricted state of the use of calculators in schools hereafter.

and become particularly pertinent when AI enters the picture. This observation is consistent with the recommendations in Wilson et al. (2014).

Speaking of shifts, see Kahan (1983) ("Mathematics written in Sand") for an early attack on teaching those lessons, in what was then a purely numerical environment. A relevant quotation from that work is

> Rather than have to copy the received word, students are entitled to experiment with mathematical phenomena, discover more of them, and then read how our predecessors discovered even more. Students need inexpensive apparatus analogous to the instruments and glassware in Physics and Chemistry laboratories, but designed to combat the drudgery that inhibits exploration. —William Kahan, p. 5 *loc cit.*

Teaching these principles in a way that the student can absorb them is a significant curricular goal, and room must be made for this goal in the mathematical syllabus. This means that some things that are in that already overfull syllabus must be jettisoned. In Corless and Jeffrey (1997) and again in Corless (2004a), some of us claim that *convergence tests for infinite series* should be amongst the first to go. Needless to say, this is a radical proposal and not likely to attain universal adoption without a significant shift in policy; nevertheless, if not this, then what else? Clearly, *something* has to go, to make room!

Curricular shifts are the norm, over time. For instance, spherical trigonometry is no longer taught as part of the standard engineering mathematics curriculum, nor are graphical techniques for solving algebraic equations (which formerly were part of the *drafting* curriculum, itself taken over by Computer Aided Design (CAD)). Special functions are now taught as a mere rump of what they were, once. Euclidean geometry has been almost completely removed from the high-school curriculum. Many of these changes happen "by accident" or for other, non-pedagogical, reasons; moreover, it seems clear that removing Euclidean geometry has had a deleterious effect on the teaching of logic and proof, which was likely unintended.

We have found (and will detail some of our evidence for this below) that teaching *programming* remediates some of these ill effects. By learning to program, the student will in effect learn how to prove. If nothing else, learning to program may motivate the student wanting to *prove the program correct*. This leads into the modern disciplines of Software Engineering and Software Validation, not to mention Uncertainty Quantification. Of course, there are some truly difficult problems hiding in this innocent-seeming suggestion: but there are advantages and benefits even to such intractable problems.

We will begin by discussing the teaching of Numerical Analysis and programming, in what is almost a traditional curriculum. We will see some seeds of curriculum change in response to computational tools already in this pre-AI subject.

2 Introduction to Numerical Analysis

The related disciplines of "Numerical Methods", "Numerical Analysis", "Scientific Computing", and "Computational Science" need little introduction or justification nowadays (they could perhaps benefit from disambiguation). Many undergraduate science and engineering degrees will grudgingly leave room for one programming course if it is called by one of those names. Since this is typically the first course where the student has to actually *use* the mathematical disciplines of linear algebra and calculus (and use them *together*), there really isn't much room in such a course to teach good programming. Indeed, many students are appalled to learn that the techniques of *real analysis*, itself a feared course, make numerical analysis intelligible. [Remarkably, taking numerical analysis at the same time can make real analysis more intelligible.] In this minimal environment (at Western, the course occupies 13 weeks, with 3 hours of lecture and 1^4 hour of tutorial per week), we endeavoured to teach the following:

1. The basics of numerical analysis: *backward error* and *conditioning*;
2. How to write simple computer programs: conditionals, loops, vectors, and recursion;
3. How to debug computer programs;
4. The elements of programming style: readability, good naming conventions, and the use of comments;
5. Several important numerical algorithms: matrix factoring, polynomial approximation; solving an Initial Value Problem (IVP) for Ordinary Differential Equations (ODEs);
6. How to work in teams and to *communicate mathematics*.

As for what constitutes good programming, it is useful to relate these to the eight Best Practices for Scientific Computing (Wilson et al., 2014). By emphasizing readability of code, communication, and working in teams, we encourage students to "Write programs for people", to "Document design and purpose" and to "Collaborate". The nature of exercises is to incrementally develop codes to target problems, and we encourage modularization through functions and recursion. While we do consider debugging, time limitations mean that we are not able to teach testing or introduce version control. We will return to these as they are important omissions, with significance for mathematics and AI, particularly in relation to reproducibility and ethics.

The students also had some things to *unlearn*: about the worth of exact answers, or about the worth of some algorithms that they had been taught to perform by hand, such as Cramer's rule for solving linear systems of equations, for instance.

Western has good students, with an entering average amongst the highest in the country. By and large, the students did well on these tasks. But they had to work, in order to do well. The trick was to get them to do the work.

---

[4] Students were enrolled in one of three tutorial hours, but often went to all three hours.

2.1 Choice of Programming Language

We used Matlab. This choice was controversial: some of our colleagues wanted us to teach C or C++ because, ultimately for large Computational Science applications, the speed of these compiled languages is necessary. However, for the goals listed above, we think that Matlab is quite suitable; moreover, Matlab is a useful scientific language in its own right because *development time* is minimized by programming in a high-level language first (Wilson et al., 2014), and because of that Matlab is very widely used.

Other colleagues wanted us to use an open-source language such as Python. This is quite attractive and Python may indeed eventually displace Matlab in this teaching role. Unlike Matlab or Maple, Python is free to download and supports a very large collection of standard and community-provided libraries. Indeed, Python often features in the "Top 5" most popular languages.[5] Another reason for changing is that Python is the language of choice for Machine Learning and AI applications, with all the popular Machine Learning libraries offering a Python interface. But as of this moment in time, Matlab retains some advantages in installed methods for solving several important problems and in particular its sparse matrix methods are very hard to beat.

We also used the computer algebra language Maple, on occasion: for comparison with exact numerical results, and for program generation. Matlab's Symbolic Toolbox is quite capable, but we preferred to separate symbolic computation from numeric computation for the purposes of the course.

2.2 Pedagogical Methods

We used *Active Learning*, of course.[6] By now, the evidence in its favour is so strong as to indicate that *not* using active learning is academically irresponsible (Handelsman et al., 2004; Freeman et al., 2014). However, using active learning techniques in an 8:30am lecture hall for 90 or so students in a course that is overfull of material is a challenge. To take only the simplest techniques up here, we first talk about *Reading Memos* (Smith & Taylor, 1995).

---

[5] Often, these lists rely on fairly baseless claims or are derived from the number of Internet searches for a language; here, we base this claim on proportion of code on GitHub as measured by https://madnight.github.io/githut/ visited on 2021-02-09.

[6] Active learning is defined, for instance, in a well-known teaching and learning website at Queen's University, Kingston, Ontario: "Active learning is an approach to instruction that involves actively engaging students with the course material through discussions, problem solving, case studies, role plays and other methods. Active learning approaches place a greater degree of responsibility on the learner than passive approaches such as lectures, but instructor guidance is still crucial in the active learning classroom. Active learning activities may range in length from a couple of minutes to whole class sessions or may take place over multiple class sessions.".

We gave credit—five per cent of the student's final mark—for simply *handing in a Reading Memo*, that is, a short description of what they had read so far or which videos they had watched, with each programmatic assignment. Marks were "perfect" (for handing one in) or "zero" (for not handing one in). Of course, this is a blatant bribe to get the students to read the textbook (or watch the course videos). Many students initially thought of these as trivial "free marks" and of course they could use them in that way. But the majority learned that these memos were a way to get detailed responses back, usually from the instructor or Teaching Assistant (TA) but sometimes from other students. They learned that the more they put into a Reading Memo, the more they got back. The feedback to the instructor was also directly valuable for things like pacing. Out of all the techniques we used, this one—the simplest—was the most valuable.

The other simple technique we used was discussion time. Provocative, nearly paradoxical questions were the best for this. For instance, consider the following classical paradox of the arrow, attributed to Zeno, interpreted in floating point (actually, this was one of their exam questions this year):

```
% Zeno's paradox, but updated for floating-point
% initial position of the arrow is s=0, target is
    s=1
s = 0
i = 0; % Number of times through the loop
% format hex
while s < 1,
    i = i+1;
    s = s + (1-s)/2 ;
end
fprintf( 'Arrow reached the target in %d steps\n',
    i)
```

Listing 1 Zeno's paradox in floating point arithmetic

In the original paradox, the arrow must first pass through the half-way point, and then the point half-way between there and the target, and so on, *ad infinitum*. The question for class discussion was, would the program terminate, and if so, what would it output? Would it help to uncomment the format hex statement in line 5? Students could (and did) type the program in and try it, in class; the results were quite surprising for the majority of the class.

Another lovely problem originates from one posed by Nick Higham: take an input number, x. Take its square root, and then the square root of that, and so on 52 times. Now take the final result and square it. Then square that, and again so on 52 times. One expects that we would simply return to x. But (most of the time) we do *not*, and instead return to another number. By plotting the results for many x on the interval $1/10 \le x \le 10$ (say), we see in Fig. 1, in fact, horizontal lines. The students were asked to explain this. This is not a trivial problem, and indeed in discussing this problem amongst the present authors, JH Davenport was able to teach RM Corless (who has used this problem for years in class) something new about it.

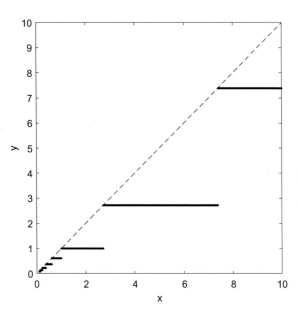

Fig. 1 The function $y = \text{Higham}(x) = (x^{1/2^{52}})^{2^{52}}$, i.e., take 52 square roots, and then square the result 52 times, plotted for 2021 points on $0.1 \leq x \leq 10$, carried out in IEEE double precision. Students are asked to identify the numerical values that y takes on, and then to explain the result. See Sect. 1.12.2 of Higham (2002) and also Exercise 3.11 of that same book, and Kahan (1980b)

We will *not* give "the answers" to these questions here. They are, after all, for discussion. [A useful hint for the repeated square root/repeated squaring one is to plot $\ln(y)$ against x.] We encourage you instead to try these examples in your favourite computer language and see what happens (so long as you are using floating-point arithmetic, or perhaps rounded rational arithmetic as in Derive)!

We will discuss, however, Kahan's proof of the impossibility of numerical integration (Kahan, 1980a) here, as an instance of discussing the limits of technology. This lesson must be done carefully: too much scepticism of numerical methods does much more harm than good, and before convincing students that they should be careful of the numerics they must believe that (at least sometimes) computation is very useful. So, before we teach numerical integration, we teach them that symbolic integration is itself limited, especially if the vocabulary of functions is limited to elementary[7] antiderivatives. As a simple instance, consider

$$E = \int_1^\infty \frac{e^{-y^2/2}}{y+1}, \tag{1}$$

which occurs in the study of the distribution of condition numbers of random matrices (Edelman, 1988). The author laconically states that he "knows of no simpler form" for this integral. In fact, neither do we, and neither do Maple or Mathematica: the indefinite integral is not only not elementary (provable by the methods of (Davenport, 1986)), it is right outside the reference books. Of course, the sad (?) fact

---

[7] The *elementary functions* of the calculus are not "elementary" in the sense of being simple, but instead they are "elementary" in a similar sense to the elementary particles of physics.

is, as observed in Kahan (1980a), the vast majority of integrands that occur in "real life" must be dealt with numerically. This motivates learning numerical quadrature methods.

However, it is a useful thing for a budding numerical analyst to learn that numerical techniques are not infallible, either. Consider the following very harmless function: Aphra(x) := 0. That is, whatever x is input, the Aphra function returns 0. However, Aphra is named for *Aphra Behn*, the celebrated playwright and spy for King Charles.[8] The function is written in Matlab in such a way as to *record its inputs x*.

```
function [ y ] = Aphra( x )
%APHRA A harmless function, that simply returns 0
%
    global KingCharles;
    global KingCharlesIndex;
    n = length(x);

    KingCharles(KingCharlesIndex:KingCharlesIndex+n-1)
        = x(:);
    KingCharlesIndex = KingCharlesIndex + n;
    y = zeros(size(x));
end
```

Listing 2 A function named for Aphra Benn

If we ask Matlab's *integral* command to find the area under the curve defined by Aphra(x) on, say, $-1 \leq x \leq 1$, it very quickly returns the correct answer of zero. However, now we introduce another function, called Benedict:

```
function [ y ] = Benedict( x )
%BENEDICT Another harmless function
%   But this function is not zero.
    global KingCharles;
    global KingCharlesIndex;
    global Big;
    s = ones(size(x));
    for i=1:length(KingCharles),
        s = s.*(x-KingCharles(i)).^2;
    end
    y = Big*s;
end
```

Listing 3 A routine using information gathered by Aphra

This function is defined to be zero exactly at the points reported by Aphra, but strictly positive everywhere else: indeed the "Big" constant can be chosen arbitrarily large. If we choose Big equal to 10^{87}, then after calling Aphra with integral(@Aphra, -1, 1) first, we find the function plotted in Fig. 2. It is clearly not zero, and indeed clearly has a positive area under the curve on the interval $-1 \leq x \leq 1$.

However, asking the Matlab built-in function integral to compute the area under Benedict(x) on the interval $-1 \leq x \leq 1$ gives the incorrect result 0 because the

---

[8] Aphra Behn 1640–1689 has one of the most interesting, if only dubiously accurate, biographies that we have read.

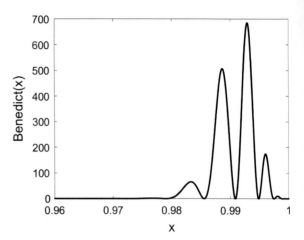

Fig. 2 The function Benedict$(x) = K \prod_{i=1}^{150}(x - s_i)^2$ where the s_i are the 150 sample points in the interval $-1 \leq x \leq 1$ reported by the function Aphra(x), and with $K = 10^{87}$. We plot only an interesting region near the right endpoint of the interval. We see that the area under this curve is not zero

deterministic routine `integral` samples its functions adaptively but here by design the function Benedict traitorously behaves as if it was Aphra at the sample points (and only at the sample points). This seems like cheating, but it really isn't: finding the good places to sample an integrand is remarkably difficult (and more so in higher dimensions). One virtue of Kahan's impossibility proof is that it works for arbitrary deterministic numerical integration functions. Without further assumptions (such as that the derivative of the integrand is bounded by a modest constant), numerical integration really is impossible.

The students *do not like* this exercise. They dislike learning that all that time they spent learning antidifferentiation tricks was largely wasted, and they dislike learning that computers can give wrong answers without warning. Still, we feel that it is irresponsible to pretend otherwise.

Finally, the course was officially designated as an "essay" course. This was in part recognition for the essay-like qualities of the lab reports, but was also explicitly in recognition of the similarities between a good computer program and a good essay: logical construction, clear division of labour, and good style. It is our contention that programming and proving and explaining all share many attributes. As Ambrose Bierce[9] said, "Good writing is clear thinking made visible."

We also not only allowed but actively encouraged collaboration amongst the students. They merely had to give credit to the other student group members who helped them, or to give us the name of the website they found their hints or answers on (frequently Stack Exchange[10] but also Chegg[11] and others). Many students could

---

[9] Ambrose Bierce 1844–1914(?) was an American satirist, critic, and journalist, perhaps most famous for his collection of definitions published as "The Devil's Dictionary".

[10] https://stackexchange.com/ Stack Exchange is a network of websites for communities where contributors ask and answer questions and then vote on responses. Stack Overflow is for general programming but specialist communities exist, e.g., for Maths, AI, and Data Science. While generally good, the quality does vary, and is poor for computer security (Fischer et al., 2017).

[11] https://www.chegg.com/ Chegg is a homework help website.

not believe that they were being allowed to do this. The rationale is that in order to *teach* something, the student had to know it very well. By helping their fellow students, they were helping themselves more.

But modern programming or use of computers is *not* individual heroic use: nearly everyone asks questions of the Web these days (indeed, to answer some LaTeX questions for the writing of this chapter, we found the LaTeX FaQ on the Help Page on Wikibooks[12] Useful, and this even though the authors of this present chapter have *decades* of LaTeX experience). We do not serve our students well if we blankly ban collaborative tools. We feel that it is important to teach our students to properly *acknowledge* aid, as part of modern scientific practice.

2.3 Assessment

But we did not allow collaboration on the midterm exam, which tested the students' individual use of Matlab on computers locked down so that only Matlab (and its help system) could be used. Examination is already stressful: an exam where the student is at the mercy of computer failure or of trivial syntax errors is quite a bit more stressful yet. To mitigate this, we gave *practice exams* (a disguised form of active learning) which were quite similar to the actual exam. The students were grateful for the practice exams, *and moreover found them to be useful methods to learn*.

Exam stress—assessment stress in general—unfortunately seems to be necessary[13]: if the students *could* pass the course without learning to program Matlab, they *would* do so, and thereafter hope that for the rest of their lives they could get other people to do the programming. Students are being rational, here: if they were only assessed on mathematical knowledge and not on programming, then they should study mathematics and leave programming for another day. So we must assess their individual programming prowess.

In contrast, the students were greatly relieved to have a final exam that "merely" asked them to (in part) write pencil-and-paper programs for the instructor to read and grade. In that case, trivial errors—which could derail a machine exam—could be excused. On the other hand, the instructor could (and did) ask for explanations of results, not merely for recitations of ways to produce them.

---

[12] https://en.wikibooks.org/wiki/LaTeX.

[13] Given the economic constraints of the large class model, we mean. Even then, there may be alternatives, such as so-called "mastery grading" (Armacost & Pet-Armacost, 2003). We look forward to trying that out. Exam stress is often counterproductive, and the current university assessment structures do encourage and reward successful cheating. We would like a way out of this, especially now in COVID times.

3 Computational Discovery/Experimental Mathematics

The courses that we describe in this section are described more fully elsewhere (Chan & Corless, 2017b, 2021). Here, we only sketch the outlines and talk about the use of active learning techniques with (generally) introverted mathematics students. The major purpose of these related courses (a first-year course and a graduate course, both in Experimental Mathematics, taught together) was to bring the students as quickly as possible to the forefront of mathematics.

> Short is the distance between the elementary and the most sophisticated results, which brings rank beginners close to certain current concerns of the specialists.
> —(Mandelbrot & Frame, 2002)

In this we were successful. For example, one student solved a problem that was believed at the time to be open (and she actually solved it *in-class*); although we were unaware at the time, it turned out to have actually been solved previously and published in 2012, but nonetheless we were able to get a further publication out of it, namely (Li & Corless, 2019), having taken the solution further. There were other successes. Some of the projects became Masters' theses, and led to further publications such as (Chan & Corless, 2017a), for example.

The course was also *visually* successful: the students generated many publication-quality images, some of which were from new Bohemian Matrix classes. Indeed some of the images on that website were produced by students in the course.

3.1 Choice of Programming Language

We used Maple for this course, because its symbolic, numerical, and visual tools make it eminently suited to experimental mathematics and computational discovery; because it was free for the students (Western has a site licence), and because of instructor expertise (Corless, 2004b). For instance, Maple allows us to produce the plot shown in Fig. 3 of all the eigenvalues of a particular class of matrices. This figure resembles others produced by students in the class, but we made this one specifically for this chapter. There are 4096 matrices in this set, each of dimension 7. However, there are only 2038 distinct characteristic polynomials of these matrices because some are repeated. Getting the students to try to answer questions such as "how many distinct eigenvalues are there" is a beginning (this is not obvious, because again there are repeats: the only way we know how to answer this is to compute the Greatest Common Divisor (GCD)-free basis of the set of 2038 degree 7 polynomials, in fact). A bigger goal—in fact, the main goal of the course—was getting the students to come up with their own questions. It helped that the students were encouraged to invent their own classes of matrices (and they came up with some quite remarkably imaginative ones).

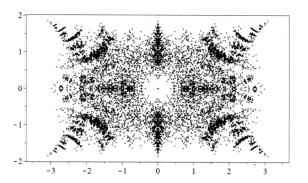

Fig. 3 All the complex eigenvalues of all the 7-dimensional symmetric tridiagonal (but with zero diagonal) matrices with population $\{-5/3 - i, -5/3 + i, 5/3 + i, 5/3 - i\}$. There are $4^6 = 4096$ such matrices, but only about half as many distinct characteristic polynomials in the set

3.2 Pedagogical Methods

This course was designed wholly with active learning in mind. It took place in the Western Active Learning Space, which was divided into six tables called Pods, each of which could seat about seven students; the tables were equipped with technology which allowed students to wirelessly use the common screens to display materials to each other. The smartboards were (in principle) usable in quite sophisticated technological ways; in practice, the varieties of whiteboards with simple coloured pens were just as useful.

Students enrolled in the first-year course were grouped with students enrolled in the graduate course. Each group benefitted from the presence of the other: the presence of the senior students was a calming factor, while the junior students provided significant amounts of energy. The grad student course also had an extra lecture hour per week where more advanced topics were covered in a lecture format.

Active learning techniques run from the obvious (get students to choose their own examples, and share) through the eccentric (interrupt students while programming similar but different programs and have them trade computers and problems) to the flaky (get them to do an interpretive dance or improvisational skit about their question). We tried to avoid the extremely flaky, but we did mention such, so that these introverted science students knew that this was within the realm of possibility.

The simplest activity was typing Maple programs that were handwritten on a whiteboard into a computer: this was simple but helpful because students learned the importance of precision, and had *immediate* help from their fellow students and from the TA.

Next in complexity was interactive programming exercises (integrated into the problems). Mathematicians tend to under-value the difficulty of learning syntax and semantics simultaneously. The amplification of human intelligence by coupling it with computer algebra tools was a central aspect of this course.

We describe our one foray into eccentricity. The paper Strange Series and High Precision Fraud by Borwein and Borwein (1992) has six similar sums. We had six teams program each sum, at a stage in their learning where this was difficult (closer to the start of the course). After letting the teams work for 20 minutes, we forced one member of each team to join a new team; each team had to explain their program (none were working at this stage) to the new member. This exercise was most instructive. The lessons learned included

- people approach similar problems very differently.
- explaining what you are doing is as hard as doing it (maybe harder).
- basic software engineering (good variable names, clear structure, and economy of thought) is important.
- designing on paper first might be a good idea (nobody believed this, really, even after).
- social skills matter (including listening skills).

3.3 Assessment

The students were assessed in part *by each other*: we used peer assessment on class presentations. The instructor informed the students that he would take their assessments and *average them with his own* because peer assessment is frequently too harsh on other students; they found this reassuring. The main mark was on an individual project, which took the full term to complete. They had to present intermediate progress at a little past the half-way point. Marks were also given for class participation.

Collaboration was encouraged. The students merely had to make proper academic attribution. While, technically, cheating might have been possible—one might imagine a plagiarized project—there was absolutely no difficulty in practice. The students were extremely pleased to be treated as honourable academics.

4 Programming and Discrete Mathematics

This course described in this section is also more fully explained elsewhere; see Betteridge et al. (2019). We restrict ourselves here to an outline of the goals and methods.

The course XX10190, Programming and Discrete Mathematics, at the University of Bath is both similar and dissimilar to the Western courses described above. One of the big differences is that it was designed specifically for the purpose of teaching programming to mathematical scientists by using mathematics as the proving ground. The course was designed after significant consultation and a Whole Course Review in 2008/2009. In contrast, the Western course designs were driven mostly by the individual vision of the instructor. The Bath course therefore has a larger base of support and is moreover supported by the recommendation from Bond (2018) that "every mathematician learn to program". As such, it is much more likely to have a long lifespan and to influence more than a few cohorts of students; indeed, since it has been running for 10 years, it already has.[14] Now that RM Corless has retired from Western and the numerical analysis course has been taken over by a different instructor, the course there is already different. Conversely, all the Bath authors have moved on from XX10190, but the course is much the same. This is the differential effect of institutional memory.

Another big difference is that the course is in the first year, not the second year; moreover, it runs throughout the first year, instead of only being a 13-week course. This gives significant scope for its integrated curriculum, and significant time for the students to absorb the lessons.

However, there are similarities. The focus on discrete mathematics makes it similar to the Experimental Mathematics courses discussed above, with respect to the flavour of mathematics. Indeed, perhaps the text (Eilers & Johansen, 2017) might contain some topics of interest for the course at Bath. Although the focus is on discrete mathematics, some floating-point topics are covered and so the course is similar to the Numerical Analysis course above as well. But the main similarity is the overall goal: to use mathematical topics to teach programming to mathematical scientists, and simultaneously to use programming to teach mathematics to the same students. This synergistic goal is eminently practical: learning to program is an effective way to learn to do mathematics.

Another similarity is respect for the practical *craft* of programming: the papers Davenport et al. (2016) and Wilson (2006) discuss this further. To this end, the instructors use Live Programming (Rubin, 2013), defined in Paxton (2002) as "the process of designing and implementing a [coding] project in front of class during lecture period". This is in contrast to a counter-principle, namely the approach called "Never touch the keyboard" (during instruction), which can also have benefits. For the Western courses, an accidental rediscovery of this counter-principle proved valuable: the instructor was for several years discouraged from using keyboards owing to a repetitive strain injury, and as a consequence took to writing code on the whiteboard. This had unexpected benefits when the students would ask him to debug their code, and he would do so in a Socratic manner by asking the students to relay error messages. In doing so, the students frequently found their own solutions. However, in spite of this success, one of the most common requests from students was for live demonstrations (supplied at Western by the Tutorial Assistant): there is no question that live programming techniques can be valuable. At this moment, it is not clear which approach is more valuable, if either. We talk about some nuances of "Never touch the keyboard" in Sect. 4.2.

4.1 Choice of Programming Language

A major similarity to the Western course is the choice of programming language: Matlab. As with the Western course, Matlab may eventually be displaced by Python,

---

[14] One British citizen in 25,000 is a graduate of XX10190.

but is an admirable first language to learn for mathematical scientists. This choice came with several unanticipated benefits, as described in Betteridge et al. (2019): for instance, the statisticians teaching subsequent courses found it simpler to teach R to students who had a working knowledge of the similar language Matlab.

4.2 Pedagogical Methods

The course is fifty per cent Programming and fifty per cent Discrete Mathematics. The course is team-taught, with Lecturers and Tutors. The whole cohort have one programming lecture, one Discrete Mathematics lecture, and one Examples class per week. The roughly 300 students are divided up into tutorial groups of size roughly 10, and there is one Discrete Math tutorial per week (when budgets allow: some years this has been financially impossible, and some years these have been in groups of 20) and one Programming Lab on Fridays, after the whole-cohort classes (this apparently minor timetabling point is pedagogically very helpful, as we discovered after the first year, when we didn't have it). Management of this relatively large staff with its hierarchical structure repays attention, and the instructors have found it valuable to provide tools such as a separate mailing list for tutors. The course uses Moodle and its various electronic delivery tools.

The Lab physically holds 75 seats, divided into five tables with 15 computers each. There is one tutor for approximately ten students: students and tutors are assigned to specific groups of seats. This division allows greater and more sustained personal contact, and more active learning.

Tutors must take great care helping students in labs. The student is not just learning a language but a new logical structure, while instructors are proficient coders. When a student asks for help, it is far too easy for a tutor to "fix" the code for them, particularly when one is new to teaching. While this is the path of least resistance, because the student's priority is working code, for many not only does this do little for learning but also in fact this can be detrimental to learning. If a tutor rewrites code with no sympathy for the student's approach, this can just alienate and destroy confidence.

The philosophy of "never touch the keyboard", alluded to previously, embodies our approach. As one practices, this approach reveals subtler layers. [We have also noted that with remote teaching, although one is physically removed, practising the method is more difficult!] The philosophy applies to both instructor and student. It really means not telling students the difficulty with their draft code, but rather discovering it with them. One method is to ask what the student is trying to do, read their code with them, and try to nurture *their* creativity. It can be time- intensive, and is not easy. One needs to react to the student, taking care not to add to the student's pain by repeating the same question[15]; methods like pseudocode and flow diagrams can be useful for withdrawing from the screen. Any suffering (on both sides) is justified when the students "get it" and the sparks of understanding light in their eyes.

4.3 Assessment

Similar to the "Reading Memos" of the Western courses, the Bath course has what is called a "tickable". These are short exercises—gradually increasing in difficulty throughout the year—which are graded only on a Yes/No basis. A tickable therefore differs from a Reading Memo in that it requests some well-defined activity, whereas a Reading Memo is less well-defined and more open-ended. The similarity is in their assessment and in their encouragement of continual work throughout the course.

For instance, one tickable from midway through the first semester is given here:

Tickable: Write a recursive Matlab function, in the file `myexpt.m`, which will compute A^n (via the call `myexpt(A,n)`) using Eq. (2), for any square matrix A.

$$x^n = \begin{cases} 1 & \text{if } n = 0 \\ (x \cdot x)^{n/2} & \text{if } n \text{ is even} \\ x \cdot (x \cdot x)^{(n-1)/2} & \text{if } n \text{ is odd} \end{cases} \qquad (2)$$

This tickable is then used to write another small program for quickly calculating the nth Fibonacci number. During lab sessions, a tutor (who has approximately 7–10 students assigned to be their tutees for the whole semester, or ideally a year) walks around the computer terminals offering help with the mathematical or programming aspects of the exercise. Students who successfully get this code running can also re-use this routine for parts of the coursework at the end of the semester.

An insufficient number (fewer than 80% of the total) of tickables marked "Yes" results in a pro rata reduction in the summative mark. This is widely perceived as fair, because there is general agreement that doing the work as you go along helps you to learn the material.

Otherwise, there is significant use of automatic assessment tools via Moodle, with tutors providing more detailed feedback on programming style.

---

[15] Although it's true that, sometimes, simply reading a question aloud can be surprisingly useful, of course tone matters, here. Reading the question aloud as if it were a reminder to the *instructor* can be less painful for the student.

5 Artificial Intelligence and Programming

The relationship between "Artificial Intelligence" and Programming is clearly deep: all forms of AI as we know it depend on programming computers. The details depend on the sort of AI being considered: much Good Old-Fashioned Artificial Intelligence (GOFAI) is programmed in LISP, Prolog, or languages based on these. The emphasis is on the manipulation of complex, often irregular, data structures containing symbolic data.

Conversely, the currently fashionable (and extremely useful in, for example, image and speech recognition) field of Machine Learning (which used to be known, perhaps more correctly, as Pattern Recognition) depends on large amounts of, generally structured, data, and a worryingly large amount[16] of numerical computation to compute the appropriate weights (anthropomorphically described as "training the network").

It is common to believe that the major languages for Machine Learning are R and Python, and in terms of the user interfaces provided, that is generally true. But we pointed out that Machine Learning is basically vast amounts of numerical computation, at which neither are particularly efficient. If one "peeks under the hood", as contributors to Croucher (2020) did, one sees a very different picture (Fig. 4). It is at this level that techniques like the use of GPUs or the more recent Tensor cores are actually deployed.

There are significant initiatives to use AI for programming now, as well; one such initiative goes by the name "AI-enabled software development". This makes the connection go both ways, just as the connection between AI and mathematics goes both ways, and just as the connection between AI and mathematics education is now beginning to go both ways. One particular area is that of "safety-critical" software. Formal proofs of safety-critical software generate vast numbers of "verification conditions" (essentially lemmas on the way to "Theorem: this software is safe") to be proved. One example (the U.K.'s National Air Traffic Services iFACTS system) is documented in Chapman and Schanda (2014, Sect. 2.6)—see Fig. 5. The 74 thousand lines of SPARK contracts (8% of the total software) generated 152,927 verification conditions, of which at the time 151,026 were proved automatically, by a combination of "Good Old-fashioned AI" and SAT/SMT checkers. A 98.76% automatic rate looks impressive, but that leaves 1901 for manual proof. Manual proof is both time-consuming and error-prone, hence the aim is to eliminate, or at least reduce, it. In automated proof theory, a relatively recent development (e.g., Kühlwein et al., 2013) is the use of Machine Learning to select the "relevant" subset of already-proved results to give the theorem-prover as a basis.

---

[16] Strubell et al. (2019) report that training a big model with neural architecture search can generate as much CO_2 as five cars during their lifetime, including fuel.

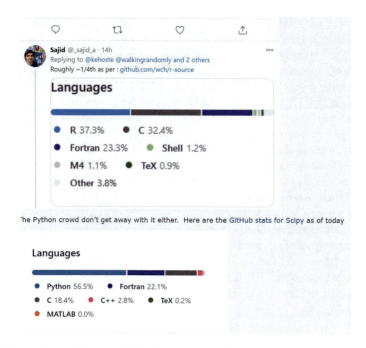

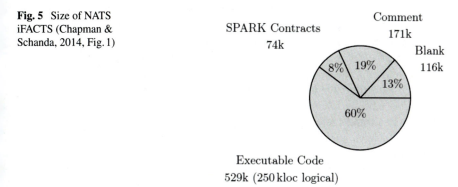

Fig. 4 Fortran Lives! (Croucher, 2020) Of course, no SciPy means you also can't have anything that depends on SciPy including things like Keras or scikit-learn. The up-to-date figures are at https://github.com/wch/r-source and https://github.com/scipy/scipy

Fig. 5 Size of NATS iFACTS (Chapman & Schanda, 2014, Fig. 1)

6 Outcomes

In both the Western and the Bath cases, the student surveys showed great satisfaction. For instance, the TA for the Western Numerical Analysis course twice won the "TA of the Year" award from the Society of Graduate Students. True measurement of the effectiveness of these courses is naturally difficult, but the indications pointed out in Betteridge et al. (2019), which include superior outcomes in downstream courses, seem quite solid. There are also excellent, though again informal, pieces of evidence at Bath, which is largely an industrial placement ("sandwich" in the U.K., "co-op" in Canada) university, even in Mathematics. While this chapter was being written, one of the authors paid an industrial placement visit to a former XX10190 student spending a year as a Data Analyst at an I.T. staffing company. The student said "Programming (XX10190/MATLAB) has been really useful in my Placement year, since Power BI also requires expressing formulae to a computer system". That author had in fact never heard of Power BI before, which demonstrates the wide applicability of these skills. The company representative said "it would be great to hire another of your brilliant students". The exceptionally high employment statistics of Bath mathematics students was part of the evidence for the recommendation in Bond (2018) that "All mathematics students should acquire a working knowledge of at least one programming language".

However, since no controlled experiments were made about teaching methods—in neither case was there a control group, where different methods were used—this kind of qualitative good feeling about outcomes may be the best indication of the success that we can obtain. This clearly touches on the ethics of testing different methods of teaching, and we take this up briefly in the next section.

7 Ethics, Teaching, and Eudaemonia

Much published research on teaching just barely skirts rules about experimentation on humans. The "out" that is most frequently used is the *belief* on the part of the teachers that what they are doing is "best practice". It is rare to have a proper statistical design with control groups to compare the effects of innovation with mere placebo change over the status quo. The previously mentioned research on Active Learning includes some that meets this stringent standard, and as previously mentioned the evidence is so strong that it is now known to be *unethical* not to use active learning. Still, active learning is labour-intensive (on everyone's part—it's a lot simpler for a student to sit and pretend to listen in class, and then cram for an exam in the traditional "academic bulimia" model) and not everyone is willing to pay the price for good ethics.

Another significant piece of active learning is the social aspect. Humans are social animals and teaching and learning is part of how we interact in person. University students appear to value *personal contact* above nearly anything else (Seymour & Hewitt, 1997). Working against that, economics of scale means that universities want to provide certificates of learning by using only small numbers of teachers for many students; this impersonal model is already unsatisfactory for many students. This time of social isolation due to COVID-19 is making this worse, of course, in part because teaching and learning are becoming even more impersonal. One response to this pressure—and this was happening before COVID—is to try to let computers help, and to use AI to personalize instruction and especially assessment.

There is an even deeper ethical question at work, however. A teacher who taught lies[17] would be properly viewed as being unethical, even as being evil. A teacher who hid important facts from the students would be scarcely less unethical. This observation seems to be culturally universal (with perhaps some exceptions, where secret knowledge was jealously guarded, but valued all the more because of its exclusiveness). Yet, aside from idealism, what are the motivations for the teacher to tell the truth, the whole truth, and nothing but the truth?

When humans are the teachers, this is one question. We collectively know to be sceptical of the motives of people: who benefits from this action, and why are they doing this? Teaching pays, and not only in money: perhaps the most important part of our pay is the respect of those that we respect. Most of us understand that the best teachers do their jobs for the love of watching their students understand, especially seeing "light-bulb moments". But when the teacher is an app on your smartphone, the questions become different. We will take as an example the popular language app Duolingo (Von Ahn, 2013). The goals of a company that sells (or gives away—Duolingo is free by default, supported by advertising) an app to teach you something may very well be different from the goals of a human teacher. Indeed, and there is nothing hidden or nefarious about this, one of the goals of the maker of Duolingo is to *provide low-cost translation services*, essentially by distributing the translation tasks to (relatively) trusted human computers. It is an ingenious idea: make the skilled app user pay for the service of learning a new language by providing some services, more as the learning progresses, that others want. The question then becomes not "what does my teacher gain?" but rather "what does the creator of this service gain?"; more insidiously, if a teaching app became truly viral, it might be "what reproductive value does this app gain?".

The modern university system has evolved from its religious roots to provide the desired service today—namely access to the scholarship of the world—to anyone who can find a way to access the University system. We (mostly) share a belief that access to education is one of the great benefits, and provides the key to a better life, a good life, the best life possible (indeed to *eudaemonia,* in Aristotle's term,[18] although people still argue about what exactly he meant by that). It is not at all clear to us that an artificially intelligent teacher (even if humans are in the loop, as with Duolingo) would necessarily share this belief. The benefits to such a "teacher" of actively *discouraging* critical thinking are unfortunately extremely clear: one only has to look at the unbearable success of lies on social media to see the problem.

It seems clear to us that we as teachers should pay attention to the ethics of teaching by or with the help of AIs.

---

[17] Except as an important stepping stone to the real truth—see the entry "Lies to Children" in Wikipedia. Sometimes a simplistic story is the right first step.

[18] Aristotle may have done us a disservice by looking down on crafts and craftspeople; the term Software Carpentry is not likely to induce respect for the discipline in academia, for instance. We lament this prejudice.

8 Concluding Remarks

> Instead there must be a more serious concern with the significant ways in which computational resources can be used to improve not so much the **delivery** but rather the **content** of university courses.
> —(Abelson, 1976)

The content of mathematics courses has changed over the past few decades (this has been noted in many places, but see, for example, Corless & Jeffrey, 1997). Some of that change has been forced by the increasing number of students and their necessarily more diverse backgrounds and interests; some of that change has been deliberate abandonment of no-longer-useful techniques; and some of that change has been driven by the introduction of new tools, some needed for new applications such as AI, which in turn is being applied to teaching. We claim that an important part of that curricular change was in fact AI; remember that *symbolic computation was the first form of AI that worked*. One might consider compilers as early thin AI, or even numerical computation as AI (after all, machine computation replaced human intellectual labour)! But symbolic computation came out of the AI labs at MIT. Getting a computer to find an antiderivative was considered to be fundamentally harder than finding a derivative; that was a watershed. Symbolic computation has been affecting the mathematics curriculum for more than 60 years now. We don't think it's finished affecting the curriculum.

These tools synergistically affect one another: the development of AI changes mathematics teaching, mathematics, and programming; mathematics teaching changes mathematics, programming, and AI; and of course, advances in programming change AI, mathematics, and mathematics teaching. The cycle is dizzying.

One new tool that we have not yet talked about is WolframAlpha. This is nearly universally available, free, almost uses natural language input—it's pretty accepting, and the students find it simple to use—and produces for the most part very legible, neat, and useful answers to problems at roughly the first year university level. Its language recognition features are indeed a kind of AI. We believe that its use (or the use of similar tools) should not only be allowed in class but also encouraged. The students will still be *responsible* for the answers, and it helps to give examples where WolframAlpha's answers are wrong or not very useful, but it is irresponsible of us to ignore it. Matlab, Maple, Mathematica, Python, NumPy, and SymPy provide other tools for mathematical thinking, on a larger scale. We believe that it is incumbent on us as educators to teach our students the kinds of mathematics that they can do when using those tools.

Developing AI further depends upon mathematicians who understand the numerical methods that underpin machine learning, reinforcement learning, etc., as well as how to program. But good programming practices are not just about what code is written but the way in which it is created. While theorems need proofs and the requirement for correctness and reproducibility in mathematical sciences is paramount, academia has been slow to apply this rigorously to its codes. In software development, this is achieved with testing, validation, and version control. While comparison with expectation and (better) analytic proof is adequate for validation, we have not formally taught testing or version control in our undergraduate programmes. The time pressure on curriculum cannot excuse this omission much longer. The value of adopting professional practices goes beyond those who will work as software engineers. They are vital tools for working efficiently, contributing to open software, for data scientists and AI engineers to manage data and to ensure trust in the methods that they develop and apply in their careers. These enable students to use their computational tools responsibly.

We also have not talked here about GeoGebra, which is probably now the most popular computational thinking tool for mathematics in the world. This is because we are "old guard" (well, some of us are) and GeoGebra is a newer tool, one that we have not yet used. However, it seems clear to us that the same principles that we have been using for our other tools also apply here: the students should be aware of the program's limitations; the students should know when the answer is correct and when it is not, and the students should be responsible for the answers.

And we have not talked at all about perhaps the most significant tool: the Internet itself. Collectively, we have vastly more capable memory now than we did before—mathematical definitions can be looked up instantly (indeed, Wikipedia is quite reliable for mathematics, though not perfect)—and we have nearly universal access to papers, projects, course materials, and more. The growth in mathematical videos on YouTube is astonishing: Tom Crawford ("The Naked Mathematician"), Bobby Seagull, Numberphile, Online Kyne, and many more provide wonderfully accessible mathematics. Is that AI? Given that many videos are automatically close-captioned, we would say that there is at least some AI involved. Then there are the algorithms for finding such videos automatically for you.

With the advent of modern AI tools like these for mathematics education, more questions arise. We believe that amongst the most important questions for AIEd will be about the *ethics* underlying the tools. We are not merely talking about the debates over using homework-assistance websites to cheat for class. We all know now that machine learning can easily copy our biases and prejudices, without us intending, or unintentionally alter the content of what we are conveying.[19] We also know that the goals of developers of AIEd tools may well be different than the goals of good teachers. We forbear from trying to define what it means to be a "good teacher'; this has been the subject of considerable debate since forever, in all cultures, but see, for instance, (Boynton, 1950). We take from that short, well-written article, out of many possible choice quotations, "Good teaching must be based on an unequivocal, sincere interest in the human individual." Achieving this alone will be a challenge for AIEd.

The ethics of AIEd is beginning to be studied intensively (see, for instance, the works (Aiken & Epstein, 2000; Sijing & Lan, 2018)), but clearly we are only just

---

[19] See also Bradford et al. (2009), which shows that computational tools can affect the basic meaning of equality: pedagogical equality is not the same as mathematical equality. It is perfectly possible for two expressions to be mathematically equal, but only one expression to be the desired student response.

scratching the surface of the issues, which include some very deep classical philosophical problems, including how to live a good life (achieve eudaemonia). The amplified human life, when humans use computers to increase their thinking capability, clearly also needs philosophical study, as we are touching directly on very classical problems. Not only philosophers, but cognitive scientists, as well as computer scientist experts in AI, will be needed to properly develop these tools.

Plus ça change, plus c'est la même chose.—Jean Baptiste Alphonse Karr, 1849

Acknowledgements RMC thanks the Isaac Newton Institute for Mathematical Sciences and the staff of both the University Library and the Betty and Gordon Moore Library at Cambridge for support and hospitality during the programme Complex Analysis: Tools, techniques, and applications, by EPSRC Grant # EP/R014604/1 when some of the work on this project was undertaken. RMC likewise thanks the University of Bath for an invitation to visit Bath, at which this project was started. EYSC and RMC also thank Western University for a grant to work on the project *Computational Discovery on Jupyter*, some of whose results are discussed here.

References

Abelson, H. (1976). Computation in the undergraduate curriculum. *International Journal of Mathematical Education in Science and Technology, 7*(2), 127–131.

Aiken, R. M., & Epstein, R. G. (2000). Ethical guidelines for AI in education: Starting a conversation. *International Journal of Artificial Intelligence in Education, 11,* 163–176.

Armacost, R. L., & Pet-Armacost, J. (2003). Using mastery-based grading to facilitate learning. In *33rd Annual Frontiers in Education, 2003. FIE 2003* (Vol. 1, pp. T3A–20). IEEE.

Betteridge, J., Davenport, J. H., Freitag, M., Heijltjes, W., Kynaston, S., Sankaran, G., & Traustason, G. (2019). Teaching of computing to mathematics students: Programming and discrete mathematics. In *Proceedings of the 3rd Conference on Computing Education Practice* (pp. 1–4).

Bond, P. (2018). The era of mathematics–review findings on knowledge exchange in the mathematical sciences. Engineering and physical sciences research council and the knowledge transfer network. https://epsrc.ukri.org/newsevents/news/mathsciencereview/.

Borwein, J., & Devlin, K. (2009). *The computer as crucible: An introduction to experimental mathematics*. The Australian Mathematical Society (p. 208).

Borwein, J. M., & Borwein, P. B. (1992). Strange series and high precision fraud. *The American Mathematical Monthly, 99*(7), 622–640.

Boynton, P. L. (1950). What constitutes good teaching? *Peabody Journal of Education, 28*(2), 67–73.

Bradford, R., Davenport, J., & Sangwin, C. (2009). A comparison of equality in computer algebra and correctness in mathematical pedagogy. In J. Carette et al. (Eds.), *Proceedings intelligent computer mathematics* (pp. 75–89).

Broussard, M. (2018). *Artificial unintelligence: How computers misunderstand the world*. MIT Press.

Camargos Couto, A., Moreno Maza, M., Linder, D., Jeffrey, D., & Corless, R.M. (2020). *Comprehensive LU factors of polynomial matrices*. Mathematical Aspects of Computer and Information Sciences MACIS (pp. 80–88).

Chan, E. Y. S., & Corless, R. M. (2017a). A new kind of companion matrix. *The Electronic Journal of Linear Algebra, 32,* 335–342.

Chan, E. Y. S., & Corless, R. M. (2017b). A random walk through experimental mathematics. In J. M. Borwein (Ed.), *Commemorative Conference* (pp. 203–226). Springer.

Chan, E. Y. S., & Corless, R. M. (2021). *Computational discovery on Jupyter* (In preparation).
Chapman, R., & Schanda, F. (2014). Are we there yet? 20 years of industrial theorem proving with SPARK. In *Proceedings of Interactive Theorem Proving* (pp. 17–26).
Corless, R. M. (2004a). Computer-mediated thinking. *Proceedings of Technology in Mathematics Education.* https://github.com/rcorless/rcorless.github.io/blob/main/CMTpaper.pdf.
Corless, R. M. (2004b). *Essential Maple: An introduction for scientific programmers*, 2nd ed. Springer Science & Business Media.
Corless, R. M., & Jeffrey, D. J. (1997). Scientific computing: One part of the revolution. *Journal of Symbolic Computation, 23*(5), 485–495.
Croucher, M. (2020). No Fortran? No data science in R and Python! https://walkingrandomly.com/?p=6696.
Davenport, J. (1986). On the Risch differential equation problem. *SIAM Journal on Computing, 15*, 903–918.
Davenport, J. (2018). Methodologies of symbolic computation. In: *Proceedings AISC* (pp. 19–33).
Davenport, J. H., Hayes, A., Hourizi, R., & Crick, T. (2016). Innovative pedagogical practices in the craft of computing. In *2016 International Conference on Learning and Teaching in Computing and Engineering (LaTICE)* (pp. 115–119). IEEE.
Edelman, A. (1988). Eigenvalues and condition numbers of random matrices. *SIAM Journal on Matrix Analysis and Applications, 9*(4), 543–560.
Eilers, S., & Johansen, R. (2017). *Introduction to experimental mathematics*. Cambridge University Press.
Fischer, F., Böttinger, K., Xiao, H., Stransky, C., Acar, Y., Backes, M., & Fahl, S. (2017). Stack overflow considered harmful? The impact of copy& paste on android application security. In *38th IEEE Symposium on Security and Privacy (SP)* (pp. 121–136).
Flanagan, O. (2009). *The really hard problem: Meaning in a material world*. MIT Press.
Freeman, S., Eddy, S. L., McDonough, M., Smith, M. K., Okoroafor, N., Jordt, H., & Wenderoth, M. P. (2014). Active learning increases student performance in science, engineering, and mathematics. *Proceedings of the National Academy of Sciences, 111*(23), 8410–8415.
Handelsman, J., Ebert-May, D., Beichner, R., Bruns, P., Chang, A., DeHaan, R., et al. (2004). Scientific teaching. *Science, 304*(5670), 521–522.
Hegedus, S., Laborde, C., Brady, C., Dalton, S., Siller, H.-S., Tabach, M., Trgalova, J., & Moreno-Armella, L. (2017). *Uses of technology in upper secondary mathematics education*. Springer Nature.
Higham, N. J. (2002). *Accuracy and stability of numerical algorithms*, 2nd ed. SIAM, Philadelphia.
Kahan, W. M. (1980a). Handheld calculator evaluates integrals. *Hewlett-Packard Journal, 31*(8), 23–32.
Kahan, W. M. (1980b). Interval arithmetic options in the proposed IEEE floating point arithmetic standard. In *Interval Mathematics 1980* (pp. 99–128). Elsevier.
Kahan, W. M. (1983). Mathematics written in sand. In *Proceeding of the Joint Statistical Meetings of the American Statistical Association* (pp. 12–26). http://people.eecs.berkeley.edu/~wkahan/MathSand.pdf.
Kovács, Z., Recio, T., Richard, P. R., & Vélez, M. P. (2017). GeoGebra automated reasoning tools: A tutorial with examples. In *Proceedings of the 13th International Conference on Technology in Mathematics Teaching* (pp. 400–404).
Kühlwein, D., Blanchette, J., Kaliszyk, C., & Urban, J. (2013). MaSh: Machine learning for Sledgehammer. In *International Conference on Interactive Theorem Proving* (pp. 35–50).
Li, A., & Corless, R. M. (2019). Revisiting Gilbert Strang's "A chaotic search for i." *ACM Communications in Computer Algebra,53*(1), 1–22.
Luckin, R., Holmes, W., Griffiths, M., & Forcier, L. B. (2016). *Intelligence unleashed: An argument for AI in education*. Pearson Education.
Mandelbrot, B. B., & Frame, M. (2002). Some reasons for the effectiveness of fractals in mathematics education. In *Fractals, graphics, & mathematics education* (pp. 3–9).
Monaghan, J., Trouche, L., & Borwein, J. M. (2016). *Tools and mathematics*. Springer.

Papert, S. (1993). *Mindstorms*, 2nd ed. Basic Books.
Paxton, J. (2002). Live programming as a lecture technique. *Journal of Computing Sciences in Colleges, 18*(2), 51–56.
Richard, P. R., Venant, F., & Gagnon, M. (2019). Issues and challenges in instrumental proof. In *Proof Technology in Mathematics Research and Teaching* (pp. 139–172). Springer.
Rosati, P. A., Corless, R. M., Essex, G. C., & Sullivan, P. J. (1992). An evaluation of the HP28S calculator in calculus. *Australian Journal of Engineering Education, 3*(1), 79–88.
Rubin, M. J. (2013). The effectiveness of live-coding to teach introductory programming. In *Proceeding of the 44th ACM Technical Symposium on Computer Science Education* (pp. 651–656).
Seymour, E., & Hewitt, N. M. (1997). *Talking about leaving*. Boulder, CO: Westview Press.
Sijing, L., & Lan, W. (2018). Artificial intelligence education ethical problems and solutions. In *2018 13th International Conference on Computer Science & Education (ICCSE)* (pp. 1–5). IEEE.
Slagle, J. (1963). A heuristic program that solves symbolic integration problems in freshman calculus. *Journal of the ACM, 10,* 507–520.
Smith, R. C., & Taylor, E. F. (1995). Teaching physics on line. *American Journal of Physics, 63*(12), 1090–1096.
Strang, G. (2001). Too much calculus. http://www-math.mit.edu/~gs/papers/essay.pdf.
Strubell, E., Ganesh, A., & McCallum, A. (2019). Energy and policy considerations for deep learning in NLP. In *Proceedings of the 57th Annual Meeting of the Association for Computational Linguistics* (pp. 3645–3650).
Tepylo, D. H., & Floyd, L. (2016). Learning math through coding. https://researchideas.ca/mc/learning-math-through-coding/.
Von Ahn, L. (2013). Duolingo: Learn a language for free while helping to translate the web. In *Proceedings of the 2013 International Conference on Intelligent User Interfaces* (pp. 1–2).
Wilson, G. (2006). Software carpentry: Getting scientists to write better code by making them more productive. *Computing in Science & Engineering, 8*(6), 66–69.
Wilson, G., Aruliah, D. A., Brown, C. T., Hong, N. P. C., Davis, M., Guy, R. T., et al. (2014). Best practices for scientific computing. *PLoS Biology, 12*(1), e1001745.

The present and future of AI in ME: Insight from empirical research

Introduction to Section 3 by the Coordinator Jean-baptiste Lagrange and the Editors

Introduction to Part III
The present and future of AI in ME: Insight from empirical research and a little more
Jean-baptiste Lagrange
LDAR, University of Paris, France

Common to all chapters in this part is a shared focus on empirical research in realistic settings, to which we include a more exploratory contribution that remains closer to the philosophical essay. As a *last chapter* in the middle, Freiman and Volkov's contribution may come as a surprise compared to the five chapters presented below. This is an historical study of a case of "rearrangement method" along centuries and in various cultures, proposing to discuss visual and dynamic approaches. This is certainly a theme of interest with regard to intelligent models of learner interaction, and further studies will bring empirical evidence about the potentialities of these approaches. From an empirical research, Jarvis et al. consider the use of computer algebra systems (CAS) at the tertiary level, taking advantage of rich data from the actual practice, developed over several years, of a university mathematics instructor. Through seven iterations of an Optimization course, significant progress is observed regarding most factors pertaining to technology integration, particularly that of CAS-focused assessment practices. Emprin's chapter introduces interactive computer simulation as a means of educating teachers and tutors (i.e., teacher educators). Part of the work is in computer science: designing expert systems able to recreate plausible classroom situations and tutor-teacher interaction like in (Paneque, J. J., Cobo, P. & Fortuny, J. M., 2017). This work could not be carried out without data and a theoretical framework issued from years of empirical research about teacher education. Moreover, investigating the full potential of these systems implies using them in realistic settings and collecting more data. Rodríguez offers a consideration of Virtual Reality use in the classroom. Potentialities for visualization and manipulation can be inferred from current uses in other domains, and Rodríguez' empirical study, carried out during the pandemic, aims to confront these potentialities with the reality of mathematics teaching/learning at the tertiary level. Insight is obtained from student feedback considering different positions of the learner (i.e., observing or manipulating) and the use of different personal hardware, and specific benefits for the teacher's practice are emphasized.

Diego-Mantecon and Trgalová do not focus on a particular software environment, but rather investigate potentialities of digital technology in a broad sense, and in connection with questions about their contribution to mathematics teaching and learning. The question addressed by Diego-Mantecon deals with the benefits of classroom STEAM projects carried out jointly by mathematics and technology teachers, with teams of 14- and 15-year-old students within an empirical study. Digital technologies were used by the teams according to their project, including 3-D printing and programming environments. Visual/Spatial competences as well as reasoning were stimulated, but mathematics teachers tended to overlook learnings opportunities made possible through connections between mathematics and technology: teachers with a mathematical background privileged curricular content while the others avoided introducing mathematics in the project. Regarding empirical research, Trgalová builds upon forty years of investigation of classroom use of Dynamic Geometry (DG) to address the diversity of ways in which digital technology can be implemented. A rigorous classification of tasks in four levels is proposed, from tasks where DG is simply substituted for paper/pencil constructions through to tasks especially designed to take advantage of DG potentialities. Since students' engagement and cognitive activity develop better in the high-level abilities of the classification, Trgalová insists on not considering technology as a vector of transformation, but rather looking more closely at how it is used in educational settings.

Drawing from these chapters, it is possible to address central issues raised in this book. What is AI, and how does it pertain to math education research? How can math education progress further, taking advantage of AI's potentialities? The ambition of this introductory text is to focus on significant periods within that development. In a first such period, beginning with Turing (1950), computer scientists became aware of capabilities of digital machines to perform symbolic treatments by way of specific representations and algorithms. At this time, computers were thought to be able to achieve tasks generally considered to involve 'intelligence', or able to behave in an 'intelligent' manner. Computer algebra systems (CAS), can be viewed in this perspective since they are basically symbol manipulators operating on representations of algebraic and calculus entities. CAS was introduced in the seventies and eighties as means to ease the work of mathematicians, but also with a broader vision of mathematical practices, including teaching and learning. At the time of the first ICMI study (Hodgson & Muller, 1992) CAS was the technology that attracted the most attention. Because CAS technology facilitated, up to a certain point, algebraic manipulations, proponents remained hopeful that it would serve to reorient teaching practices towards more conceptual issues. According to Hodgson and Muller (ibid), there was *"no doubt that Symbolic Manipulation Systems must be introduced into the Mathematics curriculum ... for compelling change in secondary and university math education in the near future."*

We are now 40 years beyond the introduction of CAS, and efforts have been made by designers to develop capabilities and user-friendly interfaces, as well as software environments intended to help teachers and students to take better advantage of CAS technology (Lagrange, 2010). In their chapter, Jarvis et al. present a comprehensive analysis of a current example of CAS use for both teaching and assessment at the tertiary level. Throughout the years, CAS has been like a double-faced coin: one side representing the many opportunities it offers for teaching/learning, and the other side the difficult aspects of integration into actual teaching/learning. Studies like Jarvis et al. serve as a great contribution in showing how these two faces can be reconciled through years of thoughtful practice. In applying a technology integration taxonomy, it was found that the only component that did not significantly progress throughout the seven course iterations was that of "Staff Facility." While the integration of the CAS-based teaching and assessment practices were not intended to be a departmental undertaking in this context, this finding is nonetheless indicative of a larger, common truth: an instructor's often solitary journey towards suitable integration reflects the deep links that exist between practices involving technology and one's personal conception of mathematics and of mathematics learning.

In a second historic period, one initiated by Carbonell (1970), computer scientists were interested in the use of AI techniques to create new systems for computer-aided teaching and learning. Existing systems, based upon a behaviorist approach of learning, relied on multiple choice questions organized in rigid learning paths. In contrast, the new systems were based on a semantic network, an information structure which organizes facts, concepts, and procedures relevant within a domain. From this structure, systems were able to propose to the learner a more flexible approach to knowledge. In mathematics education, it appeared that building a suitable semantic network for a given content implied a deep reflection on epistemological and cognitive aspects, as well as on knowledge about situations that favor learning. Then computer scientists specialized in AI for computer-aided teaching and learning, and didacticians of mathematics initiated a close collaboration (Balacheff & Vivet, 1994). In further development, this orientation was amended to allow for a renewed focus on the learning process. Tchounikine (2002) pointed out that the naive vision of a transfer of knowledge contained in the system to the student is now largely abandoned, and it is now considered that the heart of the learning process is the interaction between the learner and the system. This means that semantic networks should not be just models of knowledge; they should model the interaction between the learner and an environment in which knowledge makes sense.

From this perspective, Emprin presents two related systems. In the first one, the learner is a tutor, and in the second one, the learner is a teacher. In both systems, the domain of knowledge is the classroom use of DG. Emprin's case studies illustrate the necessity and the power of models of interaction based on didactical knowledge at several levels—ranging from DG use by students through to tutor/trainee interaction. The case studies shed light on three important characteristics of these models. First, they represent a functional synthesis of didactical research at each level. Second, they are not rigidly fixed: feedback from users helps the researcher to adapt the model; this has a practical effect, increasing the system's efficiency, and contributes to didactical research at the different levels. Third, for the user, the interaction with the system includes a reflective component: while the system gives the user feedback regarding their professional knowledge, he/she is also encouraged to discuss this feedback, as well as the underlying model.

In the chapter by Diego-Mantecon, the Rubiks cube project involves designing and building a device able to manipulate the cube, and then programming the resolution in Scratch, a contemporary evolution of the Logo software (Papert and Harel, 1985). In this activity, intelligence does not preexist in the programming environment. Students had to conceive a matrix representation of the cube, reflect about different 'intelligent' solving methods, and think of these methods as algorithms. As noted, before, Diego-Mantecon observed that the teachers privileged activities in terms of their own area of expertise. When programming was undertaken outside of this expertise, then the potential of Scratch as a micro-world seems to be missed. In a recent study of a large-scale project involving programming in Scratch, Noss et al. (in press) observed a major obstacle in the absence of teacher understanding regarding the situations proposed by the researchers.

In a micro-world, the meaning that learners construct is shaped by their understanding of the feedback and by their own actions. That is why Balacheff and Kaput (1997) insist on the necessity of a description of the domain of phenomenology, and of the type of feedback it allows. For Rodriguez, the domain is one of perception and construction of 3-D objects in relationship with their mathematical definition, and his chapter provides an in-depth examination of the phenomenology offered by virtual reality systems, especially when compared with 3-D, screen-based software, in the light of students' appreciation of their experience of learning mathematics in virtual reality. Trgalová's analysis privileges the functionality of dragging in DG, a technology often considered as a typical micro-world. However, her analysis goes beyond DG, and that classifying tasks at four levels (substitution, augmentation, modification, redefinition) is relevant for any digital environment being used for teaching and learning. The distinction between AI-aided systems and micro-words reflects two differing viewpoints on technology: the first is centered on internal computational models of the knowledge and of the learner; the second focuses on the learning opportunities offered to users for action and perception. Both viewpoints can be productive: micro-words could benefit from an explicit consideration of internal models; AI-aided systems could take more into consideration the user experience at the interface.

A third period of significant AI evolution is what is happening now with the development of machine learning, i.e., algorithms that are able to exploit a huge volume of data in order to automatically identify significant patterns. Emprin suggests machine learning as a means of obtaining accurate models of interaction from data that is collected empirically. Certainly, educational empirical research does produce a large amount of data. This data is often tedious to analyze manually, and despite the help provided by inferential statistics, the researcher is often left to his/her own intuition for building models. Thus, machine learning could provide help, especially in the domain of technology where it is possible to get automatically pre-formatted data.

By focusing on both symbolic and material milieus, Freiman and Volkov's present an exploratory work that attempts to bring symbolic artefacts closer to material artefacts (Flores-Salazar, Gaona and Richard, 2022), to prepare the modelling of the mathematical work that AI will be able to feed by better understanding the means of human learning. As Nicolas Balacheff reminds us in the preface, there are epistemological and semiotic tensions between knowledge and its representations, but also discursive ones between the art of expressing ideas and the codified expression of discourse. We know that mathematics is already fond of multi-referenced expressions and multimodal processing. However, the resurgence of artefacts in mathematics education and the new possibilities of visualization are changing traditional relationships. With the growing interest in interactive representation systems, there is a need to revisit the genesis of the past to look at how knowledge can be developed with today's means of instrumental representation. This is an inexhaustible source of inspiration, because it was at this point that we already struggled to bring knowledge to life on a medium as plain as paper.

A final remark is that most chapters conceptualize the relationship of the learner to the knowledge mediated by technology. This is productive, notably when Trgalová investigates the influence of tasks on students' engagement and cognitive activity. However, in an educational context tasks do not exist in isolation, and thus further analysis is needed to understand how they can be developed within school and tertiary level institutions. That is why Lagrange and Richard (forthcoming) propose the consideration of instrumental genesis within mathematical working spaces using two complementary viewpoints: a psychological and institutional viewpoint dealing with instrumented techniques, and how they emerge within school institutions that are characterized by more usual paper/pencil techniques.

Empirical research on mathematics education and technology is a long but exciting journey, always open to new perspectives. The chapters in this part will no doubt become very valuable landmarks within this journey.

References

Balacheff, N., & Kaput, J. J. (1997). Computer-based learning environments in mathematics. In A. Bishop (Ed.), *International handbook in mathematics education* (pp. 469–501). Kluwer Academic.

Balacheff, N., & Vivet, M. (1994). Introduction. *Recherches en didactique des mathématiques.* 14.(1–2). 5–8.

Carbonell, J. R. (1970). AI in CAI: An artificial-intelligence approach to computer-assisted instruction. IEEE Transactions on Man-Machine Systems, *11*(4). 190–202.

Flores-Salazar, J. V., Gaona, J. & Richard, P. R. (expected early 2022). Mathematical work in the digital age. Variety of tools and the role of genesis. In Kuzniak, A., Montoya-Delgadillo, E. & Richard, P. R. (Eds) *Mathematics Education in the Age of Artificial Intelligence. How artificial intelligence can serve the mathematical human learning.* Springer International Publisher.

Hodgson, B. R., & Muller, E. R. (1992). The impact of symbolic mathematical systems on mathematics education. In B. Cornu & A. Ralston (Eds.), *The influence of computers and informatics on mathematics and its teaching*, (pp. 93–107). UNESCO (Science and Technology Education Document Series 44). Paris: UNESCO.

Lagrange, J. B., (2010). Teaching and learning about functions at upper secondary level: designing and experimenting the software environment Casyopée.*International Journal of Mathematical Education in Science and Technology*, 41(2), 243–255.

Lagrange, J.-B., & Richard, P. R. (expected early 2022). Instrumental Genesis in the Theory of MWS: Insight from didactic research on digital artefacts. In Kuzniak, A., Montoya-Delgadillo, E. & Richard, P.R. (Eds) *Mathematics Education in the Age of Artificial Intelligence. How artificial intelligence can serve the mathematical human learning*. Springer International Publisher.

Noss, R., Hoyles, C., Saunders, P., Clark-Wilson, A., Benton, L., & Kalas, I. (in press). Making Constructionism work at scale: The story of ScratchMaths. In N. Holbert, M. Berland, & Y. Kafai (Eds.), *Designing Constructionist futures: The art, theory, and practice of learning designs*. MIT Press.

Paneque, J. J., Cobo, P. & Fortuny, J. M. (2017). Intelligent Tutoring and the Development of Argumentative Competence. *Technology, Knowledge and Learning*, 22, 83–104.

Tchounikine, P. (2002). Pour une ingénierie des environnements informatiques pour l'apprentissage humain. *Revue I3 - Information Interaction Intelligence*, 2(1), 59–95.

Turing, A.M. (1950). Computing machinery and intelligence, *Mind*, 49, 433–460.

CAS Use in University Mathematics Teaching and Assessment: Applying Oates' Taxonomy for Integrated Technology

Daniel Jarvis, Kirstin Dreise, Chantal Buteau, Shannon LaForm-Csordas, Charles Doran, and Andrey Novoseltsev

1 Introduction

The creation of new and innovative working spaces in post-secondary mathematics education that involve both powerful instructional technologies and more traditional paper/pencil techniques requires a number of key factors including resource accessibility, institutional support, and perhaps most importantly, a change in teacher beliefs and attitudes. Beyond mere visualization or classroom demonstration applications, instructional technology tools can and should become an integral part of the conceptualization of mathematical concepts and, further, the broadening of curricular scope. However, in order for this to become reality in post-secondary mathematics courses, the pedagogical practices adopted by instructors must ultimately also include assessment strategies that equally allow for the use of these powerful technological tools.

In this chapter, we examine the findings from a case study conducted within a university mathematics department in western Canada in which a special Teaching and Learning Enhancement Fund grant allowed for the restructuring of an existing undergraduate course on Optimization through the creative use of available CAS technology. One particular instructor increasingly incorporated Computer Algebra System (CAS)-based software into both his teaching and assessment practices within

---

D. Jarvis (✉) · S. LaForm-Csordas
Schulich School of Education, Nipissing University, North Bay, ON, Canada
e-mail: danj@nipissingu.ca

K. Dreise · C. Buteau
Faculty of Mathematics and Science, Brock University, St. Catharines, ON, Canada

C. Doran · A. Novoseltsev
Faculty of Mathematical and Statistical Sciences, University of Alberta, Edmonton, AB, Canada

© The Author(s), under exclusive license to Springer Nature Switzerland AG 2022
P. R. Richard et al. (eds.), *Mathematics Education in the Age of Artificial Intelligence*, Mathematics Education in the Digital Era 17,
https://doi.org/10.1007/978-3-030-86909-0_13

seven distinct iterations of the same Optimization course over time. In our first analysis (Jarvis et al., 2018), we aimed at providing insights into the perceived challenges and affordances relating to technology integration at the university level by examining the perceptions of key stake-holders (i.e., Project Leader, Instructors, students) regarding the implementation of the technology-enhanced Optimization course. In this chapter, we are interested in further investigating the technology integration within the Optimization course, particularly in terms of how this integration changed over time within the multiple, subsequent offerings of the course by the same instructor. We are thus interested in quantifying the degree of technology integration in the course as it evolved over time, and in identifying key aspects that seem to have significantly promoted the increased and effective use of technology within this particular course.

In this introductory section we begin by providing some background information by reviewing the literature regarding the use of instructional technology in mathematics education including issues such as teacher beliefs, teacher behaviors, and changes to curriculum and assessment practices. We then explain the university context in which funding enabled two mathematics instructors to initiate rich experimentation with the open source mathematics software known as *SageMath* (Stein & Joyner, 2005) within the two above-mentioned mathematics courses.

2 Literature Review

A growing number of international studies have shown that Computer Algebra Systems (CAS)-based instruction has the potential to positively affect the teaching and learning of mathematics at various levels of the education system, even though this has not been widely realized in secondary schools and in higher education (Artigue, 2002; Beaudin & Picard, 2010; Bossé & Nandakumar, 2004; Bray & Tangney, 2017; Kendal & Stacey, 2002; Lagrange, 2005; Lavicza, 2006; Pierce & Stacey, 2004; Richard et al., 2019; Smith Risser, 2011; Somekh, 2008). Following the 17th International Commission on Mathematical Instruction (ICMI) Study Conference, entitled *Technology Revisited* and held in Vietnam in December 2006, then ICMI President, Artigue (2009), shared her insights regarding the resistance of instructional technology:

> Making technology legitimate and mathematically useful requires modes of integration allowing a reasonable balance between the pragmatic and the epistemic power of instrumented techniques. This requires tasks and situations that are not simple adaptation of paper-and-pencil tasks, often tasks without equivalent in the paper-and-pencil environment, thus tasks not so easy to design when you enter in the technological world with your paper-and-pencil culture. (p. 467)

Clearly, the incorporation of powerful technologies requires a careful and deliberate rethinking of mathematics curriculum, learning goals, tool/software usage, pedagogical strategies, and assessment practices at all levels of education.

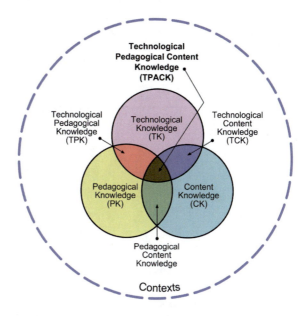

Fig. 1 Technology, Pedagogy, and Content Knowledge (TPACK) model (Koehler & Mishra, 2009)

> In addition to its computational power, modern technologies can help increase collaboration and bring about more of an emphasis on practical applications of mathematics, through modelling, visualisation, manipulation and the introduction of more complex scenarios. . . . For these reasons, the use of technology in mathematics education is becoming increasingly prioritised in international policy and curricula. (Bray and Tangley, 2017, pp. 256–25)

Building upon Shulman's (1986) work on Pedagogical Content Knowledge (PCK), Koehler and Mishra (2009) have developed their own *Technology, Pedagogy, and Content Knowledge* (TPACK) model (see Fig. 1) for the analysis of teacher instructional practice:

> TPACK is the basis of effective teaching with technology, requiring an understanding of the representation of concepts using technologies; pedagogical techniques that use technologies in constructive ways to teach content; knowledge of what makes concepts difficult or easy to learn and how technology can help redress some of the problems that students face; knowledge of students' prior knowledge and theories of epistemology; and knowledge of how technologies can be used to build on existing knowledge to develop new epistemologies or strengthen old ones. . . . Teaching successfully with technology requires continually creating, maintaining, and re-establishing a dynamic equilibrium among all components. (Koehler & Mishra, 2009, pp. 66–67)

Another informative taxonomy for understanding the different uses of technology in mathematics instruction is the Substitution Augmentation Modification Redefinition (SAMR) model created by Puentedura (2006, 2014). Although the SAMR model (see Fig. 2) has been criticized on several points (e.g., diverse interpretation/application of the model; an absence of context; an overly rigid structure; and emphasizing product over process), it remains an increasingly popular tool for practitioner reflection and planning (Hamilton et al., 2016).

Fig. 2 Substitution augmentation modification redefinition (SAMR) model (Puentedura, 2006)

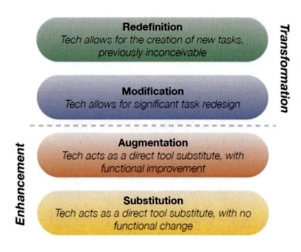

Somekh (2008) described this difficult yet required paradigm shift relating to instructional technology as follows: "The affordances of the Internet, digital photography and cyberspace are radically changing how knowledge is constructed, represented and accessed in the world outside school, and policy-makers need to acknowledge this and restructure the systems of curriculum, assessment and school organization" (p. 458).

School level studies suggest that beyond the availability of technology, teachers' beliefs and cultural influences are key factors in technology integration into mathematics teaching and learning. In categorizing barrier types as either first-order (external) or second-order (internal) in nature, Ertmer et al. (2012) have provided a helpful set of related definitions:

> First-order barriers were defined as those that were external to the teacher and included resources (both hardware and software), training, and support. Second-order barriers comprised those that were internal to the teacher and included teachers' confidence, beliefs about how students learned, as well as the perceived value of technology to the teaching/learning process. Although first-order barriers had been documented as posing significant obstacles to achieving technology integration... underlying second-order barriers were thought to pose the greater challenge. (Ertmer et al., 2012, p. 421)

Oates (2009) noted that effective integration of technology into the teaching and learning of mathematics presents a significant challenge to tertiary mathematics educators: "Assessment issues in particular are widely considered in the literature as a critical factor in technology implementation.... With respect to assessment, both pedagogical consistency and the impact of CAS on examination questions, are seen as particularly significant issues" (paragraph 1).

Teacher beliefs represent deeply held assumptions and values relating to education, and these beliefs are not easily changed. As Bray and Tangney (2017) note, "In order to achieve an environment that facilitates technology usage in an inquiry-based, constructivist manner, a change in the pedagogical approach and the learning experience of the students is required, and this is fundamentally dependent on the actions and beliefs of teachers" (p. 257).

In contrast to the growing body of research focusing on CAS technology use at the secondary school level (Connors & Snook, 2001; Fey et al., 2003; Haapasalo, 2013; Kieran & Drijvers, 2006), there is relatively little parallel research at the postsecondary level (Buteau & Muller, 2014; Buteau et al., 2014; Decker, 2011; Martinovic et al., 2013; Rosenzweig, 2007; Stewart et al., 2005; Tall, 2013; Thompson et al., 2013; Thompson, Ashbrook, & Musgrave, 2015; Tobin & Weiss, 2016). This is particularly true in the area of student *assessment*, where powerful technology tools such as CAS computer software and CAS-enabled calculators have rarely played a part in formal evaluation in undergraduate mathematics courses (Heidenberg & Huber, 2006; Sevimli, 2016).

Some have been critical of CAS use in secondary schools pointing to some of the unintended consequences of heavy dependency on these powerful tools in light of how they might be used within curriculum. Authors Jankvist et al. (2019) suggested that the following characteristics represent some of these negative effects: loss of distinctive features of concept formation, a consequential reclassification of mathematical objects, instability of CAS solutions as objects, and prevailing a posteriori reasoning on students' behalf when relying solely on CAS in their mathematical work (p. 67). Using video documentation of classes and subsequent interviews, Meagher (2012) found that college level Calculus students had mixed feelings about the powerful capabilities of CAS-enabled technologies being used within their curriculum.

> It is interesting to observe that for many students the reaction to CAS technology is very similar to that of teachers... Some are very positive about CAS but many students, including those who are comfortable with graphing calculator technology, have this sense that CAS is finally technology that does too much mathematics to be appropriate for education... . The results of this study show that students often have a very delicate relationship with technology and while they may develop mastery in using a particular technology they don't see its use as necessarily advantageous. (p. 13)

Other researchers have provided support for CAS-based technologies being effectively used at the university/college (tertiary) level. For example, Albano et al. (2005) studied the use of CAS technology in an undergraduate Statistics course at the University of Salerno and concluded that "the use of CAS for educational purpose can help students to learn how to deal with mathematics and can increase the certainty of their own knowledge and skills" (p. 2), adding that "Computer-based-education is not meant to replace the role of the teacher in the classroom... indeed it is best used in conjunction with traditional techniques" (p. 8).

Lavicza's comprehensive study (2008a, b) featured an online survey of 1100 mathematicians as well as interviews with 22 mathematicians in three countries, namely, Hungary, United Kingdom, and United States, which examined mathematicians' beliefs/conceptions regarding CAS and its instructional potential. Findings showed some similarities, but also notable differences, between university- and school-level research findings (e.g., use of CAS in one's research being the greatest factor influencing the use of CAS in one's teaching). Building on the findings from Lavicza's international work, the team of Jarvis, Buteau, and Lavicza implemented a mixed-methods research study to examine individual and systemic CAS usage in undergraduate mathematics instruction. Based on their findings, they concluded that: Instructor beliefs regarding the nature of mathematics learning, required curriculum/assessment changes, the use of technology in one's own research, and the availability of resources are among the complex set of factors that affect the degree to which technology is implemented within undergraduate university mathematics courses. The Canadian survey of over 300 participating mathematicians clearly indicated that many university instructors were using CAS in their instructional practice (69%), and also reinforced the Lavicza finding that the greatest factor influencing the use of CAS in one's post-secondary mathematics teaching was the use of CAS in one's own research (Jarvis, Lavicza, & Buteau, 2012). Buteau et al. (2010) also reviewed 326 papers relating to CAS use in teaching and noted that "to the extent that CAS was used to enhance conceptual discussions and support visualizations, it was reported as promoting greater understanding of mathematics" (p. 64). The research team concluded that "We believe that mathematicians will, at some future point, need to rethink the tertiary curriculum vis-à vis the growing ubiquity and computational power of technological learning tools" (p. 66). Further, a detailed comparative case study of two university mathematics departments (one in Canada and another in the UK) in which technology had been heavily integrated within the department over time has been documented by Jarvis et al. (2014).

Also at the university level, Sarvari et al. (2010) reported that CAS use in an undergraduate engineering could potentially provide many benefits for student learning:

> CAS can offer effective tools to enhance students' cognitive strategies. In addition, the use of CAS extends the number of the available learning strategies, [and] shifts the pragmatic, heuristic and epistemic values of mathematical schemes. Thus, CAS can be an effective tool to highlight connections between knowledge elements and to develop the knowledge representation network, so that CAS can enhance students' knowledge of mathematics. (p. 153).

For example, Zeynivandnezhad and Bates (2018) studied how CAS technology can be an effective tool for learning differential equations, and Gyöngyösi et al. (2011) designed and studied CAS-based tasks that aimed at supporting students' learning of the convergence of sequences and series as part of a Real Analysis course.

Such a shift must ultimately involve changes to *assessment practices* that reflect the use of powerful CAS-based technologies and are similar to the activities experienced during classes/labs. Pountney et al. (2002) made similar claims about assessment when studying the use of CAS in first year courses at Liverpool John Moores' University.

> The use of computer algebra systems (CAS) in the teaching and learning of mathematics is likely to have a minimal impact on the overall mathematics curriculum until the CAS becomes a legitimate tool to be used in mathematics examinations. The important issue is how the examinations can be written in a way that allows the student the appropriate use of a CAS and tests the students' understanding of the underlying mathematical concepts and problem solving strategies. . . . meaningful examinations can be constructed for testing in

a CAS laboratory or in a situation where every student has access to a CAS. . . . mathematics instruction, which fails to recognize the use of the CAS, stands the risk of becoming redundant. (pp. 15, 34)

Clearly, the use of CAS in undergraduate teaching and learning must encompass a strategy for assessment that combines the power and capabilities of the tools with a maximization of cognitive gains in both student competency and understanding. It is this kind of assessment experimentation in undergraduate CAS-based learning that lies at the heart of this current study.

In the following section we present the theoretical model used to frame our study, namely the radar model for quantifying technology integration in university mathematics courses as proposed by Oates (2009).

3 A Model for Quantifying Technology Integration in University Mathematics Courses

The radar diagram model was developed with an aim to generate "a more definite means of quantifying the degree of integration … for valid comparison between courses" which the author claims is necessary "if any inferences about the importance of technology integration are to be made" (Oates, 2009, p. 181). The Oates model was generated after a literature review and through the use of a survey completed by undergraduate instructors "to determine their use of technology, their views of what an Integrated Technology Mathematics Curriculum (ITMC) may resemble, and how it may be achieved" (Oates, 2009, p. ii). Visualization of the taxonomy with radar diagrams allows for comparison of courses, and easy pinpointing of which aspects of technology engagement within a course can be improved. Most importantly, Oates conclude that the six components of the taxonomy are not independent but rely on each other for "consistent, effective, and sustainable" (p. ii) integration. For our purposes, we chose to use Oates' taxonomy to compare iterations of the same course since it offered an explicit means of measuring and comparing the six components of the model across the iterations.

There are six components to Oates' (2009) taxonomy. Firstly, "Access" which addresses the extent to which students were able to access the technology (e.g., being able to use the software on personal computers versus only in school computer labs). Secondly, "Student Facility" describes how proficient in the technology students are, and what help is available to them. "Assessment" illustrates how technology is (or is not) used for evaluations of student learning. "Pedagogy" addresses how technology is used for learning by the educator (e.g., as a demonstration versus an exploration tool). "Curriculum" quantifies how technology impacts the content and structure of the course. Lastly, "Staff Facility" evaluates how well the instructors are capable of incorporating the technology into their teaching, and the departmental supports that are provided to them as part of this process.

Table 1 Oates' Taxonomy for Integrated Technology featuring six characteristics (Oates, 2009, p. 182).

	Characteristic	Example of question asked to examine the degree of integration for each characteristic
A	Access	To what extent do students have access, e.g. is it compulsory? Do they won their own, or access it in computer labs?
B	Student facility	How proficient are students with the use of the technology, and what assistance is provided to help them?
C	Assessment	Is technology expected and/or permitted in assessment?
D	Pedagogy	How and when do staff and students interact with the technology? For example, is it used mainly as a complex calculation device and demonstration tool, or to develop and explain concepts?
E	Curriculum	Has the course curriculum, for example content, order of teaching, changed to reflect the use of technology?
F	Staff facility	Are staff familiar with the use and capabilities of the technology, both mathematically and pedagogically?

Based on sample questions that were asked in his survey of undergraduate instructors, Table 1 describes each of the six taxonomy components found within the Oates model (Oates, 2009).

Each of the six components can be quantified using levels from 0 to 5. Although the levels are intended to be integer values, in this chapter levels of 0.5 are occasionally used. The taxonomy components and levels were originally created with respect to the use of graphing calculators, whereas this research investigates the engagement of the CAS-based software known as *Sage*. Therefore, sometimes part of a requirement for one level will be met, while another part is not, and in such cases 0.5 levels were used accordingly. Levels were assigned based on Oates (2009, pp. 307–308) quantification chart. A summary of the lowest and highest possible level is given in Table 2.

In Table 3, note that Oates (2009) did not adopt a qualitative approach to describing the Curriculum component, but rather the integration of technology to the Curriculum was considered based on the quantity of changes. We found it difficult to adopt this approach because in the context of this research some Curriculum factors seem to have held different levels of significance yet were given equal weight. For example, allowing technology use in assignments would receive a "point" for Curriculum, but this may be much less significant than an actual change in content due to technology use. For this reason, we have attempted to create a qualitative scale which we used for the analysis of the Curriculum component within this context. However, in creating this modified scale we do recognize that changes to curriculum and their perceived/actual significance will differ depending on which branch of mathematics is being considered, and therefore we recommend that this component should be

Table 2 Summary of the lowest and highest possible levels for the six taxonomy components (Oates, 2009)

Component/Score	0	5
Access	No student access; staff access only	Full access for staff and for students
Student Facility	No help provided to students	Previous experience of students and comprehensive training provided to students
Assessment	Prohibited	Allowed in all assessment
Pedagogy	No specific technology used by staff or expected of students	Expected use by both staff and students; used in class with student interaction
Curriculum	"One point for each factor considered, maximum of five. Consider for example curriculum design, course description, goals, inclusion in course materials, content changed (new, deleted topics), order of topics, changes to assessment" (Oates, 2009, p. 308)	
Staff Facility	No previous experience with software among staff, and no professional development for staff	Previous experience and ongoing training for staff in both pedagogical and mathematical applications

Table 3 The Curriculum component within the context of the Optimization course

Level	Curriculum Component Characteristics
0	• No technology use for learning course content
1	• Technology only used in instruction for demonstration of concepts • No expectation of use by students (e.g., no mention in the syllabus or within assignments)
2	• Technology use facilitates computations or is used for learning a concept (e.g., visualizing) but does not change the course content
3	• Some content is changed to reflect the technology use—topics added or removed due to technology, owing to a topic becoming less relevant (e.g., methods of calculating)
4	• Significant content changed • Something new can be done with technology that previously couldn't be done by hand, or could not be done by hand within a reasonable amount of time
5	• Course re-sequenced based on technology use • More than half of the course content changed due to technology use

reviewed whenever a new mathematics branch is being analyzed. Table 3 describes how we view this Curriculum component in our context of the Optimization course.

"Content changed" may include the depth of the topic being studied, for example, moving from simple examples to real-world applications. Our quantification focuses mainly on the change of course content. Oates (2009) mentioned "course description, goals, inclusion in course materials," but we believe that these components of Curriculum would be a result of a change to course content and therefore do not need specific mention. Additionally, we do not mention "changes to assessment" because these changes would already be reflected in the Assessment component of the model. However, a change to assessment may indicate a parallel change to course content since assignments and tests are often key learning opportunities for students and reflect what the instructor intends for the students to learn.

4 Methods

The case study research project involved 14 interviews with the following participants: Dr. Charles Doran, TLEF Project Leader (twice); Dr. Andrey Novoseltsev, the Course Instructor (twice); two other math faculty members; six students; a course grader; and a former UA student who had become a math instructor at another institution. These interviews took place both in Ontario during the Canadian Mathematics Society (CMS) Winter conference in December 2014, and (mostly) during a site visit to the University of Alberta campus in early June 2015. The interview questions were semi-structured (i.e., open-ended) and were designed and implemented according to qualitative case study standards (Creswell, 2013; Denzin & Lincoln, 2005; Yin, 2009). Follow-up email correspondence with Charles and Andrey provided further data for the study. During the site visit, artifacts were also collected regarding course outlines, software applets, course assignments, and assessment tools.

In our first analysis (Jarvis et al., 2018), the interview data was entered into *Atlas.ti* qualitative research software for the purpose of data organization and thematic coding of the interview transcripts. Thematic analysis was used with the data and this involved a process of coding which resulted in seven large code groupings: Assessment, Instructor, Learning, Optimization, Sage Software, Technology, and the TLEF Project. In this second analysis, interview data and Optimization course material (tests, syllabus, etc.) was analyzed qualitatively using Oates (2009) quantification system. This model lends itself to our research because it is designed for comparing the use of technology over time and the data was collected over the course of seven iterations of the course. During our analysis, specific focus was given to the interviews with Dr. Novoseltsev who gave description of each iteration of the course (some iterations having occurred before our case study visit and some occurring subsequent to our campus visit). This analysis resulted in six radar diagrams which visualize the changes to the course with respect to technology integration over the seven iterations of Optimization.

While the actual names of Drs. Doran and Novoseltsev are being used directly in this chapter, by permission, the names of students that were interviewed during the site visit shall be replaced with the following alphabetical order pseudonyms, as per the letter of informed consent: Akemi, Brittany, Cheng, Dawn, Ezra, Felix, and Guang. This study was approved by Nipissing University Research Ethics Board in Ontario and was similarly approved by the University of Alberta, the host institution. In the next section, we first provide some context from the case study by presenting selected relevant results from the first analysis (Jarvis et al., 2018). We then further analyze our case study data by quantifying the degree of technology integration in the Optimization course as it evolved over time, and identifying key aspects that seem to have significantly promoted the furthering of the integration throughout the course iterations.

5 Research Findings

5.1 Initial Findings Elaborating on the Context of Our Case Study

A full reporting of the initial findings of the research study, in terms of emergent themes and related discussion, can be read in the journal article entitled *Innovative CAS Technology Use in University Mathematics Teaching and Assessment: Findings from a Case Study in Alberta, Canada* (Jarvis et al., 2018).

5.1.1 TLEF Project and McCalla Professor in Science Chairship

Dr. Charles Doran, professor in the Department of Mathematical and Statistical Sciences at the University of Alberta (UA) and Site Director of the Pacific Institute for the Mathematical Sciences (PIMS), received internal funding by way of the *Teaching and Learning Enhancement Fund* (TLEF). In conjunction with the TLEF funding (2013–16), he was also named to the position of McCalla Professor of Science Chair. As part of this latter recognition, Doran had created an integrated teaching and research plan which involved the writing and delivery of a new upper level, joint (graduate and advanced undergraduate) computing and mathematics course entitled *Computing in Mathematics: Research* via *Experimentation* (MATH 497). Project funding also allowed him to focus research on an existing third year *Mathematical Programming and Optimization* (MATH 373) course in which *SageMath* (The Sage Developers, 2015), an open source mathematics software program, was being used by a Post-Doctoral Fellow, Dr. Andrey Novoseltsev, in new and creative ways in terms of mathematics teaching and assessment. A case study by Jarvis and Buteau was conducted at the University of Alberta, the results of which formed the basis of the 2018 paper, with a particular focus on what was being undertaken by Andrey in the MATH 373 course.

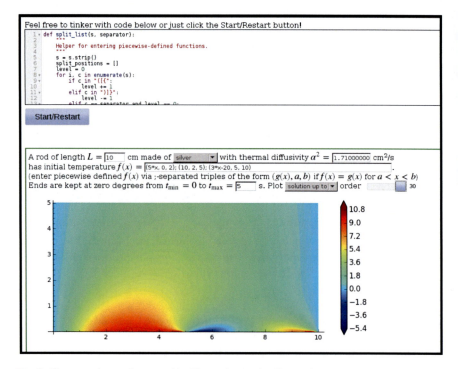

Fig. 3 Heat equation applet created by Novoseltsev using Sage software

5.1.2 SageMath Software

When asked why they chose to use open source Sage[1] software and to continue to develop their own math apps using Sage, rather than using other available commercial software and apps, Dr. Doran explained the following:

> Sage has the advantage of moving us away from proprietary systems ... The good is that we can start fresh and build from the ground up so that we can control everything about the look, and the feel, and the capacity. ... Mostly, I just think the price is right, the community is like-minded individuals who want to make everything better.

Andrey produced approximately 12 applets (i.e., very small applications, often utility programs that perform one or a few simple functions) for different mathematics courses (see Fig. 3), which represented one of their important goals for the TLEF project.

---

[1] Created by Dr. William Stein, *SageMath* (http://www.Sagemath.org/) originated at the University of Washington, but now represents an international project with many developers in dozens of different countries. *SageMath* is a freely available, open-source mathematics software system licensed under the General Public License (GPL). Since participants in this study, including the instructors that were interviewed, commonly refer to the software as simply Sage, we have used this shorter title throughout the paper for consistency.

Reflecting on the choice of Sage software, Dr. Novoseltsev explained his rationale: "I like the fact that *SageMath* software is free for students, and that they can also access it from home.... I strongly believe in the open source approach for mathematics software... you have the option to look at the code and to fix the bugs—try to make improvements." Once Sage was established as the software of preference for the TLEF initiatives, specific plans began on how to incorporate it into the new *Research via Experimentation* course and into the existing *Mathematical Optimization and Linear Programming* course in terms of both potential instructor and student use of the open source software.

5.1.3 Mathematical Programming and Optimization (MATH 373) Course

The *Mathematical Programming and Optimization* (MATH 373) course had already existed within the department and had been taught a number of times by Andrey (2011–13) before the TLEF funding began to be used for this initiative in 2014. For sake of context, what follows is the Course Description from the Andrey's syllabus from the Spring 2014 term.

> MATH 373: Introduction to optimization. Problem formulation. Linear programming. The simplex method and its variants (revised simplex method, dual simplex method). Complementary slackness and duality. Extreme points of polyhedral sets. Theory of linear inequalities (Farkas Lemma). Post-optimality analysis. Interior point methods. Applications (elementary games, transportation problems, networks, etc.).

Pre-requisites for this course were listed as first year linear algebra course, as well as any 200-level math course. In other words, this third-year level course required students to have some background in mathematics, but not any programming experience. Course objectives and expected learning outcomes for MATH 373 were described in the syllabus as follows:

> After taking this course, you should be able to formulate a linear programming problem and convert it to the standard form(s); understand the structure of dictionaries of the Simplex Method and their relation to the original problem; perform steps of the Simplex Method (and its variants) and understand why they lead to the solution; use relations between dual problems to efficiently verify optimality of solutions and to construct certificates of solutions; detect inconsistent and redundant inequalities in a system; modify optimal solutions to take into account changes in constraints and objectives. You will also develop a general sense of what optimization problems are, see "linear algebra in action," and pick up basics of using math software and typesetting mathematical expressions in LaTeX.

In interviewing Charles, Andrey, other mathematics instructors, and mathematics students at the University of Alberta, we quickly became convinced that our primary focus would be on the MATH 373 course and how Andrey had developed strategies involving technology for teaching and assessment through seven different iterations of this course. Follow-up interviews allowed us to inquire into how the MATH 373 course had changed in further offerings of the course subsequent to the TLEF funding window and to our campus site visit in Alberta.

5.1.4 Overview of Sage Software Integration Within the Optimization Course Over Time

As shown below, Table 4 provides a summary overview of the advancement of technology-related strategies and tools that the instructor, Dr. Andrey Novoseltsev, had introduced throughout the teaching of seven distinct installments of the *Mathematical Programming and Optimization* (MATH 373) course over time.

For a more elaborate recounting of the technology development within the seven iterations of the course, the reader is directed to our first paper entitled *Innovative CAS Technology Use in University Mathematics Teaching and Assessment: Findings from a Case Study in Alberta, Canada* (Jarvis et al., 2018). Examples of two MATH 373 assignments (converting a student-generated word problem into a linear programming problem, and then solving this problem using the Simplex Method) and a mid-term examination, all of which involved CAS technology components using Sage, are provided at the end of this chapter in Appendices 1–3.

5.2 Data Analysis Using the Oates Taxonomy Model

The analysis mainly involved data from the interviews with Dr. Novoseltsev who gave descriptions of each iteration of the course. It resulted in six radar diagrams which serve to visualize the changes made to the course with respect to technology integration over the seven iterations. There is limited data for some sections of the results because the data was not collected with this taxonomy in mind. The following six radar representations (Fig. 4) visually summarize the evolving degree of technology integration throughout the seven iterations of the MATH 373 class.

The area on each radar reflects the level of engagement for that iteration. Over the seven iterations several changes were made to the course that impacted how technology was used in the course. All of the changes improved technology integration, as can be seen as each radar diagram increases in area. The largest changes occurred in Access, Student Facility, and Assessment. These changes were made by the instructor for a variety of reasons, including student feedback and his own ongoing observations. Figure 5 provides another visual summary, in line graph form, of the changes taking place during the seven iterations of the MATH 373 Optimization course over time, in terms of the six integration components.

The most limited change was noted in the component of Staff Facility. In other words, although Novoseltsev made ongoing and substantive changes to his use of CAS technology in both his teaching and assessment during this period of several years, most of his math department colleagues demonstrated relatively little interest in this CAS-based approach.

Table 4 Overview of how Sage software was incorporated into the MATH 373 course over time

Installment	Sage Technology Used in Instruction	Sage Technology Used in Assessment
1. Fall Term 2011	• Wrote simple code to check his own lengthy calculations during the course • Minimal use for class demonstrations • Gave students access to it; very few used it	• Used applets to prevent exam question calculation errors in marking written papers • Grader used to mark student work
2. Fall Term 2012	• Sage applets written for most topics • Demonstration: Frequently used Sage in class lectures to incorporate student suggestions and to provide immediate feedback on overhead screen	• Sage worksheets with additional related commentary created and made available to students via the university LMS for completing assignments and for review • No grader used to mark student work; cheating noticed on submitted assignments
3. Spring Term 2013	• Student-generated linear programming problems are required, and are to be used throughout the course in various modules • Sage used to explore these problems	• Students read/respond to fellow student problems online in assigned peer groups • Sage optional for assignments (could also do by hand, or using other CAS software)
4. Fall Term 2013	• Sage moved to online platform with secure passcodes making it more accessible • Sage now required for some assignments	• Sage required for three course tests, which led to implementation of strategies to prevent cheating during test writing in computer lab • Required Sage use on final exam in lab
5. Spring Term 2014	• Extra, optional computer lab tutorials offered to small groups of students to familiarize them with Sage and LMS • Created a step-by-step video for this also	• Three term tests with Sage replaced with one mid-term examination using Sage • Mid-term/Final exams feature hand-written, optional Sage, required Sage use questions
6. Spring Term 2015	• Created more and improved online videos for LMS and Sage technical instructions	• Short exercises for students to explore included in his LMS posted lecture notes • Mid-term/Final exams feature hand-written, optional Sage, required Sage use questions
7. Fall Term 2015	• Created additional applets for demonstration and teaching purposes	• Mathematicians at other universities become interested in Simplex Method applet • Mid-term/Final exams feature hand-written, optional Sage, required Sage use questions

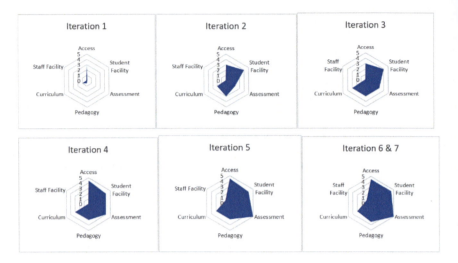

Fig. 4 Radar model representations of the seven iterations of the Optimization (Math 373) course

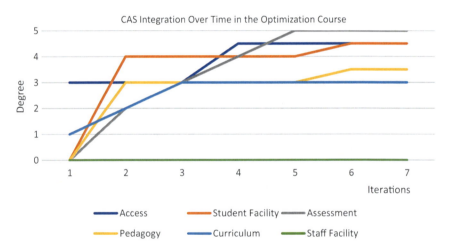

Fig. 5 Course iterations and corresponding Oates integration model component changes

5.2.1 Access

From the first iteration of the course, a free access to the technology was provided to the students by way of a username and password to the online system. Many undergraduate students owned personal laptops, and all had access to university computer labs. Therefore, students' ability to access the online software was not an obstacle to the technology integration and did not change significantly over time.

However, the frequency with which students accessed the Sage software did change noticeably.

In the first iteration of the course, students had access to the software system (Sage) that was being used by the instructor, but the students appeared to have little or no motivation to access it. Dr. Novoseltsev recognized this and reasoned that "on the one hand, once again it was usually done when we need it, and on the other hand, it was not much useful for them, and I didn't provide much help with it. So, I don't think many students used it" (Participant 2(Interview 1), page 1 of transcript; henceforth abbreviated to simply (P2(1), p. 1)). Similarly, in the second iteration the students were able to access the system and even look at, or edit, worksheets that were used for demonstration in class, but they were not required to do anything further with the system. Once more, there was limited use of the technology by the students. There was little change in the third iteration. The instructor described how he used the system for demonstration and mentioned that "after the class, I would post the version that I already had with some mathematical comments and formulas and so on" (P2(1), p. 3).

Iterations 1 through 3 are all represented by level three on the radar diagram because although students could access the system it was not mandatory for them, and they did not tend to access Sage during or following their math classes. In iteration 4, the level on the radar diagram changes to 4.5 because the students had almost full access, but classes were still not taught in a lab. So, while the instructor did use the system during class, the students did not access the system during the lectures. However, outside of class students only needed internet access to connect to the system from any computer. Iterations 5 through 7 remain at 4.5 because there is no apparent change to the factor of Student Access.

5.2.2 Student Facility

Over the seven iterations of the course, the student facility component showed one of the largest changes. Starting from a situation where no help was being offered, it developed into the facilitation of early tutorials, and then eventually online video tutorials that the students could access at any time.

In the first iteration of the course students were not expected to use Sage for their own use and therefore were not provided with assistance from the instructor. Dr. Novoseltsev mentioned that he was unsure how many students used it, but noted that "in any case, it wasn't my intention" (P2(1), p. 1). In the second iteration of the course, the component of Student Facility jumps to Level 4 because although student use was still not mandatory, it was supported for any student who was interested. The instructor provided additional commentary on worksheets that were used in class and then posted them for his students so that the mathematics behind the technology was explicit to them. Additionally, the instructor "offered a few labs for those who wanted to come" (P2(1), p. 2) so that students could be comfortable with the use of Sage. However, since use of Sage was not compulsory, he reported not "try[ing] to make sure that everybody knew how to do it" (P2(1), p. 2). There was no change to Student Facility during the third and fourth iterations of the course.

The fifth iteration of the course had a larger enrollment, which motivated the creation of video tutorials since it was no longer feasible to run in-lab tutorials at the beginning of the semester. The instructor described that he had "recorded step-by-step instructions, and also had the text instructions ... it just was a screencast showing how to do the commands and so on" (P2(1), p. 8). These videos were then uploaded online for the students to access whenever necessary. Before the sixth iteration of the course began, the instructor updated and improved the screencasts in order to "introduce them to the technical steps" (P2(2), p. 1). Students who were interviewed during the sixth iteration of the course shared specific insights about Student Facility. One student spoke to how the instructor helped them understand the use of Sage:

He actually spends a really big chunk of time on Sage, showing us how to—not just the commands, but what they mean when you're getting the output.... It wasn't as thorough for some of the other classes, [where] you sort of had to learn more on your own, and then if you had questions, then you would have to go and see the professor separately, whereas [this instructor] focuses on it in class so we don't really need that extra help. (P6, p.1).

The same student also described how helpful the video tutorials were in reviewing the new mathematics learning:

[The instructor] has really detailed instructional videos that he personally uploads on eClass, and those were really helpful because you can go back—he narrates them, and he also has the typing thing that shows what he's typing. So, it's very thorough. You can't really mess up. (P6, p.2).

Another student had a different experience and found using Sage for the first time somewhat difficult, even with the videos: "It's a little bit hard to get into the program. Like at the beginning, even my teacher gave me instructions, he has a video, on typing code, but sometimes you just feel like you've never used this technology before, so you're not really sure how to do next" (P7, p.1). Student Facility seemed to have improved since the previous iteration, especially with some in-class focus on Sage in addition to the online tutorials. However, being at a Level 5 requires students to have either "considerable previous experience and/or comprehensive training provided" (Oates, 2009), which did not seem to be the case for all students, therefore iteration 6 was given a level 4.5. There was no change to Student Facility component in iteration 7.

5.2.3 Assessment

Like Student Facility, the Assessment component represented an area of significant change during the seven iterations. However, unlike Student Facility, which jumped from 0 to 4 in one iteration, technology allowance was progressively increased in Assessment over the first five iterations of the course.

The first iteration of the course was run as a traditional class, with no technology allowed in any assessment and therefore was given a Level 0. However, the instructor did use Sage to create written assessments and to check his own solutions, which, one might argue, is an important factor in beginning to introduce students to the idea of using Sage. In iteration 2, the instructor introduced Sage to the students as a tool that they could use for their assignments but only if they wished to do so. However, the assignments were still designed to be done with pencil and paper. He noted: "It was not mandatory, in any way, and the assignment that I have used was still numerical problems which were, in principle, meant to be done by hand. And I said that, 'You can do it by hand, you can do it using Sage, if you wish'" (P2(1), p.2). Therefore, iteration 2 is given a Level 2 since "All assessment questions written so no supposed advantage from student access to technology" (Oates, 2009).

Technology use for assignments in iteration 3 became mandatory. In particular, the assignments required students to design their own problems which they would then solve using technology. Students were required to use technology by the assignment specifications, but in addition, the questions had to be made "to not be doable by hand" (P2(1), p.3). However, the instructor did "not force them to use Sage" but "let them use anything else that they were more comfortable with, like Maple, or MATLAB, or something else" (P2(1), p. 4). This iteration is best represented by a Level 3 because there was a specific technology component (i.e., the assignments), but there was also a specific component where technology was prohibited (i.e., the tests/exams). While students were allowed and encouraged to use Sage for their assignments in the third iteration, the instructor noted: "Throughout the course, nobody tried using the Sage, so for the fourth time, I removed these possibilities, and I said that, 'You have to use Sage'" (P2(1), p. 4). Requiring all students to use Sage for assignments allowed for an additional and significant change—the introduction of Sage use on tests. The reason for doing this, according to the instructor was that "One of the complaints before was that, 'We're using all this stuff, and we're using it to solve problems with Sage, but then on exams we don't use this knowledge. So, maybe on assignments we should train the skills that would help us to do well on the exams'" (P2(1), p. 5). Note that this point, which was brought up originally by a student, was recognized and reported by the instructor and thus demonstrates an important pedagogical insight and also provides evidence of his own growth as an instructor. This was the first time the instructor included a technology component in one of three tests during the semester. He set up the test with two problems so that, "One problem was material for which a computer was not particularly useful, so you really need to write your own arguments; and one was a problem where the computer was extremely useful, and I don't think that problem was doable by hand much at all" (P2(1), p. 5).

Similarly, the final exam was composed half of problems that would be done using computers and Sage, and the other half of problems that would need to be done by hand (i.e., the computer would not add value to solving the problem, although they could still be accessed). These changes increased the level of engagement during assessment to Level 4, since there were still two technology-free tests.

Iteration 5 had a larger enrolment and therefore, for logistical reasons, the three tests became one mid-term exam. The mid-term included technology and therefore there was no longer any technology-free assessment in the course. This raised the Assessment component to Level 5 in the radar model. Although for some questions on the exams, Sage might not be considered useful, students were still allowed to use technology "at any appropriate time in all assessment" (Oates, 2009). Similarly, iterations 6 and 7 allowed for Sage use in all assessment and therefore remained at a Level 5.

5.2.4 Pedagogy

Throughout the seven iterations of the course, the component of Pedagogy changes moderately, but technology did not appear, from that perspective, to ever become fully integrated.

In the first iteration of the course the instructor personally used technology for creating problems, but "didn't use it in demonstrations as much" (P2(2), p. 1). It was used in-class a few times for reviewing "material from the previous chapters" (P2(2), p. 1), but it was not intended for student use, and the instructor did not demonstrate how to use it. Therefore, the first iteration was given a Level 0.5; there was "no specific use required or expected of students", but there was "reference to specific examples in course-notes" (Oates, 2009). The second iteration of the course jumped to a Level 3. Due to the instructor's personal use of Sage in the first iteration, he was able to extend the affordance of using it to his students, and he himself used it in his teaching. He started to realize the potential of using Sage more with his students and described this increased use of Sage in his class:

> So, what I was doing was preparing parts... before we needed them in class, and then I would show the computations in class, on the one hand, much faster than if I was doing it by hand, or on the board; on the other hand, what I really liked, I was able to get input from students—asked them, "What do we do next?" And I could follow their suggestions, even if they were wrong. If their suggestions were wrong, and you spend time doing it correctly, the algorithm would show you that the end result will show you that you have done something wrong.... So, on the one hand, I do want to follow their suggestions; on the other hand, I don't want to waste time doing computations with them—with the computer it was instantaneous, so I did not worry about it. (P2(2), p. 1).

Here the instructor is describing a Level 3 pedagogy which Oates (2009) qualified as "Staff model for specific examples, but not used all the time, student use follows for specific examples or assessment component" (p. 308). There was no change in pedagogy from iterations 3 to 5.

The sixth iteration was given a Level 3.5 because the instructor used technology whenever appropriate and modeled it during the lectures. A student described his appreciation for the instructor's approach to teaching using technology:

> He actually did a sample question and showed how to do it, and he does not have to do that, but he did that. And dealing with technology difficulties, he is always answering questions and always getting questions. .... It's something I actually appreciate because he actually engages. So, it's not just, "Here's a program, go figure it out." He actually kind of walks you through it and makes sure you understand. (P6, p. 6).

However, interestingly enough, students noted that "he doesn't allow technology in class—no phones or laptops" (P6, p.6) which obviously prevented students from simultaneously participating in Sage use, which is a requirement of Level 4 in the radar model. The component of Pedagogy remained the same for iteration 7.

5.2.5 Curriculum

The Curriculum component for the Optimization course did change based on technology use but was ultimately limited by the students' technology abilities—the content that could be explored with technology required programming beyond what is realistic for the students to learn within the math course. The main change was articulated by the Chair who noted: "The kind of questions that one could ask in an optimization course and expect the students to work out either on homework or the exam went from being atypical for applications of optimization in the real-world to being typical" (P1(1), p. 4).

In the first iteration of the course, the instructor used technology only as a demonstration tool in lectures when reviewing previously covered material, but did not use it as a way to teach a new concept. Additionally, although students had access to the technology that he was using, he did not expect them to use it personally and said that "it wasn't [his] intention" (P2(1), p.1). Since there was some technology use by the instructor in class, but no change to content or student use, the first iteration had the Curriculum component represented by a Level 1. In the second iteration, the Curriculum component increases to a Level 2 as technology was incorporated into assignments. However, use of technology was not required, and the assignments were "still numerical problems which were, in principle, meant to be done by hand" (P2(1), p. 2), so the content itself was not significantly changed. In the third iteration, course content was further impacted by technology, and therefore the Curriculum component was given a Level 3. Again, the change to course content was reflected by the assignments in which "students had to each come up with their own word problem at the beginning of the course that would be convertible to a linear programming problem. There was a requirement to make the problem big enough... to not be doable by hand" (P2(1), p. 3), so they had to use some sort of technology. This requirement reflected the change that the Chair discussed about the course moving from simplified problems to more realistic problems.

In the fourth iteration, Sage became a required element and there was technology use in a test and during the final exam, but there was no further change to course content and therefore the Curriculum component remains at a Level 3. When discussing the fourth iteration of the course, the instructor elaborated more about how technology was being used for learning. In short, it was used for saving time in calculations enabling more difficult problems to be tackled or allowing previous problems to be covered in more depth. He says that before using technology he was "giving them a very simple problem with simple steps, so they could do it, or give them a little bit more complicated problem, but ask them to do just one step, or just, say, two steps" (P2(1), p. 8). However, by the fourth iteration, the questions had changed to "a bigger problem with some random coefficients, so something you couldn't do by hand, and the task was to solve it completely which usually required five or maybe even a bit more steps—and then they always had to do it to the end" (P2(1), p.8). In the instructor's mind, the idea of the questions appears to have remained the same; before technology use, students were still asked to understand the concept but to not compute it, now they were able to also fully compute it. The instructor noted that "nothing was changed officially" (P2(1), p. 10), but described how the ability to fully calculate problems did change. He noted, "I hope that I can raise the value of the method that I am using with students, and I certainly would not be able to do this computation by hand—it's crucial that we have the power of the computer" (P2(1), p. 10), showing how the content did change because of the efficiency that using technology use allowed.

In the subsequent three iterations of the course no changes were made to the course content in terms of technology use. However, it is interesting to note that the instructor did appear to have a vision for further integration of technology into the curriculum, but limited primarily by a perceived deficiency in student facility and prior knowledge. In particular, Dr. Novoseltsev described the Revised Simplex Method as follows:

> It is unquestionably more intricate, and more difficult for the students to understand and apply it.... The point of it is, of course, it is more of an efficient method, but you need to work with sparse matrices. So, on the one hand, I think the discussion of sparse matrices has to be included, and of course it does not really make sense to talk much about it unless you have some programming experience, I mean, mathematically it is the same matrix—rectangular with a bunch of zero entries, but on the programming side, it means that you can represent it in a different way... and it can be more efficient.... So, if they had some basics in understanding even what is a computer variable—what is the difference between that and a mathematical variable—that would be already useful. If they had more understanding about data structures, arrays, and separations, it would be even better. (P2(1), p. 11).

The instructor described a Level 4 or 5 Curriculum integration vis-à-vis technology integration, but recognized that curriculum can only change to the extent that students are able to understand and use technology. Therefore, because there was no pre-requisite for programming, there was a limit on how much the curriculum in this course could actually change.

5.2.6 Staff Facility

Staff Facility, across the department, was lacking throughout all iterations of the course, as the changes that were made to the course were primarily due to the instructor's own expertise and interest, and not because of any departmental motivation.

From iterations 1 through 4 the instructor was using technology according to his own preferences and based on his own related skills. In the interview he was asked: "Were you working with a technician to set that up, or did you just do it yourself because you have the understanding of that?" (P2(1), p. 5), and replied: "Yes, I did that myself. It's just a standard thing, so it's not like I have written something to do that, you just need a program that will show you this" (P2(1), p. 5). There was no evidence of "assistance or professional development for staff," and it appeared to be "largely domain-specific computational use where staff facility is assumed by virtue of their mathematical expertise" (Oates, 2009, p. 308).

Starting in the fifth iteration there was some departmental support, with the Chair noting the following:

> We have the undergraduate Associate Chair supporting us in this—actually sending out emails formally to the course heads for the courses that [the instructor] is working on, saying, "Look, you know, there's a TLEF [Teaching and Learning Enhancement Fund] project and we'd like to have you work on—set up a meeting, if there's anything I can do to help?" (P1(1), p.1).

However, the Chair felt that there was not general support from other staff members, and that even if the instructor of MATH 373 showed what could be done with Sage there were staff members who would not want to integrate technology into their own courses. Specifically, Dr. Doran described the possible related difficulties:

> We are encountering, I wouldn't say resistance, but the complexities of integrated technology in multi-session classes. As you might expect, there are some professors who are simply uncomfortable with technology. There are others who are happy to use it if it's painless and easy. We might have to have something where [the instructor] goes to the other sections and takes over the class for a day. We haven't done that yet, in any organized fashion, but I'm not sure. I'm not sure how we're going to disseminate, through the other sections, if we meet serious resistance. (P1(1), p.4).

Therefore, Iteration 5 was given a Level 1. There is some "word-of-mouth, collegial support,... but no obvious systems for training" (Oates, 2009, p. 308).

Regarding the sixth iteration of the course, the instructor described his own doubts about being able to use technology in a multi-section, multi-instructor course which involved the perceived lack of departmental support. Even with his then recent success in using Sage in a single section course he noted, "For multi-section courses, I did not try to use computers in any way, and I frankly just don't see how it could be possible, for final exams or for any other shared assessment, with more than one section [i.e., with more than one instructor]" (P2(2), p. 6). Therefore, iteration 6 saw Staff Facility remain at Level 1, and there was no apparent change to this component in iteration 7.

6 Discussion

In this section, we will discuss some of the broader issues around technology in mathematics teaching and assessment; how these novel uses of Sage software were shared with other faculty; and some suggestions for related future research. We also discuss the findings relating to the Oates taxonomy model analysis in terms of if and how the various factors changed throughout the seven course iterations, as well as highlighting potential relationships between the factors.

6.1 Technology in Mathematics Teaching, Learning, and Assessment

Clearly there is still no international consensus around the role and effectiveness of instructional technology in the learning of mathematics, particularly with regard to powerful CAS-based tools such as Sage. One of the most interesting aspects regarding the role of technology in mathematics learning is whether or not, and if so in what specific ways, the use of the technology directly impacts the understanding of existing mathematics curricula content. For some mathematicians, they appear to equate "true understanding" exclusively with the learning of traditional algorithms and hand-written solutions and proofs. To them, technology may be a helpful way to reinforce this understanding through answer checking, demonstration, or even exploration, but it is perceived as supplementary. For example, note how Zack, who had taught the Optimization course once previously, and who had included Maple software explorations in his lectures and course assignments, praises the use of powerful digital tools, yet distinguishes these from real "understanding":

> **Zack:** What's nice about using technology is that you can ask them some interesting and very difficult computational problems that you couldn't do in a reasonable amount of time with paper-and-pencil. . . . Students should learn about the technology that's available. . . . that's why this technology is fantastic. Why shut it away? They should learn how to use it, but it shouldn't be used as a replacement for understanding. . . . I think that most professors encourage students to use the technology as a checking tool and as a confidence builder, so that what you're handing in is actually correct, but it shouldn't be used to replace your understanding of the material.

In contrast, other mathematicians and educational researchers are of the opinion that the use of CAS technology not only is intrinsically tied to the understanding of existing mathematics curricula content, but that it in fact can open up whole new ways of thinking about content that were previously inaccessible, hence actually serving to expand the very possibilities of mathematical knowledge and understanding. Verillon and Rabardel's theory of *Instrumental Genesis* (Trouche, 2004) speaks to this idea of how the development of ICT, together with its usage, has led to both "instrumentation," meaning a person is able to use the instrument, and also "instrumentalisation," meaning that the tool actually shapes the actions and the character of the knowledge constructed with the tool (Haapasalo, 2013, p. 87).

Beyond just performing quicker calculations, Tobin and Weiss (2016) likewise concluded that handheld CAS-enabled calculators actually allow for expanded ways of exploring and thinking about mathematical phenomena:

> The best way to use this technology would be to write a curriculum around [the] expectation that technology like it will always be available in [the] future and that the focus on learning should shift to contexts, to applications, to learning of concepts that transfer well. It seems that we are far from being able to do this at present. (p. 40)

Dr. Novoseltsev had made significant revisions to the *Mathematical Programming and Optimization* (MATH 373) course in terms of how Sage software had been used in both his teaching style and assessment practices. Dr. Doran, the TLEF Project leader, reflected on the gradual and purposeful changes that colleague Andrey had made to the MATH 373 course:

> This is a very unique situation because Optimization is a course which has always suffered by the fact that you cannot do real world style examples by hand, you just can't. By its very nature, it involves systems of integral equations that cannot be solved in your head, and you can illustrate on the board with so few variables that it seems meaningless. On the other hand, what Andrey realized right away was that by using Sage, and also by integrating both the algebraic manipulation capacity of Sage, and the visual capacity, the actual ability to show the graphs, to show the boundary lines for the various Optimization constraints, that you could really take it to the next level. . . . So, that's what I think he did at first . . . then he created something that they could do for their homework. And then he began to realize there was a disconnect between the course that the students experienced in the classroom and on homework, and their exams. They went from having this marvelous tool that they knew how to use to having to sit there in an exam room and do everything by hand, and it just didn't make any sense. So, then he took the great leap of just trying to make the entire course Sage-based, and now it is.

Andrey had unmistakably transformed the Optimization course over time and through multiple iterations of the course, in terms of pedagogical and assessment strategies involving the purposeful use of Sage software. You can see this first affecting the way applets were used for illustrating and co-developing (with student input) key concepts in the classroom setting; then in the impact on the way the students worked with interactive applets for homework and course assignments; and, finally, through the inclusion of required Sage technology use within formal course tests and exams. One could easily argue that he had thus demonstrated all four levels of Puentedura's (2006) Substitution, Augmentation, Modification, and Redefinition (SAMR) model. Based on our relatively small participant sample, Andrey's supervisor and his students were appreciative of his math knowledge, classroom methods, supportive attitude, and novel assessment practices. To the external observer, he had also thus demonstrated key characteristics of Koehler and Mishra's (2009) Technology, Pedagogy, and Content Knowledge (TPACK) model insofar as his reflective teaching practice provided ample evidence of these qualities.

6.2 Observations from the Oates Model Analysis

Oates (2011) argued that, "The effective integration of technology into the teaching and learning of mathematics remains one of the critical challenges facing contemporary tertiary mathematics" (p. 709). The survey on which his taxonomy was based drew upon the input of 56 colleagues from international tertiary institutions involved in the teaching of undergraduate mathematics. In conclusion, he highlighted the urgent need to revisit curricular content and assessment practices:

> With respect to assessment, both pedagogical consistency, and the impact of CAS on examination questions, are seen as particularly significant issues. . . . For content, the findings reported here support the complexity of assessing the values of topics, and support the overall conclusion that a re-examination of the changing pragmatic and epistemic values of

specific topics, and the goals of mathematics education, within a rapidly evolving technological environment, remains a pressing challenge for undergraduate mathematics educators. (Oates, 2011, p. 720)

In reviewing the changes that were evident in the various iterations of the Math 373 course, the radar model (Fig. 4) assisted us in forming a number of observational opinions regarding the various model components. For example, in Iteration 2, Student Facility jumped from Level 0 to 4 with the instructor clearly expecting more from his students by way of assignments since they had experienced increasing amounts of Sage-related instruction and assistance. It would be unreasonable to make technology use mandatory (as happened in Iteration 3 and 4) without some acceptable level of Student Facility with the software. This also affected the Access component because without Student Facility (i.e., ability to confidently use the software) students did not regularly access the technology even though it was available to them. Student Facility was also limited by certain changes made to the Curriculum because some of the concepts that could be explored with technology required a higher level of computer programming than was possible for some students in the class.

Making Sage use mandatory in Iteration 4, represented a key moment because it allowed for consistency in technology use across almost all components of the Oates (2009) model. Sage was used in class demonstrations (Pedagogy), the software was made available both onsite and off-site (Access), students were all taught to use the same technology skills and therefore experienced more of a level playing field regarding Sage competency (Student Facility), students were expected to use Sage in tests and exams (Assessment), and this resulted in the ability of the instructor to alter the course content in meaningful ways with justification (Curriculum). As noted above, only the Staff Facility (i.e., beyond the instructor himself) component of the model was left generally unaffected during the seven iterations of the Optimization (Math 373) course.

One component from among the six in the model cannot easily be labeled as the most important in terms of increasing technology engagement. Oates (2009) concluded that all the components were interdependent for "consistent, effective, and sustainable" integration. Notwithstanding, some components appear to have had more of a significant impact. For example, in the Access component, having increased access without Student Facility is relatively ineffective. Curriculum change without Assessment change was also shown to be inconsistent and easily recognizable among both instructor and students. Finally, positive technology integration across components without increases in Staff Facility is ultimately unsustainable.

6.3 Future Recommendations

Dr. Novoseltsev had clearly taken bold new steps regarding the implementation of technology within his curriculum and assessment planning at the post-secondary level (Jarvis et al., 2018). While such pedagogical experiments were fraught with both first-

and second-order barriers, Andrey had indeed provided an informative example of perseverance and progress. More specifically, the adoption of personalized, revisited, and highly engaging optimization problems, along with carefully developed summative assessment tools (mid-terms/finals) that required CAS-based technology use in ways parallel to those with which his students were familiar, together demonstrate a level of sophistication and accomplishment well worth sharing.

The Oates taxonomy (2009) proved to be a very helpful model for analyzing the key components that affect the implementation of instructional technologies such as CAS-based tools within university mathematics curriculum. Significant changes in Pedagogy, Access, Curriculum, and Student Facility allowed for meaningful parallel modifications to Assessment practices (e.g., CAS use on tests and final exams). These practices were interpreted by both the instructor and his students as being relevant, timely, and justifiable in light of the independent and group learning experiences that had preceded them. In dealing with pragmatic issues involving server hosting, software uses/limitations, classroom/lab logistics, student plagiarism, plan B preparation, and tiered assessment, Dr. Novoseltsev provided meaningful guidance, by way of example, to all mathematics faculty who are willing to revisit their undergraduate mathematics teaching and assessment practices. Future research regarding the effectiveness of the Oates taxonomy in analyzing technology integration in other types of undergraduate mathematics courses, and with a number of different instructors would be beneficial, as would a closer look at departmental adoption of CAS-based assessment practices specifically in light of access and teacher belief considerations.

New and innovative working spaces that incorporate powerful instructional technologies, such as CAS-based software applications, within post-secondary mathematics learning environments require both a shift in curricular scope *and* assessment practices. Beyond mere visualization and demonstration uses, these accessible technologies can now allow university and college level instructors to rethink the types of mathematical explorations that are indeed possible for their students. Furthermore, a parallel (and sometimes painful) shift in assessment practices, such as those bravely attempted and shared by Novoseltsev, must ultimately accompany such a significant reconceptualization of mathematical inquiry.

Acknowledgements We gratefully acknowledge that this research was made possible through internal research funding support offered by the Offices of Research Services at Nipissing University, Brock University, and the University of Alberta.

Appendix 1: MATH 373 Assignment 2 (Spring 2015)

Chapter 2 Assignment: Problem Formulation

1. **Word Problem:** Compose a word problem. You should start by practicing on some other problems (e.g. from course notes) and you may use them for inspiration, but please formulate your own problem in your own words and with

your own numbers—do not look at other problems while writing yours. Taking "Dog Food" problem and replacing "Dog" with "Cat" throughout the text does NOT count at your own problem! Refrain from looking at others' submissions until you have posted your own. Your word problem should "make sense" to people who have never heard of Math 373 and optimization and be written in proper English, ask someone to proof read it, especially if English is not your native language. Points will be taken off for typos/mistakes/unclear sentences! While it may seem harsh, this is the problem that you will be working on for the rest of the term and that means your group members will have to deal with it as well—be kind to them and write accurately.

2. **Linear Programming Problem:** Formulate an LP problem corresponding to your word problem. Make sure to explicitly describe all involved decision variables: what do they represent and in what units are they measured. Explain the physical meaning of each constraint and how do you derive it (e.g., $3C + B \leq 1500$ means that the total amount of fertilizer used cannot exceed the available amount).

Once you are done, input your problem into Sage.

```
A = ([2, -1], [1, -1], [1, 0],)
b = (-1, 2, 1)
c = (1, 5)
P = InteractiveLPProblem(A, b, c)
P
```

3. **Standard Form:** Convert your problem to standard form. You are free to use Sage to do it automatically in a single step, but please explain in words what has to be done (not "use this command" but rather "multiply the second inequality by -1, replace the third equation with two inequalities, etc."). In standard form your LP problem must involve at least 5 decision variables and at least 4 constraints (not counting sign restrictions or constraints involving a single variable only). There is no upper limit: if you are so adventurous that it gets difficult to enter all the coefficient or display output, talk to me and we'll try to figure out how to deal with such a problem. If your problem has too few variables/constraints, go back to the beginning and make it more interesting! Of course, if you change your word problem, you have to adjust its conversion to an LP problem accordingly. There is no need to keep the "old" problem around. In addition, it would be nice if the numbers of constraints and variables are different for your problem. (This will help you to avoid confusion in duality theory.)

These commands have to return "True":

```
P.n_variables() >= 5
```
```
[r.nonzero_positions() >= 2 for r in P.A().rows()].count(True) >= 4
```

It is not strictly required, but it would be better if this command returns "True" as well:

```
P.n_variables() != P.n_constraints()
```

4. **Feasible Set:** Adjust your problem (both words and formulas!), if necessary, to make sure that the feasible set is non-empty (i.e. the problem is feasible) and has at least 4 vertices, i.e. the following command should give "True":

```
P.feasible_set().n_vertices() >= 4
```

5. **Solution:** Use Sage to find the optimal value and an optimal solution for your problem.

What do these numbers mean in terms of your original word problem?

Appendix 2: MATH 373 Assignment 3 (Spring 2015)

Chapter 3 Assignment: Simplex Method

1. **Word Problem:** Start with the word problem you have composed last time: your submission should include the word problem, description of decision variables, and formulation as an LP problem. (No need to keep derivation of each constraint or explanation of conversion to standard form.) If the initial dictionary of your problem is feasible, tweak the problem (both the word and formula versions so that they continue to match) a little to make the initial dictionary infeasible and force you to go through the auxiliary problem phase! You still should make sure

that your problem has a feasible set with at least four vertices, the number of decision variables is at least five, and the number of constraints involving two or more variables is at least four. Make sure also that your problem is bounded so that you do have the optimal value and at least one optimal solution!

2. **Simplex Method—Feasible Problem:** Use the Simplex Method to find ALL optimal solutions and ALL BASIC optimal solutions of your problem! (If your problem has a lot of basic optimal solutions, find at least 3 of them.) You may want to watch the "Recovering from Wrong Choices" screencast on eClass for how to "fix mistakes." It is OK to use decimal approximations if precise computations look too ugly, but you need to keep at least 5 digits for each number, i.e. you should use RealField(20) or higher. Beware of approximations issues, however! See "Approximation Issues" worksheet. You are NOT allowed to submit work invoking run_simplex_method() command and I do not recommend using it at all until you have solved the problem yourself.

If your solution uses less than 4 iterations of enter-leave-update steps (combined for auxiliary and original problem), go back to your word problem, make it more complicated, and adjust all other steps as necessary. No need to include the old "simple" version in your submission.

3. **Simplex Method—Infeasible Problem:** Take the word problem you have been working on and slightly modify its constraints/parameters in such a way that the problem becomes infeasible (e.g., for Corn and Barley the requirement to grow at least 2000 acres of corn would do the trick.). Provide below your modified word problem and its formulation as a LP problem.

You can quickly check if your modified problem is indeed infeasible via

```
P.is_feasible()
```

Apply the Simplex Method to this problem to prove that it is infeasible. (You cannot just invoke run_simplex_method().)

4. **Degenerate Dictionaries:** Have you encountered any degenerate dictionaries while working on this assignment? If yes, give a clear reference to it. If no, explain whether it is possible for your problem to have degenerate dictionaries.

Appendix 3: MATH 373 Mid-Term Examination (Spring 2015)

Instructions: Points WILL be taken off if you deviate from any of the following instructions:

1. Fill in the information above, including "Desk:" from a plaque along its top side.
2. Authenticate on the lab computer (you are not allowed to use your own device).
3. Start Mozilla Firefox web browser (not Chrome or Internet Explorer).
4. Go to [University of Alberta based url] (https:// is important!)
5. Press F11 to switch to full screen.
6. Log in to your account. You will see no worksheets—do not make any!
7. Once the test starts, make your own single copy of the published test worksheet.
8. You are not allowed to start or use any other program, access any other web site, create any other worksheets, or share/publish your test worksheet.
9. You are not allowed to use any run_... or possible_... commands.

This exam consists of 4 pages (including this title page) with 3 question(s). You can use any calculator without wireless capabilities. TURN OFF AND PUT AWAY ALL OTHER ELECTRONIC DEVICES. You can use one 2-sided sheet of notes in your own handwriting (do not submit it). You may not use any other notes or your own scratch paper. If you run out of space on the problem page, please use the back of the previous problem (which should be conveniently located on your right). Ask for more paper if it is still not enough. You must show your work on the exam paper with explanations in plain English. If a problem asks you to use a specific method, you MUST use this method. You may get zero credit for any other solution, even if it is correct. Each of the 3 questions are worth 10 marks, for a total of 30 possible marks. Good luck!

1. Consider the following LP problem with a "mystery" constraint:

$$\max \quad x_1 + x_2$$
$$3x_1 - x_2 \geq 3$$
$$-x_1 + x_2 \geq 1$$
$$Dx_1 + Ex_2 \leq F$$
$$x_1, x_2 \geq 0$$

(a) Give an example of the last constraint for which the problem is feasible, but there are no optimal solutions or explain why it does not exist.

(b) Give an example of the last constraint for which there are no feasible solutions or explain why it does not exist.

(c) Give an example of the last constraint for which (3; 5) is an optimal solution or explain why it does not exist.

2. Your company produces 4 types of fertilizer: A, B, C, and D. To produce 1 kg of fertilizer A you need 300 g of potash (P), 400 g of phosphate (H), and 300 g of nitrogen (N). To produce 1 kg of fertilizer B you need 300 g of P, 300 g of H, and 400 g of N. To produce 1 kg of fertilizer C you need 500 g of P, 200 g of H, and 300 g of N. Finally, to produce 1 kg of fertilizer D you need 400 g of P, 400 g of H, and 200 g of N. Suppliers can provide 40 kg of P, 40 kg of H, and 30 kg of N per day. Net profit is $20, $40, $50, and $30 per kilogram of A, B, C, and D respectively. Formulate an LP problem for maximizing the profit of your company. Make sure to clearly describe all decision variables, their units, and the physical meaning of each constraint. No need to simplify constraints and/or objective. You can also get 2 bonus points (but no more than 100% for the whole exam) if you find the optimal value and all optimal solutions using Simplex Method!

3. Solve the LP problem provided in the Sage worksheet using the (\Regular") Simplex Method and, based on your solution, write down the following information. Entering and leaving variables on each step:

Step	1	2	3	4	5	6	7	8	9
Entering									
Leaving									

The optimal value (or explain why it does not exist):
An optimal solution (or explain why it does not exist):
All optimal solutions (or explain why there is only one or none):

References

Albano, G., D'Apice, C., & Manzo, R. (2005, May 4). Teaching statistics with an interactive tool. *International Journal for Mathematics Teaching and Learning*. http://www.cimt.org.uk/journal/dapice.pdf

Artigue, M. (2002). Learning mathematics in a CAS environment: The genesis of a reflection about instrumentation and the dialectics between technical and conceptual work. *International Journal of Computers for Mathematical Learning, 7*(3), 245–274.

Artigue, M. (2009). The future of teaching and learning mathematics with digital technologies. In L. H. Son, N. Sinclair, J. B. Lagrange, & C. Hoyles (Eds.), *Mathematics education and technology: Rethinking the terrain—The 17th ICMI Study* (pp. 463–475). Springer.

Beaudin, M., & Picard, G. (2010). Using symbolic TI calculators in engineering mathematics: Sample tasks and reflections on a decade of practice. *International Journal for Technology in Mathematics Education, 17*(3), 69–74.

Buteau, C., Marshall, N., Jarvis, D., & Lavicza, Z. (2010). Integrating Computer Algebra Systems in post-secondary mathematics education: Preliminary results of a literature review. *International Journal for Technology in Mathematics Education, 17*(2), 57–68.

Buteau, C., & Muller, E. (2014). Teaching roles in a technology intensive core undergraduate mathematics course. In A. Clark-Wilson, O. Robutti, & N. Sinclair (Eds.), *The mathematics teacher in the digital era: An international perspective on technology focused professional development*. Springer.

Buteau, C., Jarvis, D. H., & Lavicza, Z. (2014). On the integration of Computer Algebra Systems (CAS) by Canadian mathematicians: Results of a national survey. *Canadian Journal of Science, Mathematics and Technology Education, 14*(1), 35–57.

Bossé, M. J., & Nandakumar, N. R. (2004). Computer algebra systems, pedagogy, and epistemology. *Mathematics and Computer Education, 38*(3), 298–306.

Bray, A., & Tangney, B. (2017). November). Technology usage in mathematics education research: A systematic review of recent trends. *Computers & Education, 114*, 255–273.

Connors, M. A., & Snook, K. G. (2001). A technology tale: Integrating hand held CAS into a mathematics curriculum. *Teaching Mathematics and Its Applications, 20*(4), 171–190.

Creswell, J. (2013). *Qualitative inquiry and research design: Choosing among five approaches* (3rd ed.). Sage.

Decker, R. (2011). The Mathlet Toolkit: Creating dynamic applets for differential equations and dynamical systems. *International Journal for Technology in Mathematics Education, 18*(4), 189–194.

Denzin, N. K., & Lincoln, Y. S. (Eds.). (2005). *The Sage handbook of qualitative research* (3rd ed.). Sage.

Ertmer, P. A., Ottenbreit-Leftwich, A. T., Sadik, O., Sendurur, E., & Sendurer, P. (2012). Teacher beliefs and technology integration practices: A critical relationship. *Computers & Education, 59*(2), 423–435.

Fey, J. T., Cuoco, A., Kieran, C., McMullin, L., & Zbiek, R. M. (Eds.). (2003). *Computer algebra systems in secondary school mathematics education*. National Council of Teachers of Mathematics.

Gyöngyösi, E., Solovej, J. P., & Winsløw, C. (2011). Using CAS-based work to ease the transition from calculus to real analysis. In M. Pytlak, T. Rowland, & E. Swoboda (Eds.), *Proceedings of the 7th conference of European researchers in mathematics education* (pp. 2002–2011). University of Rzeszów, Poland.

Haapasalo, L. (2013). Adapting assessment to instrumental genesis. *International Journal for Technology in Mathematics Education, 20*(3), 87–93.

Hamilton, E. R., Rosenberg, J. M., & Akcaoglu, M. (2016). The Substitution Augmentation Modification Redefinition (SAMR) model: A critical review and suggestions for its use. *TechTrends, 60*(5), 433–441.

Heidenberg, A., & Huber, M. (2006). Assessing the use of technology and using technology to assess. In L. A. Steen (Ed.), *Supporting assessment in undergraduate mathematics* (pp. 103–108). The Mathematical Association of America.

Jankvist, U. T., Misfeldt, M., & Aguilar, M. S. (2019). What happens when CAS procedures are objectified? The case of "solve" and "desolve." *Educational Studies in Mathematics, 101*(1), 67–81.

Jarvis, D. H., Buteau, C., Doran, C., & Novoseltsev, A. (2018). Innovative CAS technology use in university mathematics teaching and assessment: Findings from a case study in Alberta, Canada. *Journal of Computers in Mathematics and Science Teaching, 37*(4), 309–354.

Jarvis, D. H., Lavicza, Z., & Buteau, C. (2012, July). Computer Algebra System (CAS) usage and sustainability in university mathematics instruction: Findings from an international study. In *Proceedings of the 12th International Congress on Mathematical Education (ICME-12)*. Seoul, Korea.

Jarvis, D. H., Lavicza, Z., & Buteau, C. (2014). Systemic shifts in instructional technology: Findings of a comparative case study of two university mathematics departments. *International Journal for Technology in Mathematics Education, 21*(4), 117–142.

Kendal, M., & Stacey, K. (2002). Teachers in transition: Moving towards CAS-supported classrooms. *Zentralblatt Für Didaktik Der Mathematik (ZDM), 34*(5), 196–203.

Kieran, C., & Drijvers, P. (2006). The co-emergence of machine techniques, paper-and-pencil techniques, and theoretical reflection: A study of CAS use in secondary school algebra. *International Journal of Computers for Mathematical Learning, 11*(2), 205–263.

Koehler, M., & Mishra, P. (2009). What is Technological Pedagogical Content Knowledge (TPACK)? *Contemporary Issues in Technology and Teacher Education, 9*(1), 60–70.

Lagrange, J.-B. (2005). Using symbolic calculators to study mathematics. In D. Guin, K. Ruthven, & L. Trouche, *The didactical challenge of symbolic calculators*, Mathematics Education Library, 36. Springer-Verlag. https://hal.archives-ouvertes.fr/hal-02379891.

Lavicza, Z. (2006). The examination of Computer Algebra Systems (CAS) integration into university-level mathematics teaching. In C. Hoyles, J. B. Lagrange, L. H. Son, & N. Sinclair (Eds.), *Online proceedings for the 17th ICMI study conference* (pp. 37–44). Hanoi University of Technology.

Lavicza, Z. (2008a). The examination of Computer Algebra Systems (CAS) integration into university-level mathematics teaching. *Unpublished Ph.D. dissertation*, The University of Cambridge, Cambridge, UK.

Lavicza, Z. (2008b). Factors influencing the integration of Computer Algebra Systems into university-level mathematics education. *International Journal for Technology in Mathematics Education, 14*(3), 121–129.

Martinovic, D., Muller, E., & Buteau, C. (2013). Intelligent partnership with technology: Moving from a mathematics school curriculum to an undergraduate program. *Computers in the Schools, 30*(1–2), 76–101.

Meagher, M. (2012). Students' relationship to technology and conceptions of mathematics while learning in a Computer Algebra System environment. *International Journal for Technology in Mathematics Education, 19*(1), 3–16.

Oates, G. (2009). *Integrated technology in the undergraduate mathematics curriculum: A case study of computer algebra systems*. (Unpublished doctoral dissertation). University of Auckland, Auckland. https://researchspace.auckland.ac.nz/bitstream/handle/2292/4533/02whole.pdf?sequence=4.

Oates, G. (2011). Sustaining integrated technology in undergraduate mathematics. *International Journal of Mathematical Education in Science and Technology, 42*(6), 709–721. https://doi.org/10.1080/0020739X.2011.575238

Puentedura, R. (2006). Transformation, technology, and education [Blog post]. http://www.hippasus.com/resources/tte/.

Puentedura, R. (2014). Building transformation: An introduction to the SAMR model [Blog post]. http://www.hippasus.com/rrpweblog/archives/2014/08/22/BuildingTransformation_AnIntroductionToSAMR.pdf.

Pierce, R. L., & Stacey, K. (2004). A framework for monitoring progress and planning teaching towards the effective use of Computer Algebra Systems. *International Journal of Computers for Mathematical Learning, 9*(1), 59–93.

Pountney, D., Leinbach, C., & Etchells, T. (2002). The issue of appropriate assessment in the presence of a CAS. *International Journal of Mathematical Education in Science and Technology, 33*(1), 15–36.

Richard, P. R., Venant, F., & Gagnon, M. (2019). Issues and challenges in instrumental proof. In G. Hanna, D. Reid D., & M. de Villiers (Eds.), *Proof technology in mathematics research and teaching. Mathematics Education in the Digital Era*, Vol. 14. Springer.

Rosenzweig, M. (2007). Projects using a Computer Algebra System in first-year undergraduate mathematics. *International Journal for Technology in Mathematics Education, 14*(3), 147–149.

Sarvari, C., Lavicza, Z., & Klincsik, M. (2010). Assisting students' cognitive strategies with the use of CAS. *International Journal for Technology in Mathematics Education, 17*(3), 147–153.

Sevimli, E. (2016). Do calculus students demand technology integration into learning environment? Case of instructional differences. *International Journal of Educational Technology in Higher Education, 13*(37), 1–18. https://doi.org/10.1186/s41239-016-0038-6

Shulman, L. (1986). Those who understand: Knowledge growth in teaching. *Educational Researcher, 15*(2), 4–14.

Smith Risser, H. (2011). What are we afraid of? Arguments against teaching mathematics with technology in the professional publications of organisations for US mathematicians. *International Journal for Technology in Mathematics Education, 18*(2), 97–101.

Somekh, B. (2008). Factors affecting teachers' pedagogical adoption of ICT. In J. Voogt & G. Knezek (Eds.), *International handbook of information technology in primary and secondary education* (pp. 449–460). Springer.

Stein, W., & Joyner, D. (2005). Sage: System for algebra and geometry experimentation. *ACM SIGSAM Bulletin, 39*(2), 61–64.

Stewart, S., Thomas, M. O. J., & Hannah, J. (2005). Towards student instrumentation of computer-based algebra systems in university courses. *International Journal of Mathematical Education in Science and Technology, 36*(7), 741–750.

Tall, D. (2013). The evolution of technology and the mathematics of change and variation. In J. Roschelle & S. Hegedus (Eds.), *The Simcalc vision and contributions: Democratizing access to important mathematics* (pp. 449–561). Springer.

The Sage Developers. (2015). SageMath, the Sage Mathematics Software System (Version 6.9). [Computer Software]. https://www.sagemath.org.

Thompson, P. W., Ashbrook, M., & Musgrave, S. (2015). *Calculus: Newton meets technology.* [eBook]

Thompson, P. W., Byerley, C., & Hatfield, N. (2013). A conceptual approach to calculus made possible by technology. *Computers in the Schools, 30*(1–2), 124–147.

Tobin, P. C., & Weiss, V. (2016). Teaching undergraduate mathematics using CAS technology: Issues and prospects. *International Journal of Technology in Mathematics Education, 23*(1), 35–41.

Trouche, L. (2004). Managing the complexity of human/machine interaction in computerized learning environment: Guiding students' command process through instrumental orchestrations. *International Journal of Computers for Mathematical Learning, 9*(3), 281–307.

Yin, R. K. (2009). *Case study research: Design and methods* (4th ed.). Sage.

Zeynivandnezhad, F., & Bates, R. (2018). Explicating mathematical thinking in differential equations using a computer algebra system. *International Journal of Mathematical Education in Science and Technology, 49*(5), 680–704. https://doi.org/10.1080/0020739X.2017.1409368

Modeling Practices to Design Computer Simulators for Trainees' and Mentors' Education

Fabien Emprin

1 Introduction

Our work focuses on the simulation of human interactions for the education of mathematics teachers. We are working on three types of interactions: teacher–students in a classroom teaching situation (Emprin & Sabra, 2019), teacher–teacher in a mentoring situation, students–students in a geometric work situation (Emprin & Petitfour, 2020). Our goal, in this chapter, is to show how the work on these simulators can be upstream of the work on AI-aided educational working spaces and is articulated with the usual work without the digital tools (classroom observation), but also with digital training (use of simulators in training) and with the production and analysis of data on learning. For us, the development of AI-aided educational working spaces poses several problems: the nature of the aids that these tools can offer (on which aspects of teaching, to assist which parts of the practices), that of the use by teachers of such environments (digital technologies are struggling to integrate effectively into practices (Abboud & Emprin, 2009) and what follows, i.e., teachers preparation in such tools.

Our human interaction simulators belongs to the artificial intelligence tools in that it is an expert system that operate with pre-established rules, but our goal is to also show how they can feed deep learning algorithms. There are two main lines of work: first, human interaction simulators can be used to define and verify models that would be useful, by reducing the quantities of data to be processed; second, they generate data to power AI-aided environment.

In the first part of this chapter, we rapidly develop theoretical frameworks that guide the work of modeling human interactions in education and teacher preparation.

F. Emprin (✉)
Université de Reims Champagne Ardenne, CEREP, IREM de Reims, Reims, France
e-mail: fabien.emprin@univ-reims.fr

© The Authors(s), under exclusive license to Springer Nature Switzerland AG 2022
P. R. Richard et al. (eds.), *Mathematics Education in the Age of Artificial Intelligence*,
Mathematics Education in the Digital Era 17,
https://doi.org/10.1007/978-3-030-86909-0_14

These frameworks allow us to define teaching and training practices and to make choices for building computer models.

In a second part of this paper, we develop a model of the mentoring situation that we implement in a simulation software. This Mentoring Dialogue Simulator (MDS) is used in teacher training in France and Russia (Galiakberova et al., 2020).

The third part is dedicated to the analysis of the exploitation of a simulator to generate data that can be exploited with AI-based algorithms. For this, we take the example of the Computer Classroom Simulator (CCS). Indeed, it offers a fully controlled situation and reproducible activity. This type of tool can easily be put online or on a Learning Management System (LMS) and thus proposed to a large number of users who themselves can make several tests. The data generated are therefore quantitatively very important and comparable (since the simulated situation is the same for all). This is what we have done with the CCS by developing an online tool associated with a collection of information on the user (questionnaire that the user could fill in). The models built can then evolve according to the information collected automatically but also during training sessions.

At the end of this chapter, we discuss the use and the limits of this work for an AI-Aided environment.

2 Theoretical Framework

In a training simulator (MDS or CCS), several levels of practice are involved: that of the students facing the Dynamic Geometry Software (DGS), that of the teacher in the classroom, that of the instructor using the simulator in training. These different Layers are nested like Russian nested Matryoshkas dolls. Each layer requires the mobilization of specific theoretical frameworks: student learning, teacher practices, teacher training practices.

2.1 Choices About Student Learning

What all our simulators have in common is that they all address the issue of teaching geometry with DGS). This decision was made to limit the didactic theoretical knowledge at stake and because this is a field in which many research studies have been carried out (Laborde, 2001, Sträßer, 2001, Baccaglini-Frank & Mariotti, 2010; Jones, 2000), allowing us to anticipate the difficulties encountered by students or teachers and thus to model them. Moreover, geometry is a fundamental field of knowledge, as Sherard (1981) point out and it is a necessity for the development of both individuals and our society:

> Geometry has not died because it is essential to many other human activities and because it is so deeply embodied in how humans think. With the introduction of computers with rich graphical capacities and the recognition of multiple ways of learning, our current situation

offers an unprecedented opportunity for geometers and those who work visually (Whitely, 1999, p. 1).

The question of the teaching of geometry has moreover been the subject of several international studies such as Mammana and Villani (1998) or working groups in international conferences as CERME[1] (Jones et al., 2017).

We focus on the possibilities offered by the movement of figures in the process of conjecture and demonstration. When a drawing is made in the software using the appropriate properties, this drawing resists the dragging of its base points: a square constructed with perpendicular isometric diagonals intersecting in their middle remains a square whatever the points dragged, conversely if the properties are not used it will deform and become rectangles or simply quadrilateral. This led us to place the teaching situations in the simulators for students over 10 years old. In particular, we use two ideas of the model developed by Baccaglini-Frank and Mariotti (2010), based on the model of Arzarello et al. (1998, 2002).

First, DGS may become a potential bridge between two worlds, i.e., the mathematical world of Euclidean geometry and the phenomenological world of experience, which we will refer to as the spatial-geometry world. Therefore, DGS provides teachers with new insights and tools to overcome students' difficulties.

Second, the dragging mode in DGS can be seen as an instrument (Rabardel, 2002) that can change students' approach to the proving process.

> Dragging can help the user interpret the exploration in terms of logical dependency, in the following way [...] Such use of dragging may develop through a process of instrumental genesis (Rabardel, 2002). If we consider dragging to be an artefact and place a user in the context of solving a problem (task), it is possible to identify specific utilization schemes associated with dragging (dragging utilization schemes, in the following 'dragging schemes') [...] (Baccaglini-Frank & Mariotti, 2010, p. 228).

With regard to the proving process, we rely on the work of Arsac (1987) and Richard (2004), which highlight that the transition from the conjecture phase to the demonstration phase is complex, and on the distinction between argumentation and proof (Balacheff, 1998).

As a result, the classroom sessions that will be at the core of the simulators should highlight these elements. We are now interested in how to look at teaching practices.

2.2 Twofold Approach to Define and Model Teaching Practices

Robert and Rogalski (2005) built a theoretical framework for defining and analyzing teaching practices. This approach, which combines didactical and psychological perspectives, is known as the twofold approach didactic and ergonomic (TADE).

---

[1] Congress of the European Society for Research in Mathematics Education.

It examines practices on the basis of the following five components: a cognitive-epistemological dimension, i.e., the organization of tasks and the mathematics that is actually taught; a mediation–interaction dimension, i.e., interactions through verbal or written and direct or indirect communication; a personal dimension, i.e., whether the teacher regards mathematics as a science or merely as a subject for teaching and learning processes; a social dimension, i.e., a teacher works in a teacher's collective within which not everything is possible; and an institutional dimension, i.e., teachers are subject to academic programs and institutional injunctions.

This twofold approach to analyzing teachers' practices allows us to create a model of such practices, taking both the didactic and professional aspects of the work into account. It allows us to consider the content to be taught.

In a previous paper (Emprin, 2007, 2009), we demonstrated that the training of teachers to teach mathematics with digital technologies in France deals with very few of the dimensions referred to above. In the case of the proposed simulator, we have chosen to model the teacher's responses on four of the five components (we do not discuss the institutional component). Indeed, these four components make it possible to question the teacher's didactic choices in relation to the chosen theoretical framework (uses of the DGS for conjecturing and demonstrating) through his interactions with the pupils, his choice of learning situations, his way of seeing (from a personal point of view but also as part of a community of teachers) mathematics and in particular the place of demonstration.

In this work, the place of digital tools is important. Teachers' educators use simulation software, the students use a DGS. We need to be able to question the relationship between humans and digital tools.

2.3 Instrumental Genesis

The instrumental approach (Rabardel, 2002) is also used to analyze teacher activity when a digital artifact is used. Dragging is also an instrument for teachers, and schemes of use can be determined and implemented in a simulator.

Adopting the instrumental approach allows us to analyze the instrumental genesis of the pupils in relation to the tasks proposed by the teacher. We can then implement trajectories of these geneses in the simulator. We regard the instrument as being constructed by the teacher from an artifact, which is in this case a DGS.

The scheme of use is a central concept in the instrumental approach, especially with a view to understand the role of the artifact as a mediating tool in the construction of knowledge. The various components of the scheme have been identified as follows (Vergnaud, 1998): the goal (or subgoal) of the action, the teacher's rules of action for achieving the goal, the operating invariants that justify the rules of action, and the possibilities of inferences that allow the teacher to enrich his operational invariants according to the lived experience. According to Vergnaud (1998, p. 238), the role of mediator is the most important work of the teacher in class:

[the teacher] starts out of class where he begins to organize a complex process of implementation with several constraints, conditions and variables from one class situation to another. It continues in the classroom with a large number of decisions and mediations in a very limited time.

Vergnaud (ibid) proposes referring to the various components of the schemes to analyze these decisions and the acts of mediation. The operating invariants reflect a form of appropriation of knowledge by the teacher.

Using this approach, the virtual teacher's answers result from the analysis of the operating invariants of practices.

We combine these theoretical frameworks with specific frameworks depending on the nature of the simulation. For example, to create a mentoring dialogue simulator we have to characterize the different postures of the mentor. The role of the mentor is particularly important at a time when the question of the appropriation by teachers of AI-assisted environments arises in the same way as for other digital tools.

3 How to Create a Model of Mentor–Teacher Interactions?

The mentoring dialogue can cover several practical aspects, for example, classroom management, learning plans design, organization of the session, and proposed teaching–learning processes. All these various aspects can be treated a priori in a simulator of human interactions; however, we have chosen to focus on processes related to the learning and teaching of mathematics. We based our work on two main assumptions:

A first founding assumption is that by making mentors aware of the dilemmas and possible roles in mentoring dialogues, they will be more efficient and able to make choices. The theoretical frameworks presented in this section allow us to design an artifact that enables an academic or professional supervisor (and even a peer) to approach a dialogue to analyze and question its practice.

Our second assumption is that going beyond the teacher-centered dialogues to bring them to student learning allows the tutored teacher to question and develop his practices. Specifically, "student teacher supervision, with few exceptions, remains a teacher-centered enterprise. Typical supervision policies, procedures, and forms focus on teachers' observable behaviors." (Paris & Gespass, 2001, p. 398). It is therefore a question of considering teaching practices in several of their dimensions, in particular the cognitive-epistemological dimension. To achieve this, the simulator must make it possible for the instrumental genesis of the DGS to be questioned at two levels: that of the student and that of the teacher.

Thus, designing a Mentoring Dialogue Simulator (MDS) requires the following three elements: excerpts from a classroom session conducted by the virtual teacher, a model of current virtual teacher practices (twofold approach), and a model of the mentoring relationship. The user of the simulator will be able to view or read transcripts of the session, make choices on conducting the mentoring dialogue, and receive feedback of his choice (see Fig. 1).

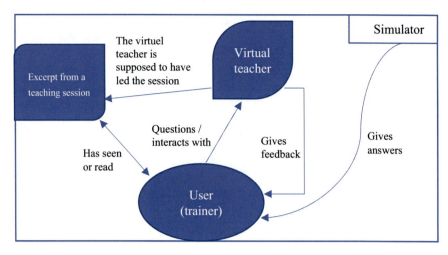

Fig. 1 Model of interactions in mentoring dialogue simulator

This MDS is designed to be used at the core of a face-to-face or training device for online trainers.

What makes our research method specific is the triple use of theoretical frameworks which make it possible to analyze the real practices observed, to design models and systems of interactions in simulators and to analyze the simulated practices of users.

We now present the choices and the design process for such a simulation.

3.1 From the Design of a Mentoring Dialogue Simulator (MDS) to a More General Model

3.1.1 Choice of Software

To simulate a dialogue of the analysis of practices first requires the existence of a teaching situation, which is the starting point of the process. This situation provides the basis for the cognitive-epistemological dimension of the practice. We must therefore provide elements that can be discussed in the interview; these are video clips showing the work of the student and verbal interactions between the teacher and the student.

Our goal is to encourage a reflection on the actions taken as well as reflexivity on the practices; we therefore opted for a part-scale simulator (Pastré, 2005). Unlike full-scale simulators, which aim to reproduce reality as faithfully as possible, part scale simulators focus on a specific part of the subject's activity. In simulated time—as opposed to in real time—it is possible to exchange views and think about an action before actually taking it. It is also possible to link several simulations quickly to test several hypotheses. To consider several practice dimensions, including social and personal dimensions, we decided to use a simulator environment with realistic conversational agents that can simulate emotions and use speech synthesis. With this choice, we dissociated ourselves from other existing simulators which use avatars with a comic/drawing appearance. This allowed us to enrich the system of interacting with the user and to increase the level of possible feedback. Studies on Embodied Conversational Agents (ECA) (Krämer & Bente, 2010) demonstrate that they have promising potential for improving the motivation and learning of users. However, a detailed discussion of these effects is beyond the scope of this chapter.

To create simulations of human interactions, we opted to use the Virtual Training Suite software (VTS—© Serious factory—Suresnes, France), which enables the generation of a realistic 3D environment with ECA to incorporate multimedia content, to build a rich scenario and to follow the answers of the users in the form of a score. Entirely customizable, the software manages the scenario as a graph. The overall scenario is divided into scenes that are depicted graphically (Fig. 2).

Within each scene, word bubbles provide elements of interaction (choice of verbal interaction, answers to a multiple-choice question, or clicking on the screen), and the branches that result are therefore the different decision paths (Fig. 3).

The scenario, once compiled, can be exported to different formats, including the web, SCORM (for LMS), iOS (iPad and iPhone), Android, Mac OS, or Windows, or deposited on a Serious Factory learning platform (the company that developed the software). We do not go into technical design details here but rather describe the conceptual aspects embedded in the simulation. For the construction of the MDS, we

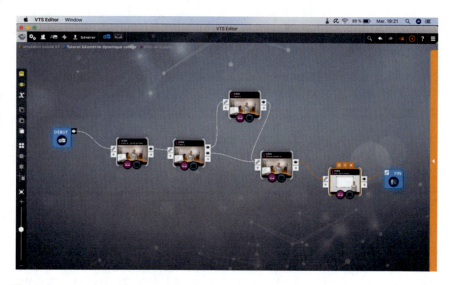

Fig. 2 Cutting the scenario in scenes

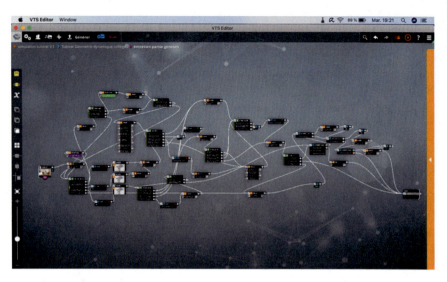

Fig. 3 Scenario graph of a scene

started from the theoretical framework MERIDs' Model and mentoring dilemmas of Brau-Antony and Mieusset (2013), with the use of didactic knowledge.

3.1.2 Combining MERID's Model and Mentoring Dilemmas

In a literature study (Hennissen et al., 2008, 2017), five key aspects were identified.

> As an answer to the first research question, from the selected empirical studies, five key aspects of mentoring dialogues emerged, which are often the focus of research: content of the dialogue, mentor teachers' style and supervisory skills, mentor teachers' input, time aspects of the dialogue and phases in the dialogue.

The authors extract/highlight three of these five aspects.

> In answer to the third research question, empirical evidence in the selected studies indicates that three key aspects connect with distinctive mentor teachers' behavior in mentoring dialogues: style/supervisory skills, input and time aspects. These three aspects are plausible candidates to constitute a conceptual framework. We connected these key aspects in the MERID model, which identifies four mentor teacher roles during mentoring dialogues: imperator, initiator, advisor, and encourager.

Mattéï-Mieusset (2013), for her part, identifies the following four dilemmas associated with this particular facet: transmitting the job or reflecting to allow the trainee to build his response; spotting the mistakes and successes of the trainee or helping them to emerge; supporting the trainee or evaluating him or her, guiding, imposing a framework, providing the trainee with tools, or leaving him or her free to make his own choices.

We combine these two frameworks which constitutes a complete theoretical framework to model the mentor/teacher relationship.

We described the didactic choices and instrumental genesis issues of the DGS implemented in the classroom scenario used in the following section. We then implemented the software in such a way as to bring out this knowledge and the knowledge of both didactic and mentoring activities in the user's possible answers and in the feedback of the software. The MDS is therefore designed in such a way that the user becomes aware of the facets and dilemmas related to the practice analysis interview activity and may identify knowledge that highlights the role of technology in the learning process.

3.1.3 Classroom Session Used as a Basis for Simulated Mentoring Dialogue

Our objectives in terms of didactic content led us to choose to focus on the use of geometry software for students aged 14–15. The students must conjecture about a property and then demonstrate it such that a triangle's three perpendicular bisectors meet at a point O, known as the circumcenter, which is also the center of the triangle's circumcircle:

> In the software. Draw a triangle. Name it ABC. Draw the perpendicular bisectors of each sides. What can you conjecture? Let O be the point of intersection of two of the perpendicular bisectors. Draw the circle with center O and radius [OA]. What can you conjecture? Then demonstrate the property.

We videotaped several teachers implementing this lesson and we transcribed some characteristic interactions of instrumental genesis problems related to DGS. From this material, we made four video clips, each of which presents a characteristic time of the lesson. We have chosen to focus on the use of DGS as part of both a conjecture and a demonstration. We rely on the research literature to identify the knowledge that could be transmitted during a training session. One key point emerges in the case of dynamic geometry, namely, the robustness of the construction of objects (Baccaglini-Frank & Mariotti, 2010; Laborde, 2000; Restrepo, 2008). The space in which students work is relevant because it brings a new system of constraints. The correct construction is one that is resistant to dragging. Moreover, this system of constraints requires an assimilation by the student that can be analyzed in terms of instrumental genesis (Rabardel, 2002).

The first video illustrates the screen of a student who built the middle point of a segment in the DGS environment by measuring the length of the segment and then placing a point on the segment and moving it to obtain the desired length (half the length of the segment). The teacher asks the student to drag the midpoint, which the student does. The resulting exchange is as follows:

Teacher (T): So, what's going on?
Student (S): It is no longer in the middle.
T: What's the problem there?

S: There's no problem.

T: So how can you say it's in the middle? You will say, "Yes, I measured and I placed it in the middle." But is the drawing something that is stable?

S: No.

T: So, is that the way to go?

Students use correct mathematical knowledge. The distance from one side of the segment to the middle is half the length of the segment, so this construction is not resistant to dragging. The teacher's intervention does not allow the student to understand the didactic rules specific to the DGS. The teacher asks the student if the drawing is stable but does not refer to the rule that "to be correct, a construction must withstand all movements." We also notice that the instrumental geneses of the DGS are poor for this student; he does not allow himself to move the point and does not use the primitive constructions implemented in the software to build resistant points. The instrumental genesis of the teacher is also weak; he does not use the software as a learning environment with feedback.

In a second excerpt, the teacher asks a student to provide a definition of the perpendicular bisector. The students answer is "a straight line that passes through the middle of a segment and is perpendicular." The teacher then begins an exchange based on the fact that the student did not use the correct article, stating that the student should not have used "a" but rather "the". This choice is debatable because it results in the student questioning an issue that is not directly related to the task at hand, even if the teacher's point is correct.

In the third video clip, when a group has correctly drawn the three perpendicular bisectors, the teacher then asks them to move a vertex. First, the student is worried that he will "have everything wrong," and then he asks himself why one of the three perpendicular bisectors does not move. The student cannot conclude anything other than the fact that there is a "base" in the triangle that results in the mediator of the base not moving. The interesting aspect here is that this example illustrates how point shifts on a resistant drawing can lead students to identify phenomena that allow them to gain a better understanding of mathematical properties. It demonstrates, once again, that the instrumental genesis of both the student and the teacher are too weak to benefit from work on DGS. The challenge of the interview will be to help the teacher not only realize the potential of the DGS but also incorporate the understanding under which he was operating. It's about following what Sträßer has highlighted: "There is no change in geometry learning and teaching without additional 'costs' in terms of investment of time and concepts to understand the software!" (Sträßer, 2001, p. 331).

The last clip extract concerns the end of the lesson, the moment when the students debate their ideas. It is a very typical situation because only the teacher speaks. He makes students guess words by means of the topaz effect (Brousseau, 1997, 2006). A question posed to students is progressively rephrased up to the point where the expected answer is ultimately generated: "If you have three lines, what is the word used to designate the point? The point of ... / con-course, and we say lines are ... // con-current. So, here you see again the difficulty of vocabulary control." The challenge is understanding why the teacher decided to share this information without interacting with the students. Is he aware of this lack? Is it a question of the length of the session? What drove him to make that choice? Could it have been social material constraints? What are the alternatives and their consequences?

We now describe the link between these classroom sessions and the system of interaction with the ECA.

3.1.4 Design of the Interview

The four video excerpts allow us to question each one a theoretical knowledge of didactics of mathematics. These four excerpts also allow us to define four entities (the scenes of the software) that make up the scenario (Fig. 4).

In each of the scenes, the mentor must therefore lead the virtual teacher to question himself on a specific problem (instrumental genesis) and help him acquire the targeted didactic knowledge. In the fifth and last scene, he must say, for each of the points, if he thinks he has succeeded, and the virtual teacher will answer him if he has indeed evolved.

In the scenes, the user (mentor) must interacting with the ECA. For this he chooses from a list the interventions and the actions he wants to perform. An example of such a list can be seen on the left in Fig. 4, which is a screenshot.

The context of the interview was chosen to correspond to the usual practices of mentors, who have indicated that they liked to settle down in a quiet place with their trainee (Brau-Antony & Mieusset, 2013).

The questions chosen by the user and the answers of the ECA are then heard. The ECA is animate and can express different feelings both by voice and by means of facial expressions, as presented in Fig. 5.

Fig. 4 Screenshot of the MDS

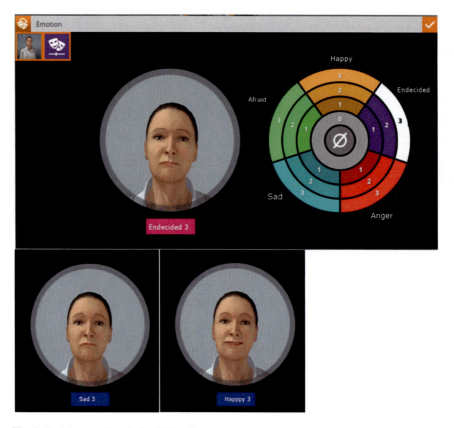

Fig. 5 Facial expressions in the VTS software

3.1.5 Choice of Interactions

The MDS simulator has been designed to allow users (i.e. future mentors) to become aware of the dilemmas and the different roles he can take in dialogues. At the end of each simulation the software must position it in relation to these different roles. To this end, each choice in the list (each possible interaction with the ECA) was assigned a weight in the form of a score for each of the axis of Fig. 7. For example, the axis "non-directive—directive" is graduated from 0 to 4. 0 for non-directive, 1 for undirective, 2 for a little bit directive, and 3 for directive. To be realistic, the sentences chosen have been extracted or inspired from the body of interviews by Mattéï-Mieusset (2013).

The intervention, "I propose that we play the video of your session to look at it together. You can stop whenever you want to, at a moment that seems interesting to you" is likely to result in the representation of the teacher emerging from a nondirective and reactive position without guiding the trainee too much. This choice therefore receives 0 on the "non-directive—directive" axis, 0 on the "reactive—active" axis, 0 on the axis (super teacher—evaluate) and 0 on the axis (leave free choice—guide).

The following intervention, however, is more guiding and comes from a nondirective and active position: "I will show you excerpt of your session and I'll ask you questions about your choices." This choice therefore receives 1 on the "non-directive—directive" axis, 3 on the "reactive—active" axis, 1 on the axis (support teacher—evaluate) and 3 on the axis (leave free choice—guide).

The following final intervention is made from an active and directive position and with a choice of guidance: "I can tell you what I saw; in fact, you asked your students the definition of the perpendicular bisector, and they told you 'we build a mediator'; you explained to them that one does not say 'a' but 'the' perpendicular bisector." This choice therefore receives 3 on the "non-directive—directive" axis, 3 on the "reactive—active" axis, 3 on the axis (support teacher—evaluate) and 3 on the axis (leave free choice—guide). All the choices made during the simulation add up and allow the user to position himself on each of the axes. For more clarity, the user receives a percentage.

The choices given to the user provide the virtual teacher to question his practices and improve his didactic knowledge, such as the following:

It is clear to me that the students who move a vertex of a triangle and wonder why one of the mediators does not move are not sufficiently familiar with either the figure or the operation of the DGS. This means that you did not spend enough time on the construction because you were guessing at something that is not at all clear to the students. If you decide to make the students work something out, you have to go to the end of the process.

An alternative element is as follows: "*First of all, I noticed that you are acting in the student's place. Some of your students do not understand the resistance of objects. They do not understand it as a criterion for checking the drawing. You do not do enough to precisely move the basic points of the drawing precisely as criteria. Finally, your last interventions on the uniqueness of the mediator changed the subject while the students were fully occupied with the construction.*" Depending on the fact that these interactions more or less allow the teacher to evolve, they are given a score, which is not visible to the user. It is used in the last scene of the scenario.

3.1.6 Feedbacks

In this last scene, the user has to answer questions about the effects of his dialogue on the virtual teacher. He must state whether he believe that his virtual trainee has identified different elements that could be at stake during the dialogue, as illustrated in the screenshot in Fig. 6. The trainee answers by stating whether he has actually identified the various elements.

The user can thus compare what he thinks he has worked and what has actually worked.

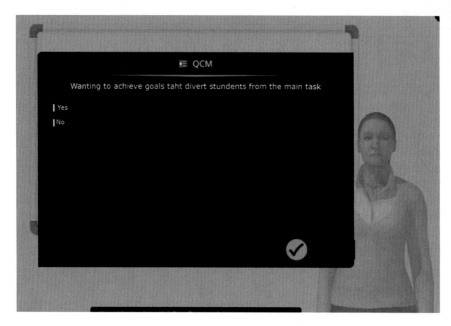

Fig. 6 Final questionnaire—determine what the ECA has identified or not

Once the work is complete, the software sends a final feedback. The user therefore receives his profile in the form of a percentage positioning on each of the axes of the model in Fig. 7. For example, the extract in Fig. 7, the user is on the transmit side of the scale (19%), which is also on the side of "point out the errors," "guide," and "evaluate." The last score concerns the appropriation of the didactic contents possible in the session. The score of 15 out of a possible 53 indicates that the trainee only worked on certain didactic concepts during this interview.

This positioning result from the score obtained in the MDS. Each choice can be interpreted according to the context, but that is precisely what makes this work interesting. In training sessions, this positioning can then be used and discussed with the mentor's educator.

Indeed, the simulator is not designed to be used alone but rather as the core of a training device, thereby allowing a collective reflection on the contents transmitted during the practice analysis period and the postures and management of the dialogue. These interpretations allow teachers to engage in a discussion during a training session. Such exchanges will relate to the institutional, personal, and social components of the practices. Our choice is therefore to generate discussions that truly involve these components and that therefore require trainees to formulate personal elements related to their teaching contexts, their representation of the profession, mathematics, etc.

All the work on the simulator, which is, let us recall it, an expert system and not a deep learning algorithms makes it possible to test theoretical models of the

Modeling Practices to Design Computer Simulators ...

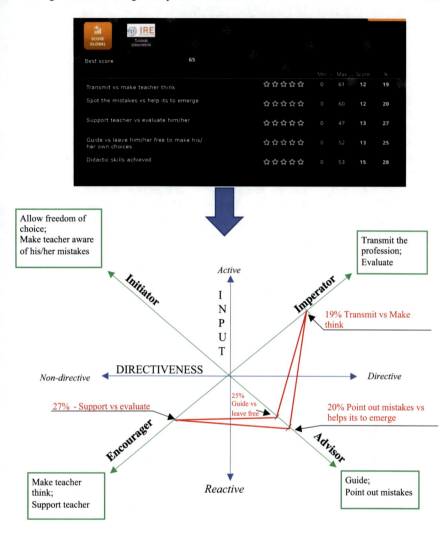

Fig. 7 Final diagram: positioning the user the theoretical model combining MERID'S model et Mattei-Mieusset's Dilemmas

mentoring activity and to check that it works, and is acceptable. The data collected are therefore qualitative but essential in our opinion to then consider systems using deep learning for example. In the last section, we will come back to the prospects of using these models. In the following section, we will look at how another simulator can be used to collect massive data.

4 Collecting Data in a Classroom Simulator

The Computer Classroom Simulator (CCS) is used to educate student (trainee) by way of "teaching" virtual student. The teacher has predetermined choices and they see the effect of their choices on student behavior. We have designed a simulator that simulates reactions centered on didactics, i.e., on the teaching-learning process. The user receives several types of feedback about the student: his mathematical work, his answers when questioned and his level of attention (represented by a VU meter). The simulated situation is a problem of conjecture in dynamic geometry software, we do not describe it in detail here, but it is presented in Emprin and Sabra (2019). To model this interaction system, we use the double approach (Robert & Rogalski, 2005) and the instrumental approach (Rabardel, 2002) presented in the theoretical part. We then operated by observing classroom situations and compiling the teacher's choices and the students' responses using the division of the activity into cognitive, meditative, social, personal, and institutional components as well as instrumental mediations. We then made statistics to recreate a situation consistent with the choice averages. The virtual teaching situation is designed accordingly to the Theory of Didactical Situation TDS (Brousseau, 1997, 2006). We implemented the CCS with the data collected and by replicating the observed behaviors.

4.1 Highlights in the Implementation Analysis

So, there are two kinds of data collected: didactic and what I will call pedagogical (in the common sense). The first concerns the problem that the student will have to solve: construction of a figure in the DGS or not (the file is provided partially or totally to the student) and the role of the drawing in the conjecture, the role of the phases of conjecture, proof, institutionalization. Here, the TDS makes it possible to measure the distance between what is proposed and the concepts of devolution, didactic situation, a-didactic phase (of a situation), antagonistic environment, institutionalization, time management of learning. The pedagogical aspects are those that are independent of the teaching situation in mathematics: choice of the conditions for giving the instruction (computers on or off), choice of the modalities of interaction with the students (collective, individual), time management from the point of view of the students' attention.

4.2 Highlights in the Design of the CSS Simulator

The first choice is to program in the simulator, teachers' decisions that have been observed in practice or in the preparations. It conditions the nature of the simulation: it simulates ordinary practices in the sense of those studied by Coulange (2012). This choice was discussed during the summer school of didactics in 2017 (Emprin et al., 2017): would it not be better to propose to the simulator user richer choices, based on didactic knowledge and therefore that he would not have thought of? It seems to me that we then have two types of classroom simulators: one that allows us to question ordinary practices and their effects, and the other that allows us to test new, richer practices.

The procedures and difficulties observed in the students were quantified and programmed in the software based on the a priori analysis. Thus, the number of students who succeeded in making a resistant construction in the DGS, in conjecturing from a right or wrong construction, in mobilizing or formulating the property of equality of the lengths of the diagonals of the rectangle corresponds to the observed proportion. The effects on student learning were quantified at three stages: at the end of the session by analyzing the students' answers to the question: are the lengths equal? But also, by the arguments written or formulated during the pooling, and finally by questioning the students a week later to find out what they had retained from the situation. The models that are programmed in CCS are based on observations: students who do not have a correct (resistant) figure are unable to conjecture. Even with a correct figure, the proof is difficult to produce if the students are not used (or encouraged) to use the DGS tools: moving points, measuring tools, adding extra segments. Moreover, in the absence of formal institutionalization during or after the session, even students who have found correct arguments remain with the superficial aspects of the problem. The only thing they retain is the answer.

The teacher's pedagogical choices are also programmed with their effects on the students. Thus, when the teacher is talking, and the students are doing something else (turning on the computers for example) only 1/3 of the students are able to repeat what has been said. On the other hand, when one of the students manages to say the sentence again, even those who did not listen the first time have understood and are then able to rephrase when they are invited to do so. Teachers who favor individual interactions spend much more time than those who interact with the group, but the latter interactions have more time consuming than individual interactions. Finally, some choices are also based on pedagogical hypotheses confirmed by numerous observations: students will not behave in the simulator when they do not understand or are unable to do the task, or when they have completed the work and have no new instructions.

The teacher's relationship to mathematics (personal component of practices in particular) is also questioned in this simulation. What is the relationship of mathematics to reality or what are the logical principles governing demonstration in mathematics? Indeed, in this problem the student must move from perception or measurement (whether paper and pencil or computer) to the demonstration of a property. Euclidean geometry, characterized by Euclid's axioms (Peyrard, 1804) corresponds to a mathematization of the real, whereas Hilbert's axiomatic (1900) gives geometry a status independent of the real. The training can thus refer to mathematical questioning. Numerous works such as those of Bkouche (1997) allow us to question and clarify teaching choices. Indeed, in simulated lesson, the question arises of the place of drawing and the figure, in particular the status of what is drawn in a DGS, of its role in the process of proof: between aid to intuition and obstacle to demonstration.

All of these elements make the simulator an artefact of the training situation that allows access to didactic concepts.

4.3 Main Concepts from the Didactics at Stake

A posteriori analysis of the simulator's design based on the observations allows us to question during a training situation several didactic aspects:

When the teachers finish the simulation session, the software output indicates that little learning has been achieved by the students. This is realistic compared to the weight of a single situation in the overall learning: what can a class really learn in 55 min? Nevertheless, this goes against the representations of teachers who declare themselves frustrated by this result. This may be indicative of their representation of learning (personal component) or of computer play. In the games, they know or serious games there is a way to win, an optimal strategy, in a simulator some choices are better than others but there is no 100% winning strategy.

Nevertheless, there are strategies that promote learning among virtual students. They are of two types: the existence of a real phase of institutionalization and a sufficient time of confrontation with the correctly constructed figure in the DGS to conjecturing, searching. With an erroneous figure or without a sufficient time, the confrontation of the student in the milieu (in the sense of the TDS) would not be effective.

Moreover, the programming of the simulation prevents the teacher from constructing the knowledge text alone. That is to say, he can only write on the blackboard what the students have said. This choice does not come from classrooms observations but is based on the researcher's desire to question the teacher's role in the institutionalization phases thanks to the simulation.

The didactic concepts of sub-figure (Mercier, 2018), drawing vs. figure (Duval, 1993; Laborde, 1985) and proof are fully evident in the simulation in that they allow for the analysis of students' difficulties or procedures: difficulties in locating rectangles in the drawing and in seeing that the second diagonal is a radius of the circle.

The status of the drawing for the students, in relation to their instrumental genesis of the DGS, remains very close to that of the paper/pencil. While they are, in fact, confronted with a class of drawings with the same properties, many of them only use it as a single representative of the figure.

Finally, the question of what forms the basis of proof in mathematics can be addressed (Mariotti, 2000) even though the software provides only a satisfactory answer. Using the measurement tool, the software indicates that the lengths are equal. Some virtual students are programmed not to offer a demonstration as soon as they have an answer given by the software.

This simulator is designed, based on the observation of practice and without introducing random parameters. Each simulated student will always react in the same way to the same series of choices. This makes it possible to obtain a perfectly controlled situation, strictly identical between each user. The only parameter which therefore varies is the user. It therefore makes it possible to collect information on teaching practices independently of the teaching–learning situation.

4.4 Collecting Data on Teachers' Practices

We have been putting online a version of the simulator for more than a year. It was associated with a form asking users to specify their gender, whether or not they are teachers, at what level they teach, their seniority, their perceived level of ability with digital technologies. Once this questionnaire was completed, the user had the choice to accept the sending of the data resulting from his simulation to the researchers. The chronological choices of the user in the simulation were collected as well as the synthesis of the students' learning (did they do the right thing? Did they learn the desired mathematical property? What do they remember after one week?).

Two hundred and sixty registered users carried out and submitted simulations which made it possible to collect 420 completed simulations. This set of data allow us to link choices with users' parameters.

We have not yet been able to use this data with deep learning algorithms for lack of a suitable partnership. We simply used statistical tools on a variety of variables (descriptive statistics, inferential statistics and multiple correspondence factor analysis) to find out whether statistical correlations emerged. The first interesting result of this correlation research is that there is none that is statistically significant. Neither age, nor gender, nor country, seniority in teaching or the perception of their level of familiarity with technology appears to be linked to the choices made. On the other hand, two main scenario models emerge. The first, where the teacher tries to have the geometric figures correctly constructed in the dynamic geometry software but does not manage to work correctly on the conjecture and the second during which the user passes very quickly on the construction to work on the conjecture. It is a central point of the training to make the user realize that the situation has two distinct objectives which are difficult to carry out simultaneously. The statistical data confirm that the simulation clearly highlights this aspect. We should now further explore the data to identify if other elements may appear, in particular elements which would not have been anticipated during the programming of the software. In addition, the data processing could also highlight actions that would be predictive of the final assessment. Finally, the processing of this data could make it possible to return personalized feedback.

5 Discussion

The two simulators that we have presented are aimed at helping teachers and mentors acquire knowledge and change their practices, more specifically regarding the use of a DGS to teach geometry.

According to Haenlein and Kaplan (2019) the initial work on AI is to be found in expert systems such as ELIZA and the General Problem Solver. These pieces of software consisted of a set of rules and a series of nodes "if–then" like our simulators.

Our approach goes in the same direction as the evolution of AI and our hypothesis is that it is an important step that allows us to lay the foundations for developing more complex systems. Indeed, the work we do advances the models of the teachers' or mentor's activity. This highlights the usual practices of teachers, the difficulties of these to develop learning situations using the potential of DGS. The experiments that we have conducted on the different simulators seem to us to be indispensable steps to consolidate the theoretical bases necessary to categorize and process data that could be collected more massively on the teaching practice.

As far as teaching practices are concerned, a first problem arises in order to move to tools using deep learning algorithms: access to data. In fact, accessing teaching practices and student learning even if we stay with the observable part of the activity, such as what the teacher says or what students respond to an exercise, requires collecting this information by filming classroom sessions. And even if we had hundreds of hours of filmed classes (or mentoring interviews) to see what students do based on what the teacher says, we would have to digitize and process this information. The models we experimented with would allow us to categorize this information and define observables. They make it possible to limit the complexity of the information to be collected.

Our goal in this chapter was to highlight two aspects that simulators can help: reduce the amount of data to be processed and the generate data usable by the AI. The models that we propose allow us to focus the data collection on the didactic dimensions of the teaching activity. For the second aspect, the quantities of data available (more than 30,000 teachers' choices associated with the simulated time in the session) or the possibility of deploying these tools on Learning Management Systems like Moodle thanks to the Scorm format opens up interesting possibilities.

There is a situation where this information on teaching practices is already digitized: this is the case of distance learning. Indeed, the exchanges between teacher and students as well as the student's activity pass through a course platform. The requested tasks and the students' results are therefore already in the form of computerized data. A first work could be based on these data, which are called learning analytics.

We have also shown that simulators can provide data that could be processed. If they are easily accessible and carried out in a completely controlled environment where the only parameter that changes is the teacher, they have a major drawback: they are not real teaching practices but simulated teaching practices. Nevertheless, it seems to us that these treatments could give information in a way 'in vitro' that could then be verified in reality. Thus, if we understand the difference between the choices on simulated and real practices, we can identify if some results, found on simulated practices, can be true in reality.

We have, in this chapter, focused our reflection on how simulators can be positioned upstream of devices using AI. However, a promising avenue seems to us to use them downstream of the AI. Indeed, deep learning algorithms that could be used to help learning will need to be tested at several levels: that of acceptability by teachers or trainers and that of effectiveness for students. The simulators can then play an important role to avoid testing, the first time, on real students with real teachers. This follows the adage that prevails for the use of simulators in the training of health personnel (e.g., doctors, dentists, nurses): "never the first time on a real (living) patient." These simulators also highlight the points of difficulty in the practices that AI-aided working spaces could attempt to resolve: the management of the assistance provided to students during the construction and conjecture phases, and the organization of the final debate phase in particular.

Simulators, as we have indicated, can integrate into LMSs such as Moodle. They thus make it possible to create new workspaces for practice, at least a certain form of practice. These new spaces offer perspectives for sharing between trainers, for collecting and processing learning data. This work is particularly the subject of the current HYPE 13[2] project. The simulators can also create new mediations between students and mathematics as in the geometrical drawing made by an embodied conversational agent.[3]

Moreover, if programs are able to offer teaching aids and to assist teaching, it will be necessary for teachers to be able to appropriate them, to understand the results provided, and also to know what their level of validity is. Simulators will then be valuable tools to interface between artificial and human intelligence.

References

Abboud-Blanchard, M., & Emprin, F. (2010). Pour mieux comprendre les pratiques des formateurs et de formations TICE. *Recherche Et Formation, 62*, 125–140.
Arsac, G. (1987). L'origine de la démonstration: Essai d'épistémologie didactique [The origin of the demonstration: Essay on didactic epistemology]. *Publications Mathématiques Et Informatique De Rennes, 5*, 1–45.
Arzarello, F., Micheletti, C., Olivero, F., Robutti, O., & Paola, D. (1998). A model for analysing the transition to formal proofs in geometry. In *PME Conference* (Vol. 2, pp. 2–24).
Arzarello, F., Olivero, F., Paola, D., & Robutti, O. (2002). A cognitive analysis of dragging practises in Cabri environments. *Zentralblatt Für Didaktik Der Mathematik, 34*(3), 66–72.
Baccaglini-Frank, A., & Mariotti, M. A. (2010). Generating conjectures in dynamic geometry: The maintaining dragging model. *International Journal of Computers for Mathematical Learning, 15*(3), 225–253. https://doi.org/10.1007/s10758-010-9169-3

---

[2] The HyPE-13 project—Hybridizing and Sharing Teachings funded by the National Research Agency (ANR) is led by the University of Pau and the Pays de l'Adour (UPPA).

[3] See this simulator at https://fabien-emprin.pagesperso-orange.fr/actioninstr/.

Balacheff, N. (1998). Contract and custom: Two registers of didactical interactions. *The Mathematics Educator, 9*(2).

Bkouche, R. (1997). Epistémologie, histoire et enseignement des mathématiques. *For the Learning of Mathematics, 17*(1), 34–42.

Brau-Antony, S., & Mieusset, C. (2013). Accompagner les enseignants stagiaires: Une activité sans véritables repères professionnels [Trainee teachers mentoring: An activity without real professional benchmarks]. *Recherche Et Formation, 72*, 27–40.

Brousseau, G. (1997). *Theory of didactical situations in mathematics*. Kluwer Academic Publishers.

Brousseau, G. (2006). *Theory of didactical situations in mathematics: Didactique des mathématiques, 1970–1990* (Vol. 19). Springer Science & Business Media.

Coulange, L. (2012). L'ordinaire dans l'enseignement des mathématiques. In *Les pratiques enseignantes et leurs effets sur les apprentissages des élèves. Mémoire d'HDR, Université Paris-Diderot*.

Duval, R. (1993). Registres de représentation sémiotique et fonctionnement cognitif de la pensée. *Annales De Didactique Et De Sciences Cognitives, 5*(1), 37–65.

Emprin, F., & Sabra, H. (2019). Les simulateurs informatiques, ressources pour la formation des enseignants de mathématiques [Computer simulators, a resource for training mathematics teachers]. *Canadian Journal of Science, Mathematics and Technology Education, 19*(2), 204–216. https://doi.org/10.1007/s42330-019-00046-w

Emprin, F. (2007). *Formation initiale et continue pour l'enseignement des mathématiques avec les TICE: Cadre d'analyse des formations et ingénierie didactique*. Doctoral dissertation, Université Paris-Diderot-Paris VII.

Emprin, F. (2009). A didactic engineering for teachers education courses in mathematics using ICT. In *Proceedings of the sixth congress of the European Society for Research in Mathematics Education (CERME6)* (pp. 1290–1299).

Emprin, F., & Petitfour, É. (2020). Using a simulator to help students with dyspraxia learn geometry. *Digital Experiences in Mathematics Education*, 1–23.

Emprin, F., & Sabra, H. (2019). Les simulateurs informatiques, ressources pour la formation des enseignants de mathématiques. *Canadian Journal of Science, Mathematics and Technology Education, 19*(2), 204–216.

Emprin, F., Maschietto, M., & Soury-Lavergne, S. (2017). *Technologies pour l'enseignement, l'apprentissage et la formation en geometrie au premier degre, in actes de la 19ème école d'été de didactique des mathématiques organisé par l'Association pour la Recherche en Didactique des Mathématiques (ARDM)*. Actes prochainement en ligne.

Галиакберова, А. А., Галямова, Э. Х. & С. Н. Матвее, Э. Х. [Galiakberova, A. A., Galyamova E. K., & Matveev S. N.] (2020). Методические основы проектирования цифрового симулятора педагогической деятельности [Methodological basis for designing a digital simulator of pedagogical activities]. *Vestnik of Minin University, 8*(3). https://doi.org/10.26795/2307-1281-2020-8-3-2

Haenlein, M., & Kaplan, A. (2019). A brief history of artificial intelligence: On the past, present, and future of artificial intelligence. *California Management Review, 61*(4), 5–14. https://doi.org/10.1177/0008125619864925

Hennissen, P., Beckers, H., & Moerkerke, G. (2017). Linking practice to theory in teacher education: A growth in cognitive structures. *Teaching and Teacher Education, 63*, 314–325.

Hennissen, P., Crasborn, F., Brouwer, N., Korthagen, F., & Bergen, T. (2008). Mapping mentor teachers' roles in mentoring dialogues. *Educational Research Review, 3*(2), 168–186.

Hilbert, D. (1900). *Les principes fondamentaux de la géométrie*, traduit par L. Laugel, Gauthier-Villard (Eds.), *Paris, Bureau des longitudes de l'école polytechnique* (114 pp.). disponible à https://gallica.bnf.fr/ark:/12148/bpt6k996866.

Houdement, C., & Kuzniak, A. (2006). Paradigmes géométriques et enseignement de la géométrie. *In Annales De Didactique Et De Sciences Cognitives, 11*, 175–193.

Jones, K. (2000). Providing a foundation for deductive reasoning: Students' interpretations when using dynamic geometry software. *Educational Studies in Mathematics, 44*(1 & 2), 55–85.

Jones, K., Maschietto, M., & Mithalal-Le Doze, J. (2017). Introduction to the papers of TWG04: Geometry education. *CERME 10*, Feb 2017, Dublin, Ireland. hal-01925528.

Krämer, N. C., & Bente, G. (2010). Personalizing e-learning. The social effects of pedagogical agents. *Educational Psychology Review*, 22(1), 71–87.

Laborde, C. (1985). Quelques problèmes d'enseignement de la géométrie dans la scolarité obligatoire. *For the Learning of Mathematics*, 5(3), 27–34.

Laborde, C. (2000). Dynamic geometry environments as a source of rich learning contexts for the complex activity of proving. *Educational Studies in Mathematics*, 44(1–2), 151–161.

Laborde, C. (2001). Integration of technology in the design of geometry tasks with cabri-geometry. *International Journal of Computers for Mathematical Learning*, 6(3), 283–317.

Mammana, C., & Villani, V. (1998). *Perspectives on the teaching of geometry for the 21st century: An ICMI study*. Kluwer.

Mariotti, M.-A. (2000). Introduction to Proof: The mediation of a dynamic software environment. *Educational Studies in Mathematics*, 44(1 & 2), 25–53.

Mattéï-Mieusset, C. (2013). Les dilemmes d'une pratique d'accompagnement et de conseil en formation. Analyze de l'activite reelle du maitre de stage dans l'enseignement secondaire [The dilemmas of a mentoring practice and training. Analysis of the actual activity of the professional supervisor in secondary education].(Doctoral dissertation, Université de Reims Champagne Ardenne).

Mercier, A. (2018). Dessin, schéma, figure. Essais sur la formation des savoirs scientifiques. In Du mot au concept, PUG, 183-203.

Pastré, P. (dir.) (2005). *Apprendre par la simulation: de l'analyse du travail aux apprentissages professionnels*. Toulouse: Octarès, 363 pp.

Paris, C., & Gespass, S. (2001). Examining the mismatch between learner-centered teaching and teacher-centered supervision. *Journal of Teacher Education*, 52(5), 398–412.

Peyrard, F. (1804). *Les éléments de géométrie d'Euclide, traduits littéralement et suivis d'un Traité du cercle, du cylindre, du cône et de la sphère, de la mesure des surfaces et des solides, avec des notes*, F. Louis: Paris, 601 pp. Disponible à l'adresse http://gallica.bnf.fr/ark:/12148/bpt6k110982q

Rabardel, P. (2002). people and technology: A cognitive approach to contemporary instruments. Translated by Wood, H., université paris 8, pp.188, hal-01020705.

Restrepo, A. M. (2008). Genèse instrumentale du déplacement en géométrie dynamique chez des élèves de 6ème [instrumental genesis of dragging in a dynamic geometry software in 6th grade] (Doctoral dissertation, Université Joseph-Fourier-Grenoble I).

Richard, P. R. (2004). L'inférence figurale: Un pas de raisonnement discursivo-graphique [Figural inference: A step of discursive-graphic reasoning]. *Educational Studies in Mathematics*, 57(2), 229–263.

Robert, A., & Rogalski, J. (2005). A cross-analysis of the mathematics teacher's activity. An example in a French 10th-grade class. *Educational Studies in Mathematics*, 59(1–3), 269–298.

Sträßer, R. (2001). Cabri-geometre: Does dynamic geometry software (DGS) change geometry and its teaching and learning? *International Journal of Computers for Mathematical Learning*, 6(3), 319–333.

Vergnaud, G. (1998). Towards a cognitive theory of practice. In *Mathematics education as a research domain: A search for identity* (pp. 227–240). Springer, Dordrecht.

Sherard, W. H. (1981). Why is geometry a basic skill? *The Mathematics Teacher*, 74(1), 19–60.

Whitely, W. (1999). The decline and rise of geometry in 20th century North America. In *Proceedings of the 1999 Conference of the Mathematics Education Study Group of Canada*. St. Catharines, Ontario: Brock University.

Exploring Dynamic Geometry Through Immersive Virtual Reality and Distance Teaching

José L. Rodríguez

1 Introduction

This article explores the benefits that a virtual reality learning environment can offer for the teaching of geometric concepts and 3D constructions. Like in Kaufmann et al. (2000) our goal is working directly in 3D space to allow better and faster comprehension of complex spatial problems and relationships than traditional teaching methods (see also Song & Lee, 2002).

The importance of 3D virtual environments for learning spatial problems is known (see, e.g., Dalgarno & Lee, 2010; Extremera et al., 2021). Head-mounted displays (HMDs) access these environments and their elements with varying degrees of interaction and immersion. They have become popular in recent years, due to their high performance at an affordable price. We refer to the HTC Vive, Oculus Rift/S/Quest, or Windows Mixed Reality headsets, which allow both room positioning and hand tracking, thus improving the feeling of total immersion. On the other hand, stereoscopic glasses like Google CardBoard are much more affordable and still offer the illusion of 3D depth of images or videos on mobile phones. Between both types of VR there is a whole range of possibilities depending on the degrees of freedom (see Derks, 2020 for a review).

One can consult (Radiantia et al., 2020) and its references on it for a detailed and systematic description of recent studies that have been carried out on virtual reality applied in education, either with stereoscopic glasses or the modern HMD headsets; see also Prodromou (2020) for update references using augmented reality (AR) in education, a complementary technology that we are not considering in this article.

In VR learning environments, like Neotrie VR, the player can manipulate a figure and interact with its elements in real time. There are other virtual environments that

J. L. Rodríguez (✉)
Universidad de Almería, ES -04120 La Cañada de San Urbano, Almería, Spain
e-mail: jlrodri@ual.es

are used to teach and learn geometry. We refer to them in Sect. 2, presented according to the level of interaction with virtual objects.

In Sect. 3 we briefly introduce Neotrie VR and how some dynamic geometry tools work. This software was designed for face-to-face classes and HDM headsets, so the adaptation to online teaching has motivated the implementation of a third person camera and the stereoscopic vision for stereoscopic glasses.

Section 4 is devoted to describing the first two sessions of our teaching experience in detail. The participants are fourth-year students of the Mathematics degree at the University of Almería, in which some feedback is collected and analyzed.

Further testimonials about the use of various supports are collected in Sect. 5, which are used to reconsider how to teach the next classes in the course, as well as fixing the bugs and improving the tools in the software.

We then provide stereoscopic glasses to students so that they could watch online lessons and videos of Neotrie in 3D. A few students could further use the HDM headsets available in the VR room of our Math department, to test the new 3D graphing calculator in Neotrie; see Sect. 6.

Section 7 advances the development plans in Neotrie: one is the multiplayer mode and another the adaptation of the software to the Oculus Quest-2 headset (or similar devices emerging in the near future) that does not require a PC but still has hand controllers, room positioning, and reasonable quality.

We finish the paper with some further discussions and conclusions of our experience.

2 Levels of Object Manipulation

Before introducing Neotrie VR, we would like to put it in context, based on the degree of interaction with virtual objects. This is based on the own experience of the author, after using several supports and software during the last years.

We next list distinct levels of how to grasp and manipulate an object; in our case one can simply consider a vertex, a geometric figure, or any of its elements. The list gets closer and closer to reality, thus increasing the feeling of actually manipulating it. We go from interactive manipulation by dragging the mouse to grabbing objects (and any of its elements) as if they were real. On this scale, Neotrie reaches a high level very close to reality, as we will illustrate in the following sections.

- A vertex on a 2D screen (PC or portable computer) and a hand with the mouse selecting and moving the point: All the 2D DGS (dynamic geometry software) usually work in this way.
- A vertex in space seen on a 2D screen and a hand with the mouse selecting and moving the point; it is not known in which direction. Most 3D design software, like Blender, Autocad, etc. incorporate three orthogonal vectors (red, green, and blue) that indicate the directions to which vertices can move, or one can use a keyboard key to set the desired direction.

Exploring Dynamic Geometry Through Immersive Virtual Reality ...

- A vertex on a touchable screen so that it can be selected and moved with a finger. This is the way it is used in the current smartphones/tablets in augmented reality. See for instance, GeoGebra-AR, and [1] Grib3D,[2] Shapes3D.[3] All of them use the finger to select and move a vertex in space. However, it does not seem easy to know in which direction the vertex (or object) is moving in space, so 3D vision is often lost. More precisely, it is easy to move a point (or object) in the current horizontal plane, but not in space (up or down). This is overcome in immersive VR by using HMD headsets with hand controllers that we discuss in the next item.
- A vertex in a VR scene: Most VR software use controllers. A ray appears to select and move the vertex or object in general, so you do not grab the object with the hand directly. The sensation when taking an object is the same as when taking a Moorish skewer. Furthermore, using the two controllers, and the two rays coming out of each hand, allows you to rotate the figure and enlarge or decrease it. This is how GeoGebra MR works, for instance. Moreover, several VR systems, like Valve Index, HTC Vive series, Windows Mixed Reality, Oculus Rift/S/Quest, use the side buttons so that when you approach an object and squeeze your hand, it gives the feeling that you are holding on, and you can move it conveniently. For instance, CalcFlow uses this approach in its interaction with objects and panels (see Fig. 1).
- A vertex in a VR scene touchable by your virtual hands: Neotrie uses controllers, which the player sees as hands. And due to the precision it requires for dynamic geometry, to select the elements, it has further a pointer (small pink sphere over each hand) so that by pressing the trigger button (this can be configured by the player), it observes how the index and thumb fingers are closed to hold the vertex (or object) and you can move it in the desired direction, which exactly matches the actual direction of space (see Figs. 2a, b.)
- Higher levels of interaction: The effort to achieve near-real interaction with virtual objects is being carried out with devices such as haptic gloves (like the new Facebook's 3D printed virtual reality gloves), leap-motions devices (the case of HandWaver (Dimmel & Bock, 2019)), or mixed reality glasses with sensors that read hand gestures, such as the Microsoft Hololens.[4] Presumably, the precision of these devices will be improved more and more, but the possibility of having controllers in the hands with the possibility of pressing different buttons seems to us the most suitable right now for dynamic geometry in VR.

The advantages that are achieved with the current virtual reality HDM headsets have been analyzed in different contexts in recent years. We can closely observe the benefits obtained in other fields of education (e.g., see applications of VR in Extremera et al. (2021) for crystallography). We highlight the usefulness of virtual

---

[1] https://www.GeoGebra.org.
[2] https://grib3d.com.
[3] https://shapes.learnteachexplore.com.
[4] See for instance the project Holo-Math (https://holo-math.org/) to application in mathematics.

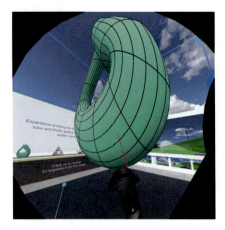 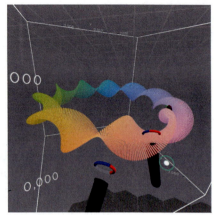

(a) A Klein bottle grabbed with the hand controllers.

(b) Strips with n-half turns grabbed with the hand controllers.

Fig. 1 Grabbing in GeoGebra MR (left) and CalcFlow (right)

 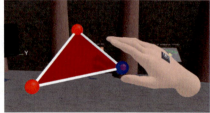

(a) Grabbing a Torus.

(b) Grabbing a selected vertex.

Fig. 2 Grabbing in Neotrie VR

reality to recreate and simulate situations that are difficult to carry out in reality. For instance, it allows us to experiment like never before with four-dimensional objects,[5] or immerse ourselves in spaces with non-Euclidean geometry (Hart et al., 2017).

3 Neotrie VR

Neotrie VR is a virtual reality software package designed for teaching and learning 3D geometry, mainly developed by Diego Cangas (CEO of the spin-off Virtual Dor of the University of Almería). It was first presented at a Congress in 2017 organized by the UNED, the Spanish distance university, where the use that virtual reality currently

---

[5] https://4dtoys.com.

has in the teaching of mathematics was already predicted; see Cangas et al. (2019a). Since then, the software has not lost its original purpose of virtually "touching" and "manipulating" geometry in a totally immersive scenario, greatly improving to the present, thanks to the collaboration of a team of teachers and researchers within an European Scientix project.[6]

Neotrie is currently used in several high schools and universities in Spain, Poland, the Netherlands, and Germany, from which we keep getting feedback and improvements to fix bugs as well as ideas to implement new tools and activities. Some case studies and classroom experiences have already been carried out with updated versions of the software; see Cangas et al. (2019b), Codina and Morales (2020), Chavil et al. (2020), Rodríguez and Romero (2019). A more in-depth study of the impact of Neotrie VR on the teaching and learning of geometry is being carried out at Codina et al. (2021).

3.1 How to Implement Neotrie in Face-to-Face and Online Teaching

In the website of Neotrie[7] one can find the minimum and recommended technical requirements for the PC, and compatible VR systems, as well as a quick installation guide.

The PC screen, which can also be projected on a giant screen, allows the students in the classroom to see what the player is doing inside the VR scene and to organize activities for small groups or the large group, even if only one person has the headset. This is how we normally work in the classroom face-to-face. The signal can also be transmitted by videoconference, sharing the full screen (Fig. 3).

A more affordable way of approaching VR is to use stereoscopic glasses for mobile phones, either for face-to-face or online teaching. Students not using the HDM headset will not have the feeling of room positioning nor interacting with the geometrical objects, but they will get 3D depth when looking at the figures and 3D explanations (Fig. 4).

The new camera gadget lets the teacher switch between the first and third person views, as well as choosing the stereoscopic option in both views (Fig. 5).

The advantage that we found in working on a 3D whiteboard in VR is that the player (with its camera in first person) can turn the head to quickly have different perspectives of the figure, helping to mentally recreate the figure on the rest of the viewers.

The new third person camera shows a fixed view of us, represented by our avatar, working in the VR scene. One can make gestures "in the air", point or serve as support in some spatial reasoning, etc. This camera can be moved to show different perspectives as well.

---

[6] http://www.scientix.eu/projects/project-detail?articleId=689498.
[7] http://www2.ual.es/neotrie.

Fig. 3 Gaming PC, Oculus Rift-S, and digital tablet Wacom used by the teacher

(a) PC screen of the teacher during the game with the HDM headset.

(b) Stereoscopic video seen on the student's glasses.

Fig. 4 Stereoscopic PC screen sent to students via video conference

From the teacher's point of view, this tool is very useful to explain concepts and reasoning in 3D. Of course, some training sessions are required to know how to use the controllers, the tools, how to import and export figures, images, or videos, how to save and load scenes, etc. On the other hand, students tend to learn very quickly, and a 10 min preparatory face-to-face session is sufficient.

Exploring Dynamic Geometry Through Immersive Virtual Reality ... 349

Fig. 5 Third person camera implemented in Neotrie VR. It allows students to get a fixed and more stable view for online teaching. Switch off button backs to the first person view, and the little glasses button activates the stereoscopic view

Fig. 6 One can select the "plus" symbol to create vertices and edges, "triangle" to create faces, "edit" to change positions of elements, "cross" to delete elements, "pencil" to free drawings, "object" to move and resize. Below there are open hand and finger touching positions, and "extrusion" modes to build prisms and pyramids

3.2 A Short Start in Neotrie VR and Their Tools

We next introduce the reader to the basic hand actions and use of various tools.

Pressing one of the buttons on the controller (button A or X in the Oculus system) opens a menu of options that are selected by passing the virtual hand over the symbols (see Fig. 6).

As it is shown in the video,[8] (see capture in Fig. 7) one can create vertices, edges, and faces by choosing the adequate hand mode and simply pressing the trigger button on the controller. For instance, the "object" mode on the hand allows you to grab any tool, and then one can select any element with each blue tip on one end of the tool. The *perpendicular tool* produces a point in the direction perpendicular to the plane formed by three vertices. The *compass* allows you to translate a distance between two points (the side), in the direction given by two other points. And the *parallel tool* produces a point on the outer parallel that passes through two others (Fig. 8a).

---

[8] https://www.youtube.com/watch?v=SVYH6pgOr10.

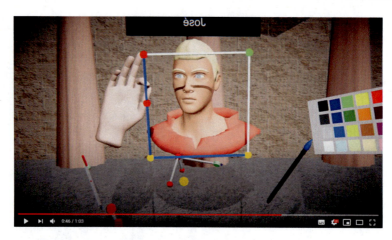

Fig. 7 Capture of the video of how to make a square in Neotrie VR. This construction allows you to move a point to change the plane, and the side of the square

(a) Tools needed to make a square: perpendicular, compass, and parallel.

(b) Further tools for dynamic geometry: slider, trace pencil, midpoint and intersection.

Fig. 8 Some tools in Neotrie VR. The red parts indicate the inputs for each tool and the green elements the outputs. The blue endings act as pointers for selecting input elements from the tools

In Fig. 8b we have other useful tools for dynamic geometry: The *slider tool* allows to make both a point moving linearly on a set of points, or on a circle. To produce a circle we have to touch first the center, then a point on the circle, and finally a perpendicular direction point above the center, to set the clockwise sense. The *trace pencil* allows drawing the trail of a moving point or line. The *midpoint tool* generates the midpoint of the selected points, and the *intersection tool* is used to make intersection of lines, planes, and line and planes, depending on what you select.

The reader can find on the website of Neotrie a detailed user guide and many other proposed activities, both for plane and spatial geometry.

4 A Real Case: Parametric Equations of Surfaces

We next describe our teaching experience with a group of students of the last year of the Math degree at the University of Almería, during the academic course 2020–21. The subject was introductory to Algebraic Topology, but we think that our methods could also be valid to teach calculus courses where curves and surfaces by parametric equations are treated, as well as other topics where 3D arguments are difficult to carry out with only 2D representations.

4.1 Contents Contextualization and Pedagogical Methodology

This course was taught for 15 weeks, from September 14 to December 21, 2020. Every week we give three theoretical-practical sessions of 1 h each and another session of 1 h to solve problems. The course was divided into two blocks. In the first one of 7 weeks we classified the compact and connected surfaces by means of its orientability, the number of boundary components, and the Euler characteristics. We followed our course notes from previous years (Gálvez & Rodríguez, 2019), complemented with classical references in the subject, like Kosniowski (1980), and extra material[9] elaborated by students from previous years. The second block of 8 weeks introduced students to the basics of homotopy theory, i.e. the fundamental group, Seifert and Van Kampen's theorem, and coverings theory, focused especially on compact surfaces, 2-complexes, and knot exterior 3-varieties.

Following Duval (2021), the course dealt with different representations of topological surfaces, the connection between them, transformations, and how and when to use each one is taken into account in the learning process. We also alternated observing and manipulating learning to increase the success of the course, as suggested in Freeman et al. (2014).

For instance, students made Möbius and cylindrical paper strips, studied their orientability according to the number of half turns, cutting them longitudinally by different parts. They saw what types of knots and links can appear, of narrow strips, orientable or not. They also explored properties of topological surfaces with the famous games by Jeff Weeks (2001). Further experiments within the standard square representations of the torus or the Klein bottle, with identified sides on the boundary can be found in Hawkins and Sinclair (2008).

Going back to the beginning of our course, one of the first preparatory sessions was to recall the parametric equations of the sphere, the torus, and the Möbius strip, to continue later with the quotient topology, group actions, varieties, etc.

---

[9] https://topologia.wordpress.com.

Table 1 Resume of relevant supports used in each session

Session	Type	Teacher's supports	Student's supports	Time
Sect. 4.3	Online lesson	Neotrie VR, HDM headset	PC	50 m
Sect. 4.4	Individual task		GeoGebra, PC	1 h

4.2 Justification of the Used Technologies

During the first weeks only half of the group of students was allowed to attend the face-to-face classes, although only about 3 or 5 people finally came. Moreover, in the week some lessons were totally online, and others mixed. Students mainly preferred to stay at home (this contrasts to 1st year courses, in which students have a greater need for face-to-face classes). Therefore, the teacher had to teach simultaneously to both groups, with a video camera, and a graphics pen drawing (our university has implemented the Blackboard Collaborate Ultra, and a Wacom digital tablet in each classroom). As a novelty this year, every session was recorded synchronously, so students could see them again at home. Extra videos were recorded, edited, and sent to students for better understanding along the course. Depending on what was required in each moment, the teacher used textbook notes, 2D whiteboard to write explanations (Google JamBoard, Windows OneNote), and also the HMD headset to explain in 3D, either in the classroom or at home. In Table 1 we resume the used supports for each session.

4.3 Observing Learning Session with Neotrie VR

This session was held on September 16, 2020. This was the first time that the teacher used Neotrie VR as a 3D whiteboard, for distance teaching. Students were all at home, and this session was recorded synchronously with the BlackBoard Collaborate Ultra video conference. The stereoscopic view was not implemented yet, but the third person camera was already available.

After a quick introduction to Neotrie and how to use the tools, the teacher started from the grid of the surface obtained from the gallery of the new 3D graphing calculator (Sánchez, 2020, see Fig. 9d).

The surface meshes were hidden to show only the grid so that the teacher's 3D annotations could be better seen. Figures 9a–c are screenshots of the deductions of the parametric equations of the sphere, the torus, and the Möbius strip. A short video was recorded, just after each explanation.[10] The reader will get first impressions of

---

[10] https://youtu.be/aTFu6RqC2ZQ.

Exploring Dynamic Geometry Through Immersive Virtual Reality ... 353

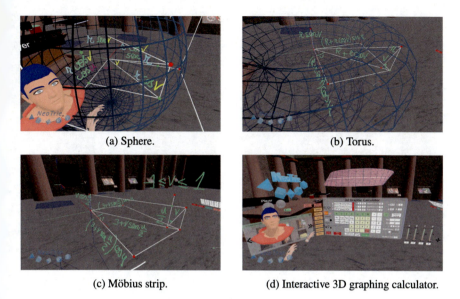

(a) Sphere. (b) Torus.

(c) Möbius strip. (d) Interactive 3D graphing calculator.

Fig. 9 Explaining parametric equations of surfaces in Neotrie VR, alternating first and third person cameras

how the students could see the different views of the figures and annotations by moving the teacher's first person camera.

In order to get the first feedback on using Neotrie, a short questionnaire was passed to the students. Anonymous responses collected from 34 students indicate a significant improvement in their perception of understanding of the deduction of the formulas, using Neotrie as a 3D whiteboard. Of course, we are not comparing this way of teaching these concepts with standard methods. It simply helps the teacher to know if the results are positive and the opinions of students; see answers in Fig. 10, questions (a) and (b).

We also asked the students what they preferred best for the teacher to use—the camera in the first person, in the third person or alternately; see answers in Fig. 10, question (c).

Due to the fact that the transmission was carried out in real time over the internet, most responded that they preferred a fixed third person camera (thus losing a bit of 3D vision). But alternating with the first camera (teacher's view) the students could get different perspectives of what is being explained in real time. In this regard, 21 students preferred the third person view (62%), 12 preferred alternating the views (33%), and only one student preferred the first camera view.

The option of alternating cameras will probably be better, once some kind of algorithmic stabilization has been implemented that slows down camera movement, even if the player moves faster.

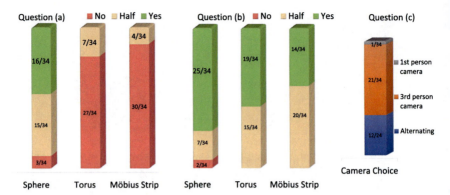

Fig. 10 Answers by students to questions **a** "Did you already know how to deduce the parametric equation of …?" Note that most may know the formulas, but not how to deduce them. **b** "The use of Neotrie VR helped you understand the parametric equation of …?" This shows a noticeable improvement in the perception of students. **c** "With which camera did you see the figures better and what we were doing in 3D?"

4.4 Manipulating Learning Task with GeoGebra

To guarantee the correct understanding of the explanations on the 3D whiteboard in Neotrie, a task with GeoGebra was then proposed. The students were asked to generalize the formula of the Möbius strip to others that made n half-turns, also varying their radius r and width a (see Fig. 11). Most of the students successfully carried out a correct parametric equation, obtaining a global mean of 8.35 on 10 points, from 29 presented (5 did not present the exercise).

5 Opinions on the Use of Various Supports

To the question "Did the use of GeoGebra help you to finish correctly deducing the parametric equation of the n half-turns strips, with variable width and radius", 75% of the students said yes, and 25% that it helped them half. Here it is evident that manipulating learning helps their understanding, since the student interacts directly with the figures and modifies them to see the changes produced. This was remarked by one of the students who wrote: "Working on the parametric equations myself, investigating how the figure changes according to the values of the sliders, made me understand how the shape of the n-half turning strips".

We conjecture that the combination of both supports, virtual reality for explaining 3D concepts and the use of GeoGebra for manipulating learning may have influenced the success of the practice in GeoGebra. For that, we asked for voluntary opinions, receiving 17 answers (see Fig. 12).

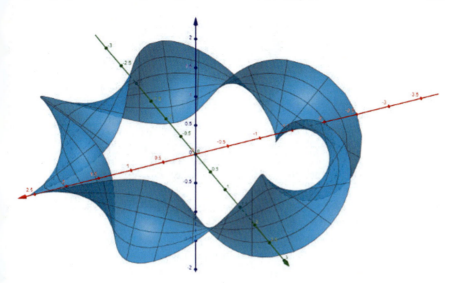

Fig. 11 Strips with n-half turns done in GeoGebra 3D by students, made with command Surface $((r + v\mathrm{Sin}(un/2))\mathrm{Cos}(u), (r + v\mathrm{Sin}(un/2))\mathrm{Sen}(u), v\mathrm{Cos}(un/2), u, 0, 2\pi, v, -a/2, a/2)$ and sliders grabbed with the mouse. The students had to figure out this formula for themselves

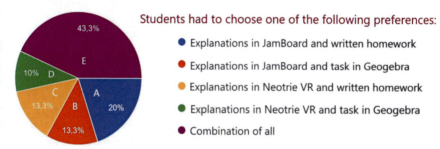

Fig. 12 Pie chart of supports chosen by students

A sample of interesting responses from students to justify their choices is collected in Table 2 (A1, A2, and A3 have chosen option A. Similarly, B1, B2, and B3 have chosen option B, etc.) (Table 2).

We can extract some interesting points from this survey:

1. Almost half of the students (43%) like E1 or E2 thought that the combined use of 2D and 3D tools in class, including GeoGebra tasks and written homework, were beneficial.
2. The combination of using Neotrie VR for explaining 3D concepts, together with GeoGebra (or similar DGS) for individual tasks, seems to be a recommendable option, well received by a majority group of students (66% chose C, D, or E).
3. However, a significant group (34% chose A or B) preferred the explanations on 2D whiteboard over 3D one. Regarding this, Student A1's argument is reasonable; it

Table 2 A sample of student's opinions on the use of different supports

Student	Opinion
A1	"Taking notes is easier and with the GeoGebra file was a bit caught"
A2	"Neotrie videos often drop low frame rate, making you lose concentration on the explanation. The GeoGebra application has given me many problems: the commands were deleted when I wanted to delete just a single character, and it was too slow"
A3	"Classic teaching methods are worth it to me, I don't notice much difference with other media. That's why I go to the classics that take up less time"
B1	"I prefer a normal explanation [2D whiteboard] whenever possible and then the use of other platforms helps to better understand these exercises"
B2	"I think that is what we are most used to".
B3	"Using several platforms creates some insecurity because in each new program or application I have to learn to move in the environment"
C1	"Doing homework with GeoGebra seems too laborious to me"
C2	"The tasks in GeoGebra took me a long time not knowing how to use it well"
D1	"Visualization is better, more comfortable"
D2	"The GeoGebra seems more useful to carry out tasks, it is more intuitive since the figure is seen as it is in 3D, while if I draw it I find it more difficult to see how the figure and perspective would look"
E1	"The explanations in 2D and the written task are the usual and as a complement, using the Neotrie helps you to clarify the doubts you have"
E2	"I think the more tools we use appropriately, the better the degree of understanding. Seeing with different perspectives helps"

is easy to take notes from 2D to 2D, as the formulas and the figures were "painted" in 3D. This could cause difficulty for some students. Others who chose A were supporters of methods close to traditional teaching, and apparently they did not need 3D software to visualize in 3D. Those who chose B agree with the use of new software but without going overboard.

4. Some students who chose option C stumbled upon GeoGebra, possibly because they weren't used to using it.

Given the variety of preferences, we encourage researchers to carry out more studies in this line with larger samples of students. We think that the combined use of several supports will benefit students with more difficulties in understanding or with less spatial, although not so much for more advanced students or those with greater spatial vision. At this point, we thought it is appropriate to implement the use of stereoscopic glasses and record 3D videos to help especially those with more difficulties.

Exploring Dynamic Geometry Through Immersive Virtual Reality … 357

Table 3 Resume of relevant supports used in each session/activity

Reference	Type	Teacher's supports	Students's supports	Time
Sect. 6.1	Videos	Neotrie VR, HDM headset	PC/phones/stereoscopic glasses	5–10 min per video
Sect. 6.2	F-to-F small groups		Neotrie VR, HDM headsets	60 m

6 Stereoscopic Videos and Extra VR Session

After carrying out the online session with Neotrie and the home task in GeoGebra 3D, we introduced the stereoscopic viewer option in the video camera that allowed students to view parts of a lesson in 3D, either broadcast or prerecorded. On the other hand, a few students were able to access the VR room. Table 3 is a resume of the supports used by the teacher and the students in each case.

6.1 Stereoscopic Videos

As in previous sessions, the camera can be placed where it is deemed best to see what you are doing, being able to switch between the first and third person camera. The added stereoscopic view would not be properly immersive virtual reality, and of course, nor would it have the high quality of HDM with hand controllers, but at least the student would have the perception of 3D depth of the figures, better than on his PC screen, as we said in Sect. 3.1.

Stereoscopic glasses for mobile phones, as in Fig. 5b, were given to 16 students so that they could watch the recorded 3D videos. (The rest did not attend face-to-face classes, then they could not pick up their glasses.)

Definitely, the 3D viewing was better done by each student individually to ensure maximum video quality, because due to connection speed problems, a 3D video broadcast in real time to the students caused a low frame update and low quality. In one of the prerecorded videos[11] we build a strip with 3 turns and in another one[12] we build again the Möbius strip (with 1 half-turn), finishing with its parametric equation. See captures in Fig. 13a, b.

We think that we have thus offered the best option to use Neotrie in the remote classroom, given the circumstances, and this is how the students have valued it:

- To the question "How do you rate that Neotrie videos can be viewed in 3D on mobile phones?", the average was 4 points out of 5.

---

[11] https://youtu.be/2irYPYOCkr4.
[12] https://youtu.be/oq4Xk99Sa64.

(a) Strip with 3 turns built with the slider, parallel, perpendicular and compass tools.

(b) Another parametric equation of the Möbius strip.

(c) Looking at torus knots, in the picture also the boundaries of n-turning half strips.

(d) Homotopy between the central double loop and the boundary in the Moebius strip.

Fig. 13 Sample of videos recorded in Neotrie VR and sent to students

- To the question "Would you like to see more 3D videos with mobile glasses in the course?", 85% answered yes.
- The time duration of this type of stereoscopic videos to choose between 5, 10, 15, and 20 min, was mainly 5 min (average 7 min).

Some students have dizziness and vision problems, so the videos for these cheap and low-quality glasses should preferably be short, and should be restricted to concrete situations, where we believe that 3D vision is beneficial to understand the figure or better follow some spatial reasoning.

We then kept recording other short videos related to the subject, in alternating stereoscopic and monoscopic views; see for instance Fig. 13c and the video,[13] and the capture Fig. 13d of the video.[14] The sliders bars were implemented out of the need to build these animations inside the VR scene.

For the interested reader, all the videos made with Neotrie on this subject are being compiled in a project entitled Introduction to Algebraic Topology on the website of Neotrie.[15]

---

[13] https://youtu.be/rlbr-s0wg4M.

[14] https://youtu.be/ALr5DdcAvYY.

[15] http://www2.ual.es/neotrie/project/introduccion-a-la-topologia-algebraica.

Exploring Dynamic Geometry Through Immersive Virtual Reality ...

 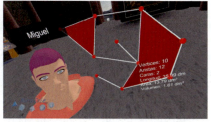

(a) Student playing with the parametric equation of the torus in Neotrie VR.

(b) Student building a figure, and showing information of it.

Fig. 14 Experimenting in the VR room of the Math department

6.2 Extra Session in the VR Room

The Math department of our university enabled at the beginning of 2019 a room with four VR headsets of Windows Mixed Reality, and the required PC gaming computers. Many students from 1st grade, 4th grade, and Master's students were able to carry out activities with previous versions of Neotrie.

This time, only a few students were able to do a practice. They could learn about Neotrie and how to use the basic hand actions and some of the tools, and in particular, they tried different options of the 3D graphing calculator, to represent some curves and surfaces (see Figs. 9d and 14a). They also investigated the changes in the Euler characteristics of 2-complexes (Fig. 14b).

According to their impressions and comments, totally immersive virtual reality would have been the best way to finish the sessions on parametric equations. An exhaustive study of the use and effectiveness of the 3D graphing calculator in the learning process should be completed and compared with the results of the cheap stereoscopic glasses.

7 Future Development Plans

It would be best if the teacher and the students were all working in the same virtual space, all with their own HMD, and interacting with each other, in a collaborative way. This could be done both in the VR room of our Math department, as well as each one at home if they had a high-quality VR system.

Neotrie had a multiplayer alpha version in mid-2019, but no study was done at that time, as the software had quite a few bugs, and it was not easy to connect with the online game.

There are already different platforms currently on the market, such as Engage,[16] but they do not incorporate tools and functions necessary to teach 3D Geometry, as Neotrie does. These are virtual reality environments where users interact, can share presentations, even 3D objects, and some allow writing in 3D too. That is, in part it could be used to teach or that students could interact. See for instance Charles Coomber's lesson on the popular video game Alix.[17]

Teachers must be attentive to the rapid development that is taking place worldwide, accelerated by the current pandemic period (see, e.g., Bambury, 2021).

Another important step in the popularization of Neotrie VR would be to adapt it to Oculus Quest series (or similar), which does not require a gaming computer and still have controllers and room positioning. All of these options will reduce costs making high-quality VR more accessible to schools and users.

An interactive web version of Neotrie would also be very useful, for students not having VR headsets or stereoscopic glasses or simply to give an alternative to those who suffer from dizziness or have vision problems to put on virtual reality glasses. An example to imitate would be the one developed by the sculptor Anton Bakker's.[18]

Augmented reality is an alternative technology increasingly used in the classroom, as it allows the student to visualize from different points of view, as one can move around the object, and would complement that offered by Neotrie VR in virtual reality. It could also be interesting to see in augmented reality what the teacher is making with his HDM headset in real time.

From now on, we will begin to see how well-known dynamic geometry software develop their own virtual reality versions, such as GeoGebra MR, or Cincerella, with its Cindy VR. It would be desirable to be able to import and export Neotrie files to each of them. Even being able to view a GeoGebra file or CalcFlow, in a Neotrie scene, and vice versa, OBJ or STL objects are right now interchangeable models but their geometric properties and interactivity with sliders, for instance, are lost.

8 Conclusions

The difficulty of current augmented reality for teaching geometry is that the dragging or handling of the points is still done on a flat screen; before it is done with the mouse on the PC screen and now with the finger on the screen of a mobile device. This difficulty is easily overcome with high-quality VR systems, which facilitate direct interaction with the 3D objects, thanks to their hand controllers. Hence, the advantage of using Neotrie with HDM headsets is clear: Objects are viewed and manipulated as if they actually existed in space and not projected onto a computer

---

[16] https://engagevr.io.
[17] https://www.youtube.com/watch?v=_T2-9MwA5JI.
[18] Geometric artwork in augmented and virtual reality https://www.antonbakker.com/momath.

screen. It is therefore much easier to build 3D figures in a virtual reality environment where we are seeing the real depth and angles, which allows us to select and handle the elements correctly and faster.

But, how to organize the lessons with the restrictions caused by the Covid-19 pandemic? In this paper, we have described a possible solution by using Neotrie as a 3D whiteboard, and also as a tool to produce monoscopic and stereoscopic videos for both the classroom and home. To this concern, we have found that students adopt different positions toward the use of VR, as an observing learning tool.

Our results show that the use of Neotrie by the teacher (observing learning by students) has favorably influenced the perception that students have of the derivation of parametric equations of surfaces. This improvement in perception is also based on exercises with GeoGebra that allow manipulation with equations. Without claiming that these results are definitive, due to the small sample, almost half of our students (43%) prefer the combined use of 2D and 3D tools in class, including GeoGebra tasks and written homework. They mainly think that 3D tools like GeoGebra or Neotrie are very good at visualizing, and Jamboard (or similar online apps) is important because it helps to take better notes. Two out of three students support the use of Neotrie as a 3D whiteboard. On the other hand, some advanced students may not need these technological tools to follow the classes, while others had technical difficulties using GeoGebra.

Reduced groups could have been later working in the VR room, but unfortunately we could not continue after the strict recommendations from our responsible for risk prevention. Our experience from previous years certainly confirms that it would have been very convenient for students.

The plan envisaged by the Neotrie developers is to adapt the software to the Oculus Quest 2 headset (or similar). This would greatly reduce costs, as this system does not require a PC, which would make virtual reality technology more accessible for many schools. On the other hand, the multiplayer mode will allow immersive collaborative activities between groups, even with members in different countries.

The personal experience of the author using VR, and particularly Neotrie, is incredible, being able to explain parametric equations better than in previous courses. It has been easier to explain 3D concepts on the 3D board than to use pen and paper, or the 2D board. It would be interesting to receive comments in this regard from other teachers who use this tool.

On the student's side, is making it so easy for them to see and manipulate 3D objects an obstacle to learning 3D geometry? Didactic activities should be designed to take full advantage of new technological support for students to explore topics beyond their scope with current systems of representation.

The need for new strategies has motivated the use of VR for distance teaching, as well as the development of new tools and improvements in Neotrie. However, there is still much to investigate and test with this software, and find the most effective strategies in each educational level, in order to improve the teaching and learning of 3D geometry by using virtual reality.

We ended up with a quote from Ng and Sinclair (2018), which refer to 3D pens; "3D drawing is dynamic and unregimented; it opens up a new, 3D territory for mathematizing that was unimaginable in the era of paper and pencil, and even in the era of the computer screen". The same could be claimed for the totally immersive DGS that is being developed in virtual reality.

Acknowledgements To Diego Cangas for the daily technical support, and to the members of the Neotrie project, who with their contributions are enriching the software to make it more useful and interesting to teach geometry. Special thanks to Isabel Romero for the accurate improvements made in all the revisions of this work, as well as the effort and substantial changes proposed by the referees. The author was partially funded by the Ministry of Science and Innovation grant PID2020-117971GB-C22 and FEDER-Junta de Andalucía grant UAL2020-SEJ-B2086.

References

Cangas, D., Crespo, D., Rodríguez, J. L., & Zarauz, A. (2019a). Neotrie VR: Nueva geometría en realidad virtual. *Pi-InnovaMath, 2*, 1–8.

Cangas, D., Morga, G., & Rodríguez, J. L. (2019b). Geometric teaching experience with Neotrie VR. *Psychology, Society, & Education, 11*(3), 355–366.

Chavil, D. Y., Rodríguez, J. L., & Romero, M. I. (2020). *Introducción al concepto de fractal en enseñanza secundaria usando realidad virtual inmersiva. Desde Sur Rev Cienc Hum Soc, 12*, 615–629.

Bambury, S. (2021). *Immersive Tech For Distance. Learning pt. 1*. https://www.virtualiteach.com/post/immersive-tech-for-distance-learning-pt-1. Cited February 17, 2021.

Codina, A., & Morales C. (2020). *Cognición y metacognición en geometría con realidad virtual utilizando Neotrie VR*. Investigación en didáctica de la matemática, homenaje a Encarnación Castro, Editorial Octaedro 2020 (pp. 155–178).

Dalgarno, B., & Lee, M. (2010). What are the learning affordances of 3-D virtual environments? *British Journal of Educational Technologies, 41*(1, Special Issue: Crossing Boundaries: Learning and Teaching in Virtual Worlds), 10–32.

Derks, L. (2020). *How to explain virtual reality*. https://ictlaurens.medium.com/how-to-explain-virtual-reality-415c2dc89277. Cited November 4th, 2020.

Dimmel, J., & Bock, C. (2019). Dynamic mathematical figures with immersive spatial displays: The case of handwaver. In G. Aldon & J. Trgalová (Eds.), *Technology in mathematics teaching*. Mathematics Education in the Digital Era (Vol. 13). Cham: Springer.

Duval, R. (2021). Registres de représentations sémiotique et fonctionnement cognitif de la pensée. In *Annales de Didactique et de Sciences Cognitives* (Vol. 5, pp. 37–65). IREM de Strasbourg, France.

Extremera, J., Vergara, D., Dávila, L. P., & Rubio, M. P. (2021). Virtual and augmented reality environments to learn the fundamentals of crystallography. *Crystals, 10*(6), 456, 1–18.

Freeman, S., Eddy, S. L., McDonough, M., Smith, M. K., Okoroafor, N., Jordt, H., & Wenderoth, M. P. (2014). Active learning increases student performance in science, engineering, and mathematics. *Proceedings of the National Academy of Sciences, 111*(23), 8410–8415.

Gálvez, J. F., & y Rodríguez, J. L. (2019). *Manipulación y clasificación de superficies compactas*. Bubok Publishing SL.

Hart, V., Hawksley, A., Matsumoto, E., & Segerman, H. (2017). Non-euclidean virtual reality I: Explorations of H^3. In *Proceedings of Bridges 2017: Mathematics, Music, Art, Architecture, Culture* (pp. 33–40) Waterloo, Canada.

Hawkins, A., & Sinclair, N. (2008). Explorations with sketchpad in topogeometry. *The International Journal of Computer Mathematics, 13*, 71–82.

Jang, S., Vitale, J. M., Jyung, R. W., & Black, J. B. (2017). March). Direct manipulation is better than passive viewing for learning anatomy in a three-dimensional virtual reality environment. *Computers & Education, 106*, 150–165.

Kaufmann, H., Schmalstieg, D., & Wagner, M. (2000). Construct3D: A virtual reality application for mathematics and geometry education. *Education and Information Technologies, 5*, 263–276.

Kaufmann, H. (2009). Virtual environments for mathematics and geometry education. *Themes in Science and Technology Education, 2*(1), 131–152.

Kosniowski, C. (1980). *A first course in algebraic topology*. Cambridge University Press, (first printed version, 1980; online publication 2010). https://doi.org/10.1017/CBO9780511569296.

Ng, O., & Sinclair, N. (2018). Drawing in space: Doing mathematics with 3D pens. In L. Ball, P. Drijvers, S. Ladel, H.-S. Siller, M. Tabach, & C. Vale (Eds.), *Uses of technology in primary and secondary mathematics education* (pp. 301–313). Cham: Springer.

Prodromou, T. (Ed.). (2020). *Augmented reality in educational settings*. Leiden Boston: Brill Sense.

Radiantia, J., Majchrzaka, T.A., Frommb, J., & Wohlgenannt, I. (2020). A systematic review of immersive virtual reality applications for higher education: Design elements, lessons learned, and research agenda. *Computer & Education, 147*.

Rodríguez J. A., & Romero, M. I. (2019). Optimización de superficies a partir de un volumen dado mediante realidad: una experiencia en 6° de Primaria. *Actas del Congreso Innovación y tecnología en contextos educativos*, UmaEditorial (pp. 583–593).

Rodríguez, J. L., Romero, I., & Codina, A. (2021). The Influence of NeoTrie VR's immersive virtual reality on the teaching and learning of geometry. *Mathematics, 9*, 2411. https://doi.org/10.3390/math9192411.

Sánchez, C. (2020). *Visualización y tratamiento de gráficas 3D en Neotrie VR*, final degree project, Universidad de Almería.

Sinclair, N., Bartolini, B., Maria, G., de Villiers, M., Jones, K., Kortenkamp, U., Leung, A., & Owens, K. (2016). Recent research on geometry education: An ICME-13 survey team report. *ZDM Mathematics Education, 48*, 691–719.

Song, K. S., & Lee, W. Y. (2002). A virtual reality application for geometry classes. *Journal of Computer Assisted Learning, 18*, 149–156.

Weeks. J. (2001). *Exploring the shape of space (2001) and the Jeffrey Weeks' geometry and topology website*. http://www.geometrygames.org/.

Historical and Didactical Roots of Visual and Dynamic Mathematical Models: The Case of "Rearrangement Method" for Calculation of the Area of a Circle

Viktor Freiman and Alexei Volkov

1 Introduction

It has been widely acknowledged that the concept of areas of plane shapes and rigorous methods of their calculation are complex and difficult to learn and to teach.[1] Among these methods, the calculation of the area of a circle is particularly challenging as it requires dealing with infinitesimal methods and an irrational number (π). At the same time, it is considered crucial for middle school curriculum being embedded into a rich context of real-life applications along with introduction of ideas and concepts which are supposed to be used in later schooling.[2] Not surprisingly, the number of resources produced to help students deal with the complexity of explanations and justifications of the formula for the area of a circle has been growing, both in the traditional form of printed textbooks and teacher's manuals[3] and in novel, digital format. The latter category includes a number of online resources with useful information (explanations and exercises, learning scenarios and suggestions) for teachers, some of which are published by official curriculum bodies (e.g., NCTM,[4]

---

[1] See, for instance, Kordaki and Potari (1998), Rejeki and Putri (2018), among others.
[2] Cavanagh (2008), O'Dell et al. (2016).
[3] See, for instance, Van de Walle et al. (2020), p. 530. For a preliminary discussion of some of them, see Freiman and Volkov (2006).
[4] NCTM (National Council of Teachers of Mathematics), see nctm.org.
[5] LearnAlberta.ca is a website providing digital learning and teaching resources correlated with the educational programs from kindergarten to Grade 12 approved by the Government of Alberta Province, Canada.

V. Freiman (✉)
Université de Moncton, Moncton, NB, Canada
e-mail: viktor.freiman@umoncton.ca

A. Volkov
National Tsing-Hua University, Hsinchu, Taiwan

© The Author(s), under exclusive license to Springer Nature Switzerland AG 2022
P. R. Richard et al. (eds.), *Mathematics Education in the Age of Artificial Intelligence*, Mathematics Education in the Digital Era 17,
https://doi.org/10.1007/978-3-030-86909-0_16

Learn Alberta,[5] and NZ Math[6]). With the raise of digital tools and environments, several applications (applets) were created allowing dynamic investigation of the formula of the area of a circle by the learners.[7]

In the present paper, we shall argue that at some point in the history the educators started adopting (sometimes with considerable modifications) methods coming from pre-modern (ancient and medieval) mathematical treatises available to them via translations and interpretations provided by historians of mathematics. The reason for this adoption usually was not explicitly stated by the educators and deserves a further study; for the time being we assume that the educators did it in hope that the "archaic" (and, therefore, presumably "simple") methods could stimulate interest and facilitate understanding of the learners.[8] It is also possible that this conviction of the educators may have been based on a tacit assumption concerning the growth of the child's understanding of geometrical concepts as mirroring the stages of their historical genesis.

In the context of teaching geometry, this vision led the educators to a transition from approaches found in Euclidean *Elements* (ca. 300 BC) and based on quasi-formal deduction system[9] to more practical, visual, intuitive, and dynamic ones.[10] However, while some of these methods indeed allowed a relatively simple visualization, the others involved concepts and ideas particularly difficult for young learners. The infinitesimal methods of calculation of the areas of a circle and its parts, as well as the methods of calculation of the volumes of a pyramid and a sphere, arguably belonged to the latter category. In the case of calculation of the area of a circle, the search for an appropriate didactical apparatus suitable for this case led some educators to the use of so-called "method of rearrangement of sectors" or, more precisely, a group of approaches based on the division of circle into sectors followed by rearrangements of different kinds and investigation of possible approximations based upon them.

Nowadays, this *rearrangement method* (we will use this term further in the chapter for the sake of convenience) seems to be the most often used by the authors of textbooks who suggest it for teaching calculation of the area of a circle; it can be

---

[6] See https://nzmaths.co.nz/, the website containing information about mathematics education in New Zealand.

[7] See, for instance, https://www.mathed.page/constructions/pi/index.html for an applet designed by Henri Picciotto. With the help of this applet a circle can be divided into $2n$ sectors, $n = 6, \ldots, 180$, and these sectors can be rearranged to form a "rectangle-like" figure.

[8] See, for instance, Francis (1976), Ernest (1998), Marshall and Rich (2000), Farmaki and Paschos (2007), Furinghetti (2007), Jankvist (2009), Mac an Bhaird (2011) (listed in chronological order). The earliest publications of this kind we are aware of were authored by D.E. Smith (1860–1944); yet, he was not the only one who claimed that history of mathematics can be used in classroom; see, for instance, Wiltshire (1930).

[9] For a recent study of the logical structure of Euclid's *Elements*, see Acerbi (2008).

[10] On this process see, for example, Montoito and Garnica (2015), Herbst et al. (2018).

43. **JUSTIFYING THEOREM 11.9** You can follow the steps below to justify the formula for the area of a circle with radius r.

Divide a circle into 16 congruent sectors. Cut out the sectors.

Rearrange the 16 sectors to form a shape resembling a parallelogram.

a. Write expressions in terms of r for the approximate height and base of the parallelogram. Then write an expression for its area.

b. *Explain* how your answers to part (a) justify Theorem 11.9.

Fig. 1 A "justification" of the formula of the area of a circle. Larson et al. (2007), p. 755

found in almost all modern textbooks for middle school mathematics and in numerous resources available online.[11]

In particular, the *rearrangement method* explores the idea of dividing the circle into a number of equal sectors, then rearranging them in a way that makes it look like a familiar figure, namely, a polygon (e.g., parallelogram) for which the area can be easily found (see Fig. 1).

When facing this seemingly attractive, "visual," and "dynamic" justification of the formula, the learners are expected to follow the implied reasoning procedure and are supposed to realize that the area of the rearranged shape is "resembling a parallelogram," and, consequently, the area of the original shape (i.e., the circle) is equal to the product of the "height" and "base" of this "parallelogram." Further, the learners are guided toward expressing the area of a parallelogram using the radius, which is supposed to be the "approximate height" of the parallelogram, and the half of the circumference, identified as the "approximate base," and thus would obtain a "justification" (apparently, problematic[12]) of the area of the circle.

Why did the authors of this and other textbooks choose this particular approach? While seemingly appealing to students' intuition, it does contain hidden assumptions

---

[11] Freiman and Volkov (2006). The printed textbooks and teaching materials available online that feature this method are too numerous to be listed here; for the most recent ones see, for example, Johnson and Mowry (2016), pp. 574–575 and the website https://www.colorado.edu/csl/2017/03/23/slices-pi containing an animation of this kind. For an analysis of the attempts to use digital technology to estimate the value of π, see de Silva et al. (2021).

[12] There are two elements that make this kind of "justification" problematic: first, the left part of the picture in Fig. 1 shows that the sectors do not seem congruent while in the right part they are obviously congruent; second, the term "approximate" used in connection with the terms "base" and "height" of the "shape resembling a parallelogram" does not help the learner understand why the area of the circle is equal *exactly* to the product of the radius and circumference. Much of the work dealing with these subtleties is thus left to the teacher.

that are far from being obvious.[13] Stacey and Vincent (2009), for example, discuss this method among other methods that can be used for introducing the area of a circle (such as a "square grid" method or an "onion layers" model) to young learners; they stress that when using such methods the educators should be aware of "a difference in the mathematical quality of the arguments between the empirical *counting squares* method and the approaches that use deductive reasoning" (ibid., p. 9). The authors stress that these methods include "very complicated limiting processes involved in formally proving the formula," and the instructors need to take these complex elements into account when presenting students with intuitive models that might help shaping their conceptual understanding (ibid., p. 8).

In line with Tamborg (2017) who refers to classical work of Vérillon and Rabardel (1995), we look into the process of growth of conceptual understanding through a view of human cognition and knowledge building as artifact-mediated activities in which "artifacts (as opposed to natural objects) possess cultural and historical dimensions because they are constructed with a particular purpose and a particular way of fulfilling this purpose in mind" (op. cit., p. 3). These "cultural and historical dimensions" are not always made explicit, in our opinion, especially when the artifacts undergo transformations in the process of transition from traditional ("usual") "mathematical working spaces" (hereafter MWS) to modern ("new") MWS (e. g., in particular, those supported by digital technology).[14]

In our view, it appears promising to adopt a *genetic* paradigm that relates *developmental process* of mathematical work and the *manner* by which it was created and took shape. This paradigm dialectically unites mathematical, didactical, and historical aspects of an artifact which further could be transformed through, for example, an "instrumental genesis" (that is, an instrumentalization and instrumentation, as result of an interplay between a subject and an artifact, as a part of "didactical situation"; see, for example, Brousseau (1998) and Trouche (2005)). In this complex and transformational (eco)system of knowledge development, we consider the *rearrangement method* as an *artifact* whose history goes back to the Antiquity and whose genesis (or even "geneses") has/have not yet been made sufficiently explicit. While the ancient history of the *method* seems to be lost in time and space, as we will show in this chapter, it was later (approximately in the mid-second millennium CE) refined by mathematicians to help them solve some particular problems (related to the calculation of the area of circle), or to conduct investigations of other infinitesimal processes before being taken over by the nineteenth–early twentieth-century educators as "visual support" for learners. At that time, the *rearrangement method* was introduced propaedeutically, as a (novel) didactical approach in introductory geometry courses; in the case of the area of a circle, it was featured as an alternative to the relatively technical approaches of Euclid and Archimedes. Apparently, the

---

[13] This "justification" is apparently based on the unstated assumption that the length of any curved line getting closer to a given straight line tends to the length of this straight line, which is mathematically incorrect.

[14] On the concept of MWS see, for instance, Kuzniak and Richard (2014) and Richard et al. (2019).

educational reforms of the late 1990s–early 2000s (which coincided with the introduction of electronic calculators and other digital tools) placed it in the foreground of modern didactical innovations, yet without a clear explanation of the history of its genesis and possible pitfalls, as well as the challenges it might bear for teachers and students.

Therefore, we begin our chapter with a brief history of the origins of the *rearrangement method* and infinitesimal discourses that accompanied some of its versions. We will then investigate some didactical ideas brought by nineteenth–early twentieth-century educators into their textbooks. We will finally explore how these works shaped modern development of innovative spaces of reflection and practice in mathematics education. At the end of the chapter, we will outline epistemological and didactical issues that modern educators might need to be aware of when using the *method* and will also discuss how a further and deeper investigation of both historical and instrumental geneses could help dealing with these issues, both in theory and in a classroom practice.

2 Historical Roots of the *Rearrangement Method*

The origin of the *rearrangement method* is unknown. In his book on the history of π originally published in 1971,[15] Petr Beckmann (1924–1993) claimed that *rearrangement method* was used by some unidentified "ancient peoples":

> The ancient peoples had rules for calculating the area of a circle. [...] They probably did it by a method of rearrangement. They calculated the area of a rectangle as length times width. To calculate the area of a parallelogram, they could construct a rectangle of equal area by rearrangement as in the figure below, and thus they found that the area of a parallelogram is given by base times height (1976, pp. 17–18),

and provided the diagrams shown in Fig. 2.

The word "probably" used by Beckmann in the above-sited quotation reveals that the author most probably *provided here his own reconstruction* that, he apparently believed, was not found in any particular extant historical documents produced by the unidentified "ancient peoples" (who, he claimed, devised this method "almost five millennia before the integral calculus was invented"). (Beckmann, 1976, p. 17).

---

[15] Even though the author himself admitted that he was "neither an historian nor a mathematician" (op. cit., p. 3), his book was praised by some of its reviewers (see, for instance, Brieske's 1977 review of its edition of 1976 and even more recent review of its edition of 1989 by Blank (2001), p. 155). However, mathematicians and historians of mathematics provided a detailed critique of this book accompanied with a list of numerous misprints and errors (some of them indeed quite surprising, such as, for example, the claim that $p! + 1$ which is not divisible by $2, 3, \ldots, p$, "is therefore a prime") that can be found in Gould 1974 who (quite justly, in our opinion) concluded that "[t]he book should be rewritten entirely and the manuscript should be examined by qualified mathematicians and historians before being committed to the printed page" (p. 327). Unfortunately, this work has never been done.

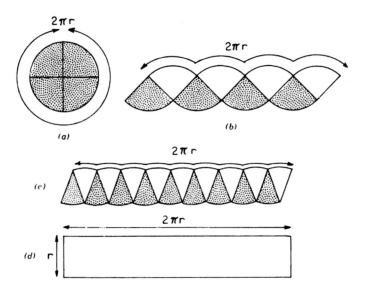

Fig. 2 Diagrams from p. 18 of Beckmann (1976) identical with the diagrams in the first edition of his book (1971). The caption reads: "Determination of the area of a circle by rearrangement. The areas of the figures (**b**), (**c**), (**d**) equal exactly double the area of circle (**a**)." The origin of the diagrams is not specified

It thus remains unknown from where Beckmann borrowed this diagram, apparently radically different from the one used by Archimedes of Syracuse (287–212 BC), to prove his statements concerning the area of a circle.[16] Some authors, when putting this method side by side with the one used by Archimedes, remind the reader that this method should be considered *less rigorous* and needs to be handled with care, for example, Bryan and Sangwin (2008, p. 145) write:

> We shall approximate the area of each sector by such a triangle. You would be well advised to retain a suspicious attitude towards the cavalier way we forge ahead with approximations such as these.

Similarly, (Casselman, 2012) states

> Roughly speaking, the reason the Theorem is true is that we can carve up both the circle and the triangle into very small regions that approximate each other closely in area.

A number of books, articles, and online resources featuring rearrangement diagrams (along with some descriptive texts) rarely contain the necessary references to the original works in which this method was introduced. Some of them mention David E. Smith's and Mikami Yoshio's 三上義夫 (1875–1950) *A History of Japanese Mathematics* (1914) where a similar diagram presumably dated of 1698 and credited to the authorship of Satō Moshun's 佐藤茂春 (whose name can also be spelled Satō Shigeharu; fl. late seventeenth century) can be found on p. 131, as

---

[16] Archimedes (1544), pp. 44–46, 1615, pp. 128–133, 1676, pp. 81–97, 1880, pp. 257–271.

well as more recent books on the history of mathematics, in particular, the monographs authored by Carl B. Boyer (1906–1976) (Boyer, 1959) and Margaret Baron (1915–1996) (Baron, 1969). In her book, M. Baron refers (p. 110) to the approach used by J. Kepler (1571–1630) whereas C. Boyer, while also pointing at Kepler's work, mentions the works of Nicholas of Cusa (1401–1464) as the possible origin of Kepler's approach (Boyer, 1959, p. 108). In turn, Beckmann (1976, p. 19) conjectures that Leonardo da Vinci (1452–1519) may have used the same method; he, however, does not provide any references to Leonardo's works.

In his summary of Nicholas of Cusa's works on mathematics, Hofmann (1971) lists several treatises of Cusanus related to the squaring of circle.[17] The earliest statements of Nicholas concerning a circle and an inscribed polygon are found in his *De Docta Ignorantia* (On learned ignorance) completed in 1440.[18] Also, in Chap. 3 of Book 1 of the treatise titled *Quod praecisa veritas sit incomprehensibilis* (The precise truth is incomprehensible) Cusanus uses the relationship between these two figures as a metaphor of the process of approaching the absolute truth. It is interesting that Cusanus makes a mathematically correct statement: at any step the area of inscribed polygon is less than the area of the circle, and two of them cannot be equal to each other unless the polygon becomes identical with the circle (he, apparently, understands that this cannot happen at any finite step of the procedure). Later Cusanus returns to the topic in his other works including *De circuli quadratura, pars theologica* (On Squaring the Circle, Theological Part), completed in 1450, *Quadratura circuli* (Squaring the Circle), 1450; *De sinibus et chordis ou Dialogus de circuli quadratura* (On sines and chords, or Dialog on the Quadrature of the Circle), 1457, and *De caesarea circuli quadratura* (On Squaring the Caesarian Circle), 1457.[19]

Pierre Duhem (1861–1916) in his study of Leonardo (1909, vol. 2) devotes a short section titled "L'infiniment grand et l'infiniment petit dans les notes de Léonard de Vinci" (pp. 49–53) to Leonardo's concepts of infinitely small and infinitely large entities. Duhem quotes Leonardo's claim that "La Géométrie est infinie parce que toute quantité continue est divisible à l'infini dans l'un et l'autre sens" (p. 50);[20] he borrows this quote from Manuscript M of the Bibliothèque de l'Institut[21] and provides his own reconstruction of Leonardo's drawing that is supposed to illustrate

---

[17] For more details, see Watanabe (2011), pp. xx–xxvi.

[18] Watanabe (2011), p. xxi.

[19] For details, see Watanabe (2011), pp. xxiv–xxv. On Cusanus' interest in mathematics and evaluation of his mathematical writings by his contemporaries, see the section "[Nicolaus von Cusa:] Die mathematischen Schriften" in Scharpff (1871), pp. 294–323; for more recent works, see Boyer (1959) (esp. see p. 91), Counet (2005) (esp. see pp. 286–289) and Vengeon 2006. Albertson (2014) on p. 172 cites a (negative) evaluation of Cusanus' geometry by Regiomontanus (1436–1476); see also idem, p. 255, n. 15 for relevant references. German translation of Cusanus' mathematical writings is available in Nikolaus von Kues (1980), while their French translation is found in Nicolas de Cues (2007).

[20] That is, "Geometry is infinite, because every continuous quantity is infinitely divisible from one direction to another" (da Vinci, 1964, p. 127). The original statement of Leonardo is found on the same page of da Vinci (1890).

[21] Da Vinci (1890), vol. 5, Ms M, fol. 18r.

this claim. Surprisingly, Duhem does not mention here Leonardo's fragment and diagrams directly related to the calculation of the area of a sector found in Manuscript E.[22] Leonardo gives his verbal formula for the area of a circle on the first page (verso) of this manuscript; Félix Ravaisson-Mollien (1813–1900) provides his French translation on the same page (1888, ibid), in our English translation it reads: "The circle is equal to a quadrilateral made of a half of the diameter of this circle, multiplied by a half of the circumference of the same circle."[23] Leonardo offers a description of his procedure accompanied by several pictures; its translation can be found in the mentioned works of C. Ravaisson-Mollien and V. Zubov.

In the context of infinitesimal (pre-calculus) debates of the sixteenth–seventeenth-century, Johannes Kepler's discussion of the circumference and area of a circle is found in his treatise *Nova Stereometria Doliorum Vinariorum* (New Stereometry of Wine Barrels) published in 1615; it makes important setback from Archimedes proof putting upfront practical utility against rigor. It opens with "Theorem 1" stating that the ratio Circumference: Diameter is close (but not equal!) to 22:7. Kepler provides his own proof using only inscribed and circumscribed regular hexagons and mentions that Archimedes found this ratio using polygons with 12, 24, and 48 sides. In his Theorem 2, Kepler states that the ratio of the area of a circle and the square built on its diameter is approximately equal to 11:14. It is surprising to find this outdated approximation in Kepler's work right after his mention of the much more precise approximation (3.141592653589793) attributed by him to Adriaan van Roomen (1561–1615), also known as Adrianus Romanus.[24] One can conjecture that for Kepler the main reason was not obtaining a good approximation of π; instead, he focused on the *method* which, as he stressed, *differed* from the method of Archimedes.[25]

3 The *Rearrangement Method* in Nineteenth–Early Twentieth-Century Western Textbooks

While logico-deductive approach to geometry based on Euclid's work was having significant influence on teaching geometry up to the modern times, alternative approaches were also making their way being grounded in its practical origins as "the art of measuring well" (Ramus, 1569, cited in Menghini, 2015). Hence, the activity of measuring, as argued by Marta Menghini (ibid.), was "preceded by a very interesting and original part of observation, a 'play' with figures" in which proofs were supported by geometric constructions and measurement activities by drawing. Later on, by the end of the eighteenth century, according to Menghini (2015, p. 568), one can observe in Western Europe an apparent shift from teaching practical geometry to teaching geometry "practically" (e.g., using problem-based approach). To

---

[22] Da Vinci (1888), vol. 3, Ms E, folios 24r-26v.
[23] For a Russian translation, see Zubov (1935), vol. 1, pp. 68–69, Zubov (1955), p. 73.
[24] Romanus (1593).
[25] For a Russian translation of these two theorems, see Kepler (1935), pp. 111–117.

support her claim, Menghini takes as example a textbook authored by Alexis Claude Clairaut (1713–1765):

> In 1741, again in France, Alexis Clairaut wrote his *Éléments* (sic.- V.F. & A.V.) *de Géométrie*. His first chapter is about the *measurement of fields*; nevertheless, Clairaut was *not* interested in teaching a practical geometry. With Clairaut we see a shift from measurement as a goal to measurement as *a means* to teach geometry via problems. This is seen by the fact that the part about measurement *doesn't contain numbers*; there is only a hint at the necessity for a comparison with a known measure (See Clairaut, 1743).
>
> The aim of Clairaut is to solve a problem 'constructing' the elements he wants to measure. The focus is on the *process* of constructing and in a *narrative* method.[26]

In their turn, Richard et al. (2019) argue that Clairaut, when building his proofs on inductive argument, "tries to make dynamic a figure by reasoning" in such a way that "the model reader will be able to visualize the animation, while in his demonstration, a classical figure like those in Euclid's *Elements* is proposed" (pp. 141–142).

The *rearrangement method* was not the only method used by mathematicians of the past that was taken over by mathematics educators of modern times. For instance, the so-called "onion" model was used by Rabbi Abraham bar Hiyya ha-Nasi (also known as Savasorda), a Jewish philosopher, mathematician, and astronomer active in the late eleventh and early twelfth centuries,[27] to justify the formula of the area of a circle (Figs. 3 and 4). This method was later used by Emma Castelnuovo (1913–2014) in her textbook on intuitive geometry (1966) as a part of a hands-on activity for elementary students as demonstration of the formula for the area of the circle (Fig. 5).

While the possible historical and didactical sources of the *rearrangement method* need to be investigated in more detail,[28] our analysis of geometry textbooks published in the nineteenth and early twentieth centuries shows that this model was already present in some of them thus becoming a *didactical artefact*; these archaic methods were introduced, we claim, because it was assumed that they may have helped articulate certain didactical ideas of modern "innovative" approaches to teaching geometry. The educators searching for alternatives to "Euclidean" methods of teaching geometry (that is, to the methods based on quasi-formal logic used in the *Elements*) in the second part of the nineteenth century often adopted an approach based on *Anschauung* (intuition) in preliminary courses on observational geometry and

---

[26] Menghini 2015, p. 568. Italics as in the original.-V.F.&A.V. On Clairaut and his approach, see also Schubring 2011, p. 81. In this paper, the author refers to Schubring 1987, 2003 (pp, 54–58) and Glaeser 1983. For a brief description of Clairaut's *Élémen(t)s de Géométrie* (1741) and its influence on European textbooks of the nineteenth and twentieth centuries, see Barbin and Menghini (2014), p. 481. Menghini (2015) provides a more detailed description of Clairaut's textbook on pages 568–571.

[27] The dates of birth and death of Abraham bar Hiyya are unknown. Some authors claim that he died in 1136, while other scholars consider this date not sufficiently justified; for more details, see Levey (1954), p. 50, n. 1. Levey himself suggests "fl. [...] before 1136" (1981, p. 22).

[28] The results of our investigation will be published elsewhere.

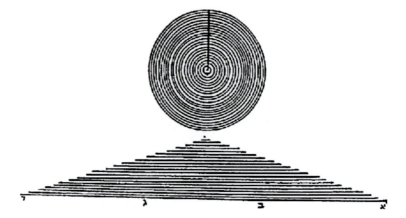

Fig. 3 The "onion model" used by Abraham bar Hiyya (Savasorda) (1116). The diagram reproduces Fig. 73 from Guttmann's (1903) edition of Savasorda's book

Fig. 4 The diagram from an earlier edition of Savasorda's treatise (Abraham, 1720). The authors thank Josep Maria Fortuny Aymemí for his suggestion to refer to Savasorda's work

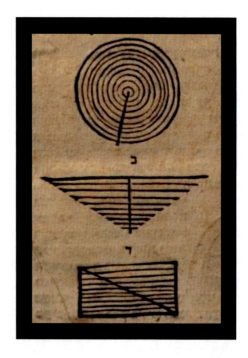

drawing; this approach was based on the ideas of the pedagogical reformers J.H. Pestalozzi (1746–1827) and J.F. Herbart (1776–1841).[29]

Indeed, several geometry textbooks published during the time period from the early nineteenth century to the mid-twentieth century by the authors from several

---

[29] For more details, see Henrichi and Treutlein (1897); Castelnuovo (1948).

Fig. 5 A hands-on activity for elementary students learning the formula of the area of a circle (Castelnuovo, 1966, p. 93)

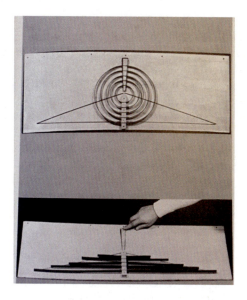

Western countries, namely, England, Austrian-Hungarian Empire, Germany, United States, and Russia, included various versions of the *rearrangement method*. In particular, its popularity might be attributed to the works of Dionysius Lardner (1793–1859), Peter Joseph Treutlein (1845–1912),[30] and Emma Castelnuovo,[31] among others.

We begin our analysis with Lardner's *The First Principles of Arithmetic and Geometry; Explained in a Series of Familiar Dialogues* (1835) to pursue with other textbooks that used the *method*, the latest of them was Zaitseva's teacher's manual (1952); none of these textbooks provides a reference to earlier sources (prior to the nineteenth century). The authors of these textbooks dealing with elementary geometry stressed various didactical aspects that were in the heart of innovative approaches at their time. In particular, the following elements were stated explicitly as the goals: the popularization of scientific (especially, mathematical) knowledge (clearly expressed by Lardner, 1835), development of the learners' intuition (Castelnuovo, 1948; Henrici & Treutlein, 1897), emphasizing "practical approach" (Stern & Topham, 1913),[32] and "hands-on experiments" (Willis, 1922). For instance, Treutlein's (1911) program contains three main principles which arguably influenced modern teaching: (1) to consider intuitive geometry as preliminary step to more formal, deductive instruction;

---

[30] On the reform of school mathematics education designed by P. Treutlein and the textbooks he co-authored with J. Henrici, see Becker (1994) and Weiss (2019).

[31] On Emma Castelnuovo's work, see Furinghetti and Menghini (2014); esp. see pp. 2–3, see also Castelnuovo (1977), pp. 42–44 on her interest in A.C. Clairaut's (1713–1765) didactical ideas.

[32] When dealing with the area of a circle (pp. 70–71), Stern and Topham describe the "matching sectors" method they borrowed from Earl (1894), p. 89–90 (see their footnote on p. 70). In turn, Earl does not provide any reference to his sources. It should be stressed that Earl's textbook was devoted to "lessons in *physical* measurement" (italics ours.- V.F. and A.V.).

(2) to conduct an implementation of modern ideas of projective and transformation geometry; and (3) to focus on spatial imagination and the fusion of spatial and plain geometry (Weiss, 2019, p. 112). The *method* (of rearrangement of the sectors) seems to align with this didactical vision.

The two books by Lardner that contain the method are *The First Principles of Arithmetic and Geometry; Explained in a Series of Familiar Dialogues* (1835) and *Lardner's Cabinet Cyclopedia* (1840). According to Peckham (1951), the audience of *Cyclopedia* were "those who seek that portion of information respecting technical and professional subjects which is generally expected from well-educated persons." The front page of the *First Principles...* ("Adapted for Preparatory Schools and Domestic Instruction; with Copious Examples and Illustrations") states that it was written as a mathematics textbook for the beginners. Lardner claimed that he considered geometry as part of public instruction having two objectives:

> First, it may be regarded as an exercise by which the faculty of thinking and reasoning may be strengthened and sharpened. It is peculiarly fitted for this purpose by the accuracy and clearness of which its investigations are susceptible, and the very high certitude which attends its conclusions. Secondly, it is the immediate and only instrument by which almost the whole range of physical investigation can be conducted; without it we could not advance a step beyond the surface of the earth in our knowledge of the universe; without it we could obtain no knowledge of the figure or dimensions of the earth itself, nor of the mutual mechanical operation, or influence of bodies upon it. (Lardner, 1840, pp. 2–3)

After having stated these two goals, Lardner complained that

> In the course of instruction followed by the great mass of students in our universities, geometry has been regarded almost exclusively as a system of intellectual gymnastics; while, on the other hand, owing to the very stinted portion of instruction attainable by those who are engaged in the useful arts, the science is with them almost degraded to a mass of rules, without reasons, and dicta, the truth of which is expected to be received on the authority of the writer, and of which the reader is not put in a condition to judge. (Lardner, 1840, p. 3)

Therefore, in his textbook, he uses a form of a (Socratic) dialogue between a student (Henry) and the teacher ("Mr. L.," with the capital "L" apparently, referring to Lardner himself). Let us consider one of such dialogues devoted to the area of a circle. It begins with the student's questioning:

> Henry (Student): [...] How shall I compute the number of inches enclosed within the circle when I know the length of its diameter?[33]

Mr. L. presents then a picture of a circle divided into 16 equal sectors while mentioning that by further dividing the circle into many even smaller sectors the arches will be made "exceedingly" short so that their curvature would not perceptible. The student says that in this case, the sectors would look as "narrow triangles." The conversation arrives then at the point at which the teacher mentions that the space enclosed within the circle would be the same as within a rectangle formed by the circumference (length) and the radius (height). Further, the teacher brings even a more "evident" way of rearrangement cutting the two identical circles into sectors-triangles and then putting them one against the other (Fig. 6).[34]

---

[33] Lardner, 1835, p. 112.

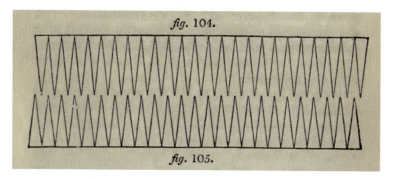

Fig. 6 Figures 104 and 105 from Lardner (1840), p. 101 originally published as Figs. 81–82 in Lardner (1835), p. 114. The author states that the formula for the area of a circle can be "made still more evident" with the help of this picture (1835, p. 113; 1840, pp. 100–101)

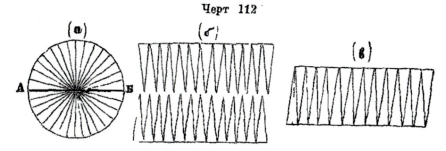

Fig. 7 Diagram 112 from Boryshkevich (1893), p. 68

A Russian textbook by Boryshkevich (?–1906) published in (1893), expressed a similar idea of using "saw-teeth," but this time with only one circle divided into 32 equal pieces; then from two half-circles, two "saws" with 16 teeth each were formed to be rearranged into a "parallelogram," see Fig. 7.[35]

The use of "rearrangement method" for the area of a circle was certainly a part of the larger didactical agenda focusing on "visualization" of geometrical statements. For instance, Max (Maximilian) Simon's (1844–1918) book (1889) provided some suggestions for teachers who intended to teach introductory geometry "intuitively." In the case of the area of the circle, he explains that one could imagine a circle divided into many triangles with a common vertex in the center of the circle and the bases on

---

[34] The regular polygon inscribed in the circle that Larnder used is rather unusual: it has 25 sides. It thus differs from the polygons of all the other authors whose works we inspected. It is possible that the division of the central angle into 25 equal parts was somehow related to the reform of measure units that took place in France after the French Revolution of 1789; according to the new system, the measure of the circumference equaled to 400°.

[35] The diagram reproduced in Fig. 7 is not clear: while the circle is divided into 32 sectors, the second (central) figure shows only $13 + 13 = 26$ triangles, while the right figure (composed of the upper and lower parts of the central diagram) contains only 24 triangles.

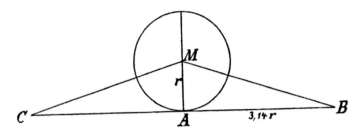

Fig. 8 Simon (1889), p. 19, Fig. 31

the circumference, so all of them, when combined, would form one big triangle with the base equal to the circumference and the height equal to the radius; see Fig. 8. So, its area should be equal to area of the circle.

Interestingly, a similar description of the method can be found in the abovementioned modern book by Bryan and Sangwin (2008), although in a slightly different manner: the authors subdivide the procedure into three "steps": (1) the circle is divided into 18 sectors; (2) the circle unfolds into a "teeth row" (making not very obvious the curvilinear shape of the "base" of each "tooth"); and then (3) suggesting a transformation of the "teeth row" into a set of triangles having equal bases and the same height (their vertices are joined together thus forming a big isosceles triangle, that is, a rectilinear figure). A warning accompanies this transition from a curvilinear to a rectilinear shape; it says that the sectors are arranged "along a straight line" whose length can be (when the number of sectors is large) "approximated very well by the circumference of the circle" (p. 145) and the height is "practically" equal to the radius of the circle, and, finally, that the "displacement of the apex of each triangle does not alter the area" (ibid.). The reader is thus supposed to be convinced that the circle was transformed into the large triangle, so the area of the circle equals to the area of the triangle.

Following the same line of thought, P. Treutlein's *Intuitive Geometry* (1911) introduces the process of calculation of the area of a circle by means of an even more sophisticated visualization passing through two consecutive transformations, one based on the division of the circle into six pieces and on the rearrangement ("gluing") of three of them on top of three others. Then a second circle, identical with the first one, would be divided into 12 pieces which would be similarly rearranged into two groups of six pieces completing each other. The student then is asked to make an observation concerning the differences between the surfaces of both shapes, and, moreover, is asked what would happen if the number of pieces grows larger. So, gradually the student would eventually come to the conclusion that the rearranged shape would become a parallelogram with the base equal to circumference, and the height, to the radius; see Fig. 9. All this serves to derive the formula for the area of the circle using the student's "imagination" (Treutlein, 1911; for a discussion, see Fujita et al., 2004).

In England, a book on practical mathematics by Stern and Topham (1913) focused on graphical and experimental processes in mathematics; the projected readership

Historical and Didactical Roots of Visual and Dynamic ...

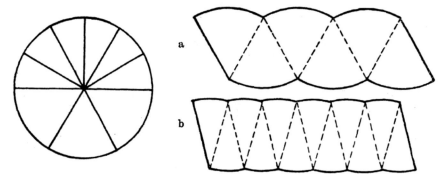

Fig. 9 Treutlein (1911), p. 186, Fig. 86[36]

was the militaries, but the book would also have been useful for more general audience. The authors called their method "approximate" (p. 70) while making reference to an earlier work by Earl (1894) who also suggested dividing a circle into a large number of equal sectors so that the obtained pieces, when put together, would form a figure whose area could be found approximately by treating it as a rectangle (Fig. 10).

As general introduction to his approach, Earl claimed that

> Public Schools should not be devoted simply to instilling into boys a certain amount of technical knowledge, but should rather train them to observe accurately, to reason rightly, and to front nature with an open and inquiring mind (Earl, 1894, p. v).

In the US, also in attempt to help teaching practical mathematics, C.I. Palmer (1871–1931) suggested in his *Practical mathematics for home study* (1919) a similar approach to approximation of the area of the circle using its division into 16 pieces. The illustration in his book reproduced below (Fig. 11) presents, however, a half of a circle divided into 8 pieces.

The ideas of laboratory methods of studying geometry, popular in the first half of the twentieth century, brought C.A. Willis (1869–?) to design an experiment-based lesson to help students "discovering" this formula using a model that suggested manipulations with sectors (1922, pp. 264–265); see Fig. 12. In his textbook, the formulas for areas of plane figures were supposed to be obtained as results of "experiments," and the entire chapter 16 titled "Measurement of areas. Principles determined experimentally" was devoted to manipulations with paper models of geometrical figures.

The calculation of the area of a circle is preceded by a definition of π and a theorem stating that "The value of π can be calculated with as close an approximation as desired" (p. 249). His discussion of the matter reads as follows:

---

[36] Note that in this diagram the lower part of the circle shows its division into six equal sectors, while the upper part shows its further division into 12 sectors. Meanwhile, the results of the recomposition of the sectors shown on the right side are inversed: the upper figure shows the result of recombination of the six sectors, while the lower figure features 12 sectors.

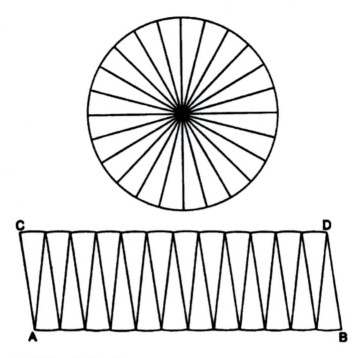

Fig. 10 Earl (1894), p. 89, Fig. 44

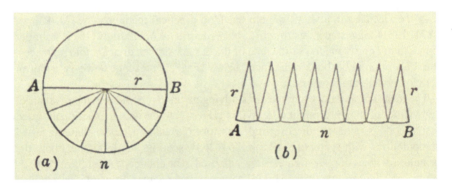

Fig. 11 Palmer (1919), p. 152, Fig. 90

356. Experiment V. The area of a circle.

Draw a circle about 4 inches in diameter and cut it out. Cut it on a diameter AB. Cut each semicircle into the same number of equal sectors; say into 8 sectors. Place the sectors as shown in Fig. 2. How may the resulting figure be made to approximate more and more closely to a rectangle as in Fig. 3? What are the base and altitude of this rectangle? What is therefore the area of a circle? (Willis, 1922, p. 264)

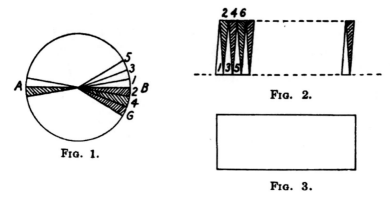

Fig. 12 Figures 1–3 from Willis (1922), p. 264[37]

The diagrams provided by Willis are quite remarkable. Firstly, they do not show the presumed division of the circle into 16 sectors, but into a larger number of sectors only 8 of which are shown. Our exploration of the diagram suggests that the diagrams shown in Fig. 12 represent a division of the circle into 32 sectors; this number is not specified in the text. Secondly, and this is even more interesting, the figure obtained as combination of sectors into which the circle is subdivided ("Fig. 2" of the diagram) is designed according to some particular strategy: the sectors from the lower half of the circle are coupled with the symmetric sectors from the upper half, for example, sector 1 is coupled with sector 2, sector 3 with sector 4, and so on. Apparently, the author of this diagram suggested that each sector from the upper half of part 1 is to be paired up with the symmetric sector from the lower half, and this operation should go from right to left to produce the couples placed in part 2 from left to right. Thirdly, the object shown in "Fig. 2" of the diagram has straight lines as its upper and lower sides, which is actually an approximation, since it is obtained as collection of sectors. So, using these diagrams would require some additional mathematical and didactical work from the teacher to avoid possible misconceptions.

Several Russian and Soviet authors used the method of rearrangement in the textbooks over the first half of the twentieth century often referring to Western sources (e.g., Treutlein's work which was translated into Russian). Some ideas of the above-mentioned experimental approach could be found, for instance, in Zaitseva (1952). On the one hand, the author used a diagram (Fig. 13) visually close to that used by Earl (1894) (however, the number of sectors, 16, was different). On the other hand, she suggested that students, after dividing a circle into 16 sectors, should have cut them out and superpose one half of them over the other half to form a figure resembling a parallelogram.

While visually convincing and eventually accessible to relatively young learners (the textbook was supposed to be used in grade 5 of Soviet school, that is, by students

---

[37] The lettering in Fig. 1 of the diagram contains a misprint: instead of number "6" the capital letter "G" is written to mark a sector, while Fig. 2 shows the same sector correctly marked with digit "6".

Fig. 13 Diagram from Zaitseva (1952), p. 70, Fig. 14

of 12 years of age), the complexity of the model and especially its infinitesimal part would have remained obscure for them, unless some extra explanations were provided by teacher.

As Earl (1894), Zaitseva suggests dividing the circle into 16 equal sectors then cutting them out and putting together as shown on the right picture to get "approximately a parallelogram"; its base will be half of circumference and its height will be equal to radius; so the area of the circle is half of circumference multiplied by radius.

In another part of the world, namely, in Quebec, Canada, the authors of a 1966 secondary school textbook on "new mathematics" (*Mathématiques Nouvelles*, Hamel et al., 1966) also were seemingly inspired by the "teeth model." On pages 203–204 of the textbook, we find first a regular hexagon divided into six triangles with the common vertex in the center of the circle. At the next step, this hexagon (which can be inscribed in a circle) is shown as decomposed into a chain of triangles; the area of one such triangle multiplied by 6 would give the area of the hexagon (Fig. 14). Then it is said that the same technique can be applied to a regular polygon with a very large number of sides, so it would *"practically coincide"* (highlighted by us. - V. F. & A.V.) with the circle. According to the authors, in the case of a circle, the base of each "triangle" (interestingly, the illustration shows figures looking rather like sectors) is "practically" a segment of a straight line and the height (of the triangles) is "practically" the radius (Fig. 15). This leads to the formula for the area of the circle.

Overall, in the 1950–60s, the bases of these "novel" approaches to initiation to geometry in the middle grades (starting from the age of 12) have been discussed,

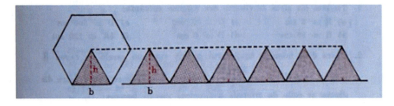

Fig. 14 Hamel et al. (1966), p. 203

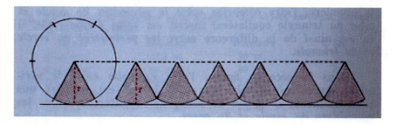

Fig. 15 Hamel et al. (1966), p. 204

according to Tessier and Beaugrand (1961) at the Congress of the International Commission for Mathematics Instruction (ICMI) in 1958 where the appropriate methods of teaching at entry level (first year of secondary school) where identified as intuitive, experimental, and empirical, based on observation and investigation, and providing an access to knowledge in a clear, direct, and immediate manner of perceiving the truth intuitively (without using deductive reasoning). The learners' access to rigorous modes of reasoning was therefore supposed to be based on experiment, and they were supposed to discover by themselves new properties of already familiar shapes as result of a process of active research (Tessier and Beaugrand (1961), see the section titled *Présentation*). We consider this essence of didactical genesis as representing an accumulation of the "innovative" elements of the didactical knowledge produced over the 150 years of reforms.

Not surprisingly, in their textbook, the authors of the "innovative approach" presented the familiar old rearrangement methods to introduce the area of the circle by suggesting to decompose the circle into "a big number of sectors, or figures that resemble to triangles" (the provided illustration shows 32 sectors) and then to operate with triangles in order to calculate the area of a sector (as base times height divided by 2 or, in other words, arc times radius divided by 2) which yields the area of the circle as the area of one sector multiplied by the total number of sectors. It remains however unclear why the formula of the area of the triangle can be applied to the area of the sector, and, moreover, why this *method* is believed to be helpful for transition to higher levels of rigor in later schooling?

This latter question can be considered fundamental in terms of the interplay between mathematical, historical, and didactical aspects of the development of the *method* as an artifact and its use to (presumably) support teaching and learning. In the next section, we will look at modern development to confirm that the issue appears to be well alive in twenty-first century's mathematical working spaces.

4 Looking into Modern Mathematical Working Spaces Through the Lens of Historical, Didactical, and Instrumental Genesis

From our genetic analysis of possible historical and didactical sources of *rearrangement method*, we can track a variety of ideas that reflect nineteenth–early twentieth-century innovations in teaching geometry by making it more visual, intuitive, and experimental, and therefore accessible to all learners at their early steps in mathematics. In particular, in Lardner's and Boryshkevich's textbooks, we find illustrations of a circle divided into equal sectors that are then rearranged into a new shape that "looks like" a parallelogram (or rectangle), eventually helping learner to *visualize* the process like it was done by the abovementioned authors centuries before.

An *intuitive approach* (Anschauung) which became the focus of innovative approaches at the end of the nineteenth–beginning of the twentieth centuries focused on the perception of a student who *imagines* "what happens if" a circle is divided into "many" sectors (considered as "triangles") that could be rearranged in what would look as a big "triangle" (Simon) or a "parallelogram" (Treutlein).

In the context of a *practical* (experimental) *approach*, the focus is put on *approximation* as the main tool used to grasp the concept of area of the circle by the learner (Earl; Stern and Topheim) who could also support her or his observation by physically *manipulating* with pieces of paper (like in Palmer, Willis, and Zaitseva); the latter approach was related to the idea of *mathematical laboratory* popular in the didactics of the first decades of the twentieth century.

All three approaches are apparently being used in the design of modern middle school geometry lessons which we will examine in the next section.

4.1 Rearrangement of the Circle in Modern Middle School Geometry Lessons: Possible Teaching Scenarios and Didactical Challenges

The approach based on the rearrangement of sectors apparently remains very attractive to the authors of geometry textbooks even nowadays, as we mentioned in the introductory section of our chapter (Larson et al., 2007; see Fig. 1).

In their quest for justification of the formula of the area of the circle, similar to the "approximate method" used in the textbooks of the late nineteenth and early twentieth centuries discussed above (e.g., Earl, 1894), the authors suggest an investigation based on the idea of dividing a circle into sectors, cutting them out and rearranging them into a new shape ("resembling a parallelogram"), and then ask the students to "write an expression" to calculate its area (using the radius of circle as the "approximate height of parallelogram.") Finally, students are asked to connect their "expression" to the area of the circle (eventually using the formula as "justification" of the theorem stating the mathematical expression for the area of a circle). However, the "task" does not provide any reference to the possibility of making the number of sectors grow indefinitely thus entirely omitting the infinitesimal part of the proof. Another difference with earlier textbooks is that the figure itself does not present a division of a circle into equal parts (sectors), for example, the circle in the left part of Fig. 1 looks like an orange sliced into pieces that do not look "congruent," while on the right side they, surprisingly, have the same size and shape. Again, in the case of Willis's (1922) scenario, some extra work from teacher would be needed to help students to grasp the subtlety of the process of "division" and "approximation."

In a popular in the North America manual for (pre- and in-service) teachers which claims that *teaching* should be done *developmentally* (Van de Walle et al., 2020), it is suggested to challenge students to "figure out the area formula for circles on their own." Teachers are, however, advised to give a "hint" by showing the students how to "cut a circle apart into sectors and rearrange them to look like a parallelogram " (ibid., p. 530). According to the authors, the process of approximation (ibid., p. 530, Fig. 18.19) unfolds in three steps: (1) dividing a circle into 8 equal sectors and rearranging them a "near parallelogram"; (2) doing the same with 24 sectors ("even closer to a parallelogram"); (3) as the number of sectors grows, the figure "becomes closer and closer to a rectangle (a special parallelogram)". Here, the issue that draws our attention is that no mechanism is suggested to make students think about the reason why the fact of being "closer and closer to a rectangle" ensures the equality of areas (in other words, what justifies the transition from "being close" to "being equal").

Another manual for teachers' preparation (to make them *think*, as the authors state) by Brumbaugh et al. (2006) along with the method of "squares counting" suggests an activity of "cutting a circle" into "several pie-shaped wedges" (the authors suggest, as example, that wedges should have the central angle of 30°, that is, the circle is supposed to be cut into 12 wedges), then getting it "unrolled" and "interlaced" to have a shape which "approximates a parallelogram" (ibid., p. 187, Fig. 9.13). Again, the authors only mention that, presumably, an exploration of this fact ("approximating a parallelogram") would "lead to the area of the circle" (that is, the area as obtained by multiplying half of the circumference by the radius, idem.). The question of why one fact "leads" to the other seems to remain open (or even not formulated).

Finally, Small (2018, p. 167) provides an example of dividing a circle into multiple parts ("like pieces of a pie") to help students to see that the parts can be re-united to form a parallelogram. Her illustration (p. 167) shows first a circle divided into eight pieces and a rearranged shape ("parallelogram"), to finally come back to a circle (as a whole) whose area will be "equal" to Pi × radius × radius. Interestingly, the author does not mention *approximation* but puts straightly an equality sign for the area of the "parallelogram" (which is still curvilinear on the illustration) to claim that it is equal to the area of the circle.

A large number of similar approaches can be found in online resources available to teachers and learners nowadays. For instance, the method of calculation of the area of a circle based on the procedure of division of circle into sectors can be found on the *official* website of the National Council of Teachers of Mathematics suggesting a

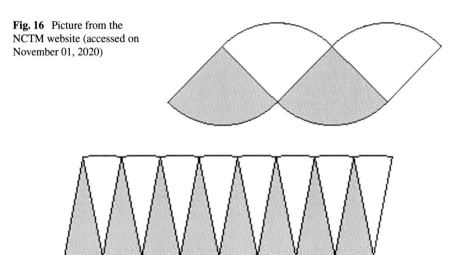

Fig. 16 Picture from the NCTM website (accessed on November 01, 2020)

Fig. 17 Picture from the NCTM website (accessed on November 1, 2020)

number of said-to-be-high-quality resources.[38] The teaching scenario for "measuring circles" begins with students working in pairs or small groups; they would "cut the circle from the sheet and divide it into four wedges." The next step would be to ask the students to rearrange the wedges in the way shown in Fig. 16 (this configuration looks similar to Treutlein's, but contains only four pieces).

The authors suggest highlighting different parts of the circle (its radius and circumference) with a color. Then the authors argue that this shape would not be recognized by students as something familiar to them. Hence, the next step would be a further division of each wedge into two thinner wedges so that there would be eight wedges; the students have to continue the process to get 16 wedges and to finally produce a shape shown in Fig. 17 (similar to what we had in the above-cited examples, such as Larson et al. (2007), and in earlier sources).

This work, facilitated by the teacher's discussion eventually helping the students, would lead them to identify the shape resembling a parallelogram, which, when being "continually divided, [...] will more closely resemble a rectangle."

Another step in the investigation (i.e., the further division of the wedges) would presumably lead the students to the conclusion that the length of the rectangle is equal to half the **circumference** of the **circle**, or πr. Additionally, students should recognize that the **height** of this rectangle is equal to the **radius** of the **circle**, r. And, similar to the previous examples, students would "try and generate a formula for **area** of this new rectangle formed by the pieces of the **circle**." Considering the **area** of this rectangle being equal to $\pi r \times r = \pi r^2$,[39] and knowing that "this rectangle is

---

[38] https://www.nctm.org/Classroom-Resources/ARCs/Measures-of-Circles/Circles-Lesson-2/.

[39] Interestingly enough, at the moment when we visited the official site of NCTM, it contained a misprint *at this crucial point*: instead of $\pi r \times r = \pi r^2$ it was typed "πr × r = πr2."

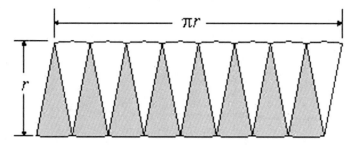

Fig. 18 Picture from the NCTM website (accessed on November 1, 2020)

equal in **area** to the original **circle**, this activity gives the **area** formula for a **circle**: A = πr^2"

The provided figure (Fig. 18) indicates dimensions of the shape to add more clarity to this conclusion (similar to Zaitseva's figure).

The final suggestion is to "have a class discussion with students explaining that total **area** is almost always an approximation."[40]

Again, as in previous examples, we can see the division of the circle into 16 pieces (yet, the authors encourage students into further exploration of situations with increasing numbers of pieces by saying that the process can be continued). The authors of this procedure also keep the idea of cutting out pieces of the circle and rearranging them into "parallelogram"; they also mention that the further division would transform the "parallelogram" into a "rectangle" without any specific suggestions of how to interpret and explain such transformation, again leaving it to the teacher's *didactical orchestration* (Drijvers, 2012). The website of NCTM does not mention any rationale of this method to explain why this activity is supposed to be relevant to teaching and learning mathematics in middle school.

4.2 Novel Approaches to Area Investigation Using Dynamic Digital Tools

The development of digital tools to support teaching and learning geometry in the late twentieth–early twenty-first centuries has brought novel didactical dimensions and tools that might be used by teachers. For instance, the Canadian website LearnAlberta provides an interactive animation which allows to increase the number of sectors (using a slider), so their rearrangement rapidly approaches the shape of a rectangle (see LearnAlberta.ca). Figures 19 and 20 show the division of the circle into 8 and 16 sectors.

---

[40] The exact meaning of this statement remains unknown. Does it refer to the area of a circle or to the area of *any* geometrical figure?

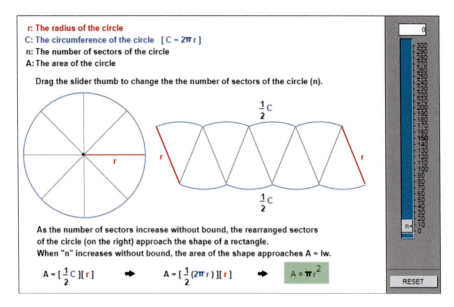

Fig. 19 LearnAlberta webpage, $n = 8$

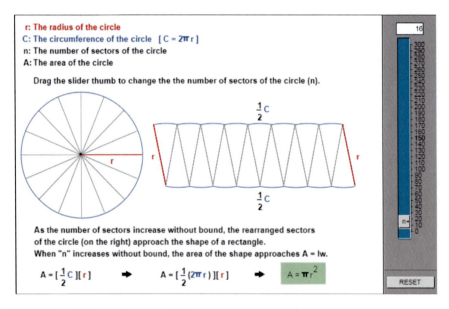

Fig. 20 LearnAlberta webpage, $n = 16$

Historical and Didactical Roots of Visual and Dynamic ...

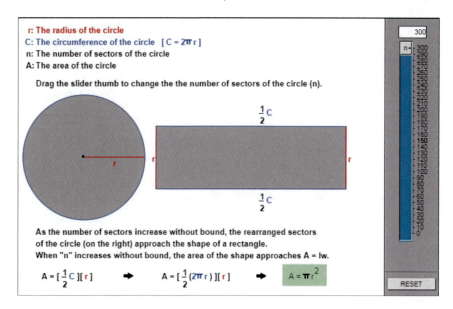

Fig. 21 LearnAlberta webpage, $n = 300$

The maximal possible number of sectors is 300, and when this number is reached, the figure composed of the sectors looks like a rectangle with the sides $C/2$ and r (Fig. 21).

The applet is accompanied by the following explanations:

The area of the interior of a rectangle is the product of the length of the rectangle (l) and the width of the rectangle (w) [$A = lw$].

When the number of sectors is large:

– the width of the rectangle (w) is approximated by the radius of the circle (r);

– the length of the rectangle (l) is approximated by half the circumference of the circle (πr).

This explanation is somewhat problematic. Firstly, at no step the figure obtained as rearrangement of sectors becomes "rectangle," so the figure referred to as "rectangle" is actually *not* a rectangle. Secondly, if l is used to note the length of a rectangle, say, drawn to approximate the "waved" figure, and w is used for its width, the relationships between these magnitudes and the half of circumference and radius are not very clear, for example, what can be l and w in Fig. 19? Thirdly, it is not very clear why the approximations suggested by the authors are valid. Again, as in above-cited textbooks, much of the work dealing with this complexity is left to the teacher.

Numerous interactive applets (like the one created using GeoGebra's affordances, see Fig. 22) allow for some more sophisticated exploration using several sliders to arrive at similar conjectures concerning the formula of area of the circle.

It can be argued that these and other similar recent computer-aided approaches to the introduction of the formula of area of the circle are largely based on the methods

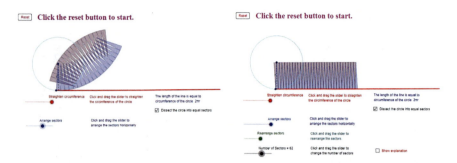

Fig. 22 GeoGebra applet for exploring the formula of area of the circle, Authored by Ooi Soo Huat. (available at https://www.geogebra.org/m/AADN5Ruq, visited on October 24, 2021)

of teaching already found in the mathematics textbooks compiled in the twentieth century or even earlier. In other words, the creators of the mentioned software tools developed their programs to "animate" the procedures that had been found in earlier textbooks, and by doing so, to transform them, presumably, into more *dynamic and interactive* MWS, and, eventually, to make them more attractive and accessible to the learners. However, this transformation (from traditional "paper-and-pencil" methods to novel ones, based on digital tools) when carried out without a substantial historical analysis of the method, especially in regard to its epistemological complexity and didactical implications, could lead to limited conceptual understanding by students. One can question whether the application of software aiming at visualization of infinitesimal procedures can indeed enhance the ability of the learner to understand the rationale of the derivation of the formal expression for the area; this question will immediately lead to an even larger question concerning the application of computer-generated simulations in teaching mathematics (in particular, geometry).

4.3 Discussion and Conclusions

In our paper, we tried to track back the history of introduction of the "matching sectors method" (or "*rearrangement method*") recently used to justify the formula for the area of a circle in school curricula and resources (printed and available online). It appears plausible to conjecture that this method was originally used by professional mathematicians of Antiquity and Middle Ages in the West,[41] and only later it was borrowed by mathematics educators and placed into school textbooks and lessons as result of didactical transposition from "knowledge to be used" to "knowledge to be taught and learned" (Chevallard, 1985; Kang & Kilpatrick, 1992). In the case of the area of a circle, the earliest sources for "rearrangement method" that we were able

---

[41] The original version of this chapter included a discussion of the similar methods found in Chinese and Japanese mathematical treatises; it was removed as not directly related to the topic discussed in the present chapter. We plan to return to the collected materials in a future publication.

to identify in the Western textbooks were produced in the first half of the nineteenth century (e.g., Lardner, 1835, 1840). What were the particular reasons for the use of this archaic method for instruction in the nineteenth and especially in the twentieth century? Was it grounded in some pedagogical or didactical traditions or was it a result of innovative efforts to improve teaching and learning comparable to other newly introduced methods?

It remains equally unclear from where did the method of "matching sectors" come to the modern school textbooks. Why was it perceived as an especially "efficient" tool for teaching? Was it because Archimedes' "exhaustion method" involved sophisticated reasoning and thus was considered difficult for beginners, while the method of "matching sectors" used in textbooks may have seemed more didactically attractive since the procedure of "matching" the sectors looked easy to understand for even relatively young learners? This simplicity, along with visual and dynamic nature of rearrangements, apparently seemed to the modern educators to be the best way to construct lessons which are hands-on, practical, investigative, and providing students with initial intuition that could lead to more complex mathematical concepts.

In this chapter, we only briefly discussed possible historical roots of "rearrangement method" while mentioning some mathematical works authored by European scientists (Cusanus, da Vinci, Kepler), among others. We used mainly the primary sources cited in modern literature, and used secondary works, such as Baron's book on the history of calculus and Beckmann's *History of Pi*,[42] as well as in online resources (e.g., the *Cut-the-Knot* educational website created by Alexander Bogomolny).[43] A more detailed historico-didactical and genetic analysis of the process of calculation of the area of a circle mainly based on primary sources will be published elsewhere.

The problem of calculation of the area of a circle solved by mathematicians of the past was of a particular nature; in certain cases, they arguably dealt with it when focused on some specific problems as, for instance, the quadrature of the circle (Cusanus) or calculation of the volumes of the wine barrels (Kepler). Yet, when one turns to the process of transmission of scientific (in this particular case, mathematical) knowledge featured in the textbooks specifically designed for learning such as Clairaut's *Elements* of 1741 (1830)[44] or, later, the geometry textbooks of the nineteenth and twentieth centuries, then the task of transmission of theoretical and practical knowledge became essentially a didactical issue, and new considerations took place.

In the case of the area of the circle, the epistemic issue of dealing with infinity interacted with the need for visualization (in particular, using diagrams in textbooks), induction (based on intuitive approach to infinitely small and large entities), and approximation (dictated by practical considerations). In particular, we found two

---

[42] Although several authors cite Beckmann's work, especially the paragraphs concerning the procedures of calculation of the area of a circle, we find that certain parts of the book are not sufficiently elaborated and reliable.

[43] http://cut-the-knot.org/.

[44] On Clairaut's didactical ideas see, for example, Sander (1982) and Glaeser (1983).

different interpretations of the "making of a circumference a piece of straight line" (with the length equal to the circumference): one related to the category of methods based on the idea of "unfolding a circle" (with sectors becoming "teeth-like" row, as, for example, in the diagram constructed by Casselman, 2012 or in the one found in Beckmann, 1976 (Fig. 2)); the other is more like "cutting out the sectors of a circle" (each sector is to be moved and rearranged), when even the notion of "being equal" seems to be sacrificed to the goal of making a new arrangement of parts fitting into a rectangular shape (like in Lardner, 1835 and Willis, 1922).

We argue that the division of a circle into a number of equal sectors and their further rearrangement thus producing "teeth-like" diagrams in school textbooks was done mainly for the sake of visualization, even though minor differences between variants of this procedure can be identified. There is an aspect that needs deeper reflection as far as the "visual" part of the process is concerned: it is related to the number of sectors into which the circle is dissected. More specifically, there are two patterns that can be identified, the first one is based on a regular hexagon inscribed in the circle and then having its sides doubled (6–12–24…), the second one features powers of 2 (4–8–16…) and is apparently based on a square inscribed in the circle and then having its sides doubled. Both visual representations serve to convince the reader (learner) that the area of a given circle can indeed be calculated in the same way as the areas of all other circles, via an approximation by the polygons.

At some point, the circle itself is considered (or should one say "defined"?) as a polygon with an infinite number of sides. Then comes a "rearrangement part" arguably based on the assumption that the operations of division, decomposition, and re-composition of a shape do not change its area; several models that we investigated above reflect this dynamic process. Then comes even more complex (and complicated) issue which remains mathematically and didactically challenging: How to deal with the infinity? Some sources we analyzed provide a rather static explanation (providing a "sufficiently large" number of sectors) saying that the new shape "looks like" a triangle, or a rectangle, or a parallelogram. Others add the word "approximately" to this visualization. Finally, a number of authors bring dynamic aspects in the process, explicitly or implicitly, pointing at a possibility to increase the number of sectors and, consequentially, to make it intuitively clear that the area of the rearranged shape of the circle becomes (approximately) equal to that of a triangle, or a rectangle, or a parallelogram for which it is known how to calculate the area.

This "knowing" (most often coming from the previously proved formula for the circumference) allows for further manipulations with the formulas for the area of a triangle (or rectangle, or parallelogram) and, finally, for advancing conjectures concerning the formula of the area of the circle.

Considering geometry as a complex activity embracing various processes of interaction between practical knowledge (measuring) and its theoretical codification (e.g., the Euclidean deductive system used for proving), we are inclined to see in the use of this "rearrangement" method an attempt to introduce a twofold procedure explaining to the learners *how* to calculate the area of a circle and, at the same time, offering a plausible reasoning strategy explaining *why* this formula works. While considering it as "instrumental activity" (implying both signs and tools) from its very beginnings,

Richard et al. (2019, p. 143) reflect on the complexity of distinguishing between mathematics as a science and mathematical thought in the context of a particular situation/task/activity. The authors further refer to Kuzniak and Richard's (2014) definition of mathematical work as a "progressively constructed process of bridging the epistemological and the cognitive aspects in accordance with three yet intertwined genetic developments as the *semiotic, instrumental and discursive* geneses" to introduce their model of MWS allowing to "report on mathematical activity, potential or real, during problem solving or mathematical tasks" (Richard et al., 2019, p. 144).

From the historico-didactical perspective, it appears to be difficult, or even impossible, to make direct connections between the mathematicians' work prior to eighteenth century and didactical innovations of the later period; however, some of the issues that they were dealing with certainly merit attentive look of modern educators. For instance, the idea of da Vinci of separating "the angles of the sectors from each other in such a way that the space between the vertices of these angles become equal to flattened bases of the sectors" (Ravaisson-Mollien 1888, Ms E, folios 24r-26v) might be resonating with that of transforming rearranged sectors into configurations where sectors (approximately) become triangles, and, when put together, form rectilinear shapes (triangles, rectangles, or parallelograms). The idea of Cusanus that the area of the circle may be found by the same means as that employed for any other polygon, that is, by dividing it up into a number (in this case, an infinite number) of triangles (Boyer), is also fruitful in terms of the treatment of "teeth-like" representations in later (modern) sources.[45] Kepler's work, besides following Cusanus' method of "indivisibles," provides an insight into the procedure that points at the approximation as a way of approaching the circle by polygons with a large (even infinite) number of sides.

In our study, we found that the modern authors (e.g., Boyer, Baron, Beckmann) seem to add details which were *not* found in the cited works. Indeed, neither Cusanus, nor Kepler explicitly discussed the "rearrangement" of sectors. Moreover, despite a seeming similarity of the way in which the rearrangement method was described in the nineteenth–early twentieth-century sources that we analyzed in this chapter, we noticed some substantial *differences* in representations that need to be reflected upon from the "teaching and learning" perspective. While the idea of dividing the circle into sectors is present in all models, there are differences in terms of the number of pieces, as well as in their pictorial representations. Some authors provided illustration of the whole circle divided into equal sectors (e.g., Boryshkevich, 1893; Earl, 1894). Others presented a division of a half of a circle (Palmer, 1919) or showed a part of the sectors (Willis, 1922). One model (Henrici & Treutlein, 1897) shows division of one half of a circle into three sectors and the other half into six sectors. When representing the result of a rearrangement, some authors show the final result of one transformation (Earl, 1894; Hall & Stevens, 1921; Palmer, 1919), or even leave the "matching" incomplete (like in Lardner, 1835, 1840), where two rows of "teeth" are getting close to each other but not yet stuck together (similar to the

---

[45] For discussions of Cusanus' quadrature, see Uebinger 1985, 1896, 1897, Wertz 2001, Nicolle 1996; 2001; 2020.

middle configuration in Boryshkevich's drawing). Other authors might show two consecutive transformations (like Treutlein's model with three and six pieces, or Willis's showing a part of sectors put together and a plain rectangle) or even three (similar to Boryshkevich). This variety of the representations of the one and the same model might have had an impact on its use (in terms of instrumentation) which also needs a deeper investigation.

Finally, the development of recent didactical models, especially employing the dynamic aspects of new digital technology (e.g., dynamic geometry) adds more complexity to the existing MWS while pointing at the fertility of genetic investigation of their historical and didactical antecedents.

References

A. Primary materials

Abraham Bar Hiyya. (Savasorda, 1116). *Hibbur ha-mêsihá wé-ha-tidbóret (Tractat de geometria i mesurament)*. Barcelona.

Abraham B., H. (1720). 地球の形 (*Séfer Surat ha-aretz*), 1720, p. 15. Offenbach. Biblioteca Virtual del Patrimonio Bibliográfico. https://bvpb.mcu.es/es/catalogo_imagenes/grupo.do?path=154236.

[Archimedes]. (1544). *Archimedis Syracvsani Philosophi ac Geometriae excellentissimi Opera* [Greek edition]. Basileae: Ioannes Heruagius.

[Archimedes]. (1615). *Archimēdus Panta Sōzomena Novis Demonstrationibvs Commentariisqve Illvstrata*. Paris: Morellus.

[Archimedes]. (1676). *Archimedis Siracusani. Arenarius et dimensio circuli*. Oxford: Sheldonian Theatre.

[Archimedes]. (1880). *Archimedis Opera Omnia cum Commentariis Eutocii*. Lipsiae: B.G. Teubner.

Boryshkevich, M. (1893). [Борышкевич, М[ихаил Федотович]] Kurs elementarnoi geometrii s prakticheskimi zadachami dlya gorodskih uchilishch [Курс элементарной геометрии с практическими задачами для городских училищ (A course of elementary geometry accompanied by practical problems for educational institutions located in towns). Kiev: Kryzhanosvkii and Avdyushenko (Типография И.Крыжановского и В.Авдюшенко), (second edition).

Castelnuovo, E. (1948). *Geometria intuitiva, per le scuole medie inferiori* (Intuitive geometry for lower secondary schools). Roma: Carabba [reprint 1949].

Castelnuovo, E. (1966). *The way of Mathematics: Geometry*. Florence: The New Italy.

Clairaut [Alexis Claude]. (1741). Élémens de géométrie. Paris: Lambert & Durand.

Clairaut [Alexis Claude]. (1830). Élémens de géométrie. Nouvelle édition, revue et corrigée. Paris: Bachelier.

Clairaut, A.-C. (1743). *Théorie de la figure de la terre, tirée des principes de l'hydrostatique*. Paris: David fils.

Da Vinci, L. (1888). *Les manuscrits de Leonard de Vinci*. Tome 3 [Manuscrits C, E, & K de la Bibliothèque de l'Institut]. Ed., transl., Charles Ravaisson-Mollien. Paris: Maison Quantin.

Da Vinci, L. (1890). *Les manuscrits de Leonard de Vinci*. Tome 5 [Manuscrits G, L & M de la Bibliothèque de l'Institut]. Ed., transl., Charles Ravaisson-Mollien. Paris: Maison Quantin.

Da Vinci, L. (1964). *Leonardo da Vinci on Painting. A Lost Book. Reassembled from* Codex Vaticanus Urbinas 1270 *and from the* Codex Leicester *by Carlo Pedretti*. Berkeley and Los Angeles: University of California Press.

Earl, A. G. (1894). *Practical lessons in physical measurement.* London and New York: Macmillan and Co.
Guttmann, M. (1903). *Chibbur ha-Meschicha we-ha-Tishboret.* Berlin: Mekis.e Nirdamim. https://www.hebrewbooks.org/38608.
Hamel, O., Richard, G. W., Hébert, M., Labrie, J.-M. (1966). *Mathématiques nouvelles: Cours secondaire -2.* Laprairie, Qc.: Les éditions F.I.C. La Mennais.
Hall, H. S., & Stevens, F. H. (1921). *School Geometry, Parts I, II, and III.* London: Macmillan.
Henrici, J., & Treutlein, P. (1987). *Lehrbuch der Elementar-Geometrie: Erster Teil. Gleichheit der Planimetrischen Grössen. Kongruente Abbildung in der Enene. Pesnum der Tertia.* Leipzig: Teubner.
Kepler, J. (1615). *Nova Stereometria Dolorium Vinariorum.* [New Solid Geometry of Wine Barrels] [...] *Accessit Stereometriae Archimedeae Svpplementvm* [[to which] A Supplement to the Archimedean Solid Geometry Has Been Added]. Lincii [=Linz]: J. Plancus.
Kepler, J. [= Johannes]. (1935). *Новая Стереометрия Винных Бочек* [New Stereometry of Wine Barrels, annotated Russian translation]. Translated by G.N. Sveshnikov, with an introduction by M.Ya. Vygodskiï. Moscow and Leningrad: GTTI.
Lardner, D. (1835). *The First Principles of Arithmetic and Geometry; Explained in a Series of Familiar Dialogues, Adapted for Preparatory Schools and Domestic Instruction; with Copious Examples and Illustrations.* [Part 2:] *Conversations on Geometry.* London: Longman and Taylor.
Lardner, D. (1840). *A Treatise on Geometry and its Application to the Arts.* Longman etc.
Larson, R., Boswell, L., Kanold, T. D., & Stiff, L. (2007). *Geometry.* Evanston (IL): McDougal Littell.
Nicolas de Cues (2007). *Les écrits mathématiques.* Introduction, translation and notes by Jean-Marie Nicolle. Paris: Honoré Champion.
Nikolaus von Kues (Nicolaus Cusanus) (1980). *Die mathematischen Schriften.* Josepha Hofmann (transl.). *Philosophische Bibliothek* 231. Hamburg: Felix Meiner Verlag.
Palmer, C. I. (1919). *Practical mathematics for home study, being the essentials of arithmetic, geometry, algebra and trigonometry.* New York: McGraw-Hill Book Company.
Ravaisson-Mollien, C. (ed., transl.). (1888). *Les manuscrits de Leonard de Vinci.* Tome 3 [Manuscrits C, E, & K de la Bibliothèque de l'Institut]. Paris: Maison Quantin.
Romanus, A. [= Adriaan van Roomen]. (1593). *Ideae mathematicae pars prima, sive Methodus polygonorum, qua laterum, perimetrorum & arearum cujiuscunque polygoni investigandorum ratio exactissima & certissima; unà cum circuli quadratura continentur.* Antwerpen: Ioannem Keerbergium.
Simon, Max (Maximilian). (1889). *Der erste Unterricht in der Raumlehre: ein methodischer Leitfaden für die unteren Klassen höherer Lehranstalten, sowie für die Volksschule in heuristischer Darstellung.* Berlin: Springer.
Small, M. (2018). *Grandes idées pour l'enseignement des mathématiques 9 à 14 ans.* Chenelière.
Stern, H. A., & Topham, W. H. (1913). *Practical Mathematics.* London: G. Bell and Sons (4th edition).
Tessier, G. J.-M., & Beaugrand, R. (1961). *Initiation à la géométrie: Géométrie intuitive.* Montréal: Centre de psychologie et de pédagogie.
Treutlein, P. J. (1911). *Der Geometrische Anschauungsunterricht als Unterstufe eines zweistufigen geometrischen Unterrichtes an unseren höheren Schulen.* Leipzig und Berlin: B.G. Teubner.
Van de Walle, J. A., Karp, K. S., Bay-Williams, J. M. (2020). *Elementary and middle school mathematics: Teaching developmentally.* 10th Edition. Harlow, etc.: Pearson.
Willis, C. A. (1922). *Plane Geometry: Experiment, Classification, Discovery, Application.* Philadelphia: Blakiston's Son & Company.
Zaitseva, N. Y. (1952). *Plany urokov po arifmetike v V classe. Is opyta raboty.* [Plans of arithmetic classes in the 5th grade. [Lessons] drawn from working experience.] Moscow: Uchpedgiz.
Zubov, V. P. (1935). *Леонардо да Винчи. Избранные естественнонаучные произведения.* Leonardo Da Vinci. [Selected works on natural sciences.] Translated and commented by V.P.

Zubov. Moscow: Academy of Sciences, 1935 [reproduced fac-simile by A.Lebedev' Studio Publishers (Moscow, 2010)].

Zubov, V. P. (1955). *Леонардо да Винчи. Избранные естественнонаучные произведения.* Leonardo Da Vinci. [Selected works on natural sciences.] Translated and commented by V.P. Zubov. 2 vols. Moscow: Academy of Sciences.

B. Secondary works

Acerbi, F. (2008). Disjunction and conjunction in Euclid's *Elements*. *Histoire Épistémologie Langage*, 30(1), 21–47.

Albertson, D. (2014). *Mathematical theologies: Nicholas of Cusa and the legacy of Thierry of Chartres*. Oxford University Press.

Barbin, E., & Menghini, M. (2014). History of teaching geometry. In A. Karp & G. Schubring (Eds.), *Handbook on history of mathematics education* (pp. 473–492). New York: Springer.

Baron, M. E. (1969). *The origins of the infinitesimal calculus*. Oxford etc.: Pergamon Press.

Becker, G. (1994). Das Unterrichtswerk "Lehrbuch der Elementargeometrie" von J. Henrici und P. Treutlein: Entstehungsbedingungen, Konzeption, Wirkung. In J. Schönbeck, H. Struve, & K. Volkert (Eds.), *Der Wandel im Lehren und Lernen von Mathematik und Naturwissenschaften*. Band 1, pp. 89–112. Weinheim: Deutscher Studienverlag.

Beckmann, P. (1976). *A history of π (pi)*. New York: St. Martin's Press, 1976 [originally published in 1971 by Golem Press, a publishing house established by the author himself].

Blank, B. (2001). *A history of Pi* by Petr Beckmann; *The joy of Pi* by David Blatner; *The nothing that is* by Robert Kaplan; *The story of a number* by Eli Maor; *An imaginary tale* by Paul Nahin; *Zero: The biography of a dangerous idea* by Charles Seife. *The College Mathematics Journal*, 32(2), 155–160.

Boyer, C. B. (1959). *The history of the calculus and its conceptual development (The Concepts of the Calculus)*. New York: Dover.

Brousseau, G. (1998). *Théorie des situations didactiques*. Grenoble: La Pensée Sauvage.

Brumbaugh, D. K., Ortiz, E., & Gresham, G. (2006). *Teaching Middle School Mathematics*. Mahwah, NJ: Lawrence Erlbaum.

Bryan, J., & Sangwin, C. (2008). *How round is your circle? Where engineering and mathematics meet*. Princeton University Press.

Casselman, B. (2012). Archimedes on the circumference and area of a circle. Feature column from the AMS. Posted on Internet in February 2012. http://www.ams.org/publicoutreach/feature-column/fc-2012-02.

Castelnuovo, E. (1977). L'enseignement des mathématiques. *Educational Studies in Mathematics*, 8(1), 41–50.

Cavanagh, M. (2008). Area measurement in year 7. *Reflections*, 33(1), 55–58.

Chevallard, Y. (1985). *La Transposition Didactique. Du savoir savant au savoir enseigné*. Grenoble: La Pensée Sauvage.

Counet, J.-M. (2005). Mathematics and the divine in Nicholas of Cusa. In T. Koetsier & L. Bergmans (Eds.), *Mathematics and the divine: A historical study* (pp. 273–290). Elsevier.

da Silva, R. S. R., Barbosa, L. M., Borba, M. C., & Andrejew Ferreira, A. L. (2021). The use of digital technology to estimate a value of pi: Teachers' solutions on squaring the circle in a graduate course in Brazil. *ZDM–Mathematics Education*, 23, 1–15. https://doi.org/10.1007/s11858-021-01246

Duhem, P. (1909). *Études sur Léonard de Vinci: Ceux qu'il a lus et ceux qui l'ont lu*. A. Hermann et fils.

Drijvers, P. (2012). Teachers transforming resources into orchestrations. In G. Gueudet, B. Pepin, & L. Trouche (Eds.), *From text to 'lived' resources: Mathematics curriculum materials and teacher development* (pp. 265–281). Springer.

Ernest, P. (1998). The history of mathematics in the classroom. *Mathematics in School, 27*(4), 25–31. [Reprinted in *Mathematics in School*, 40 (2011), No. 4, 14–20.]

Farmaki, V., & Paschos, T. (2007). Employing genetic "moments" in the history of mathematics in classroom activities. *Educational Studies in Mathematics, 66*(1), 83–106.

Francis, R. L. (1976). History of mathematics in the training program for elementary teachers. *The Arithmetic Teacher*, 23(4), 248–250.

Freiman, V., & Volkov, A. (2006). Infinitesimal procedures in modern and medieval mathematics textbooks. In S. Alatorre, J. L. Cortina, M. Sáiz, & A. Méndez (Eds.), *Proceedings of the twenty eighth annual meeting of the north American chapter of the international group for the psychology of mathematics education* (Vol. 2, pp. 519–520). Mérida, Mexico: Universidad Pedagógica Nacional. http://www.pmena.org/pmenaproceedings/PMENA%2028%202006%20Proceedings.pdf

Fujita, T., Jones, K., & Yamamoto, S. (2004). Geometrical intuition and the learning and teaching of geometry. Paper presented at the 10th International Congress on Mathematical Education (ICME-10). Copenhagen, Denmark, pp. 4–11 July 2004.

Furinghetti, F. (2007). Teacher education through the history of mathematics. *Educational Studies in Mathematics, 66*(2), 131–143.

Furinghetti, F., & Menghini, M. (2014). The role of concrete materials in Emma Castelnuovo's view of mathematics teaching. *Educational Studies in Mathematics, 87*(1), 1–6.

Glaeser, G. (1983). A propos de la pédagogie de Clairaut: Vers une nouvelle orientation dans l'histoire de l'éducation. *Recherches En Didactique Des Mathématiques, 4*(3), 332–344.

Gould, H.W. (1974). A History of π by Petr Beckmann [review article]. *Mathematics of Computation*, 28(125), 325–327.

Herbst, P., Cheah, U. H., Richard, P. R., & Jones, K. (2018). *International perspectives on the teaching and learning of geometry in secondary schools*. (ICME-13 Monographs). Cham: Springer.

Hofmann, J. E. (1971). Cusa, Nicholas. In C. Coulston Gillispie (Ed.), *Dictionary of scientific biography* (Vol. 3, pp. 512–516). New York: Charles Scribner's Sons.

Jankvist, U. T. (2009). A categorization of the "Whys" and "Hows" of using history in mathematics education. *Educational Studies in Mathematics, 71*(3), 235–261.

Johnson, D. B., & Mowry, T. A. (2016). *Mathematics: A practical odyssey*. Cengage Learning.

Kang, W., & Kilpatrick, J. (1992). Didactic transposition in mathematics textbooks. *For the Learning of Mathematics, 12*(1), 2–7.

Kordaki, M., & Potari, D. (1998). Children's approaches to area measurement through different contexts. *The Journal of Mathematical Behavior, 17*(3), 303–316.

Kuzniak, A., Richard, P. R. (2014). Espaces de travail mathématique. Point de vues et perspectives. *Revista latinoamericana de investigación en matemática educativa*, 17(4(I)), 5–40.

Levey, M. (1954). Abraham Savasorda and his algorism: A study in early European logistic. *Osiris, 11*, 50–64.

Levey, M. (1981). Abraham bar Hiyya. In C. Coulston Gillespie (Ed.), *Dictionary of scientific biography* (Vol. 1, pp. 22–23), New York: Charles Scribner's Sons.

Mac an Bhaird, C. (2011). The provision of mathematics support and the role of the history of mathematics. *Hermathena*, No. 191 (2011), Philosophy and Mathematics II, pp. 53–67.

Marshall, G. L., & Rich, B. S. (2000). The role of history in a mathematics class. *The Mathematics Teacher, 93*(8), 704–706.

Menghini, M. (2015). *From practical geometry to the laboratory method: The search for an alternative to Euclid in the history of teaching geometry*. In: S. J. Cho (Ed.). *Selected Regular Lectures from the 12th International Congress on Mathematical Education*, (pp. 561–587) Cham: Springer.

Montoito, R., & Garnica, A. V. M. (2015). Lewis Carroll, education and the teaching of geometry in Victorian England. *História da Educação, 19*(45) (January/April), 2015. (Online publication available at https://www.scielo.br/scielo.php?pid=S2236-34592015000100009&script=sci_arttext&tlng=en; retrieved on October 28, 2020).

Nicolle, J.-M. (1996). Les transsomptions mathématiques du Cardinal Nicolas de Cues. *Actes de l'Université d'été 95: Épistémologie et Histoire des Mathématiques (Besançon)* (pp. 359–372). I.R.E.M. de Franche-Comté.

Nicolle, J.-M. (2001). *Mathématiques et métaphysique dans l'œuvre de Nicolas de Cues*. Villeneuve d'Ascq: Presses Universitaires du Septentrion.

Nicolle, J.-M. (2020). *Le laboratoire mathématique de Nicolas de Cues*. Paris: Beauchesne.

O'Dell, J. R., Rupnow, T., Cullen, C. J., Barrett, J. E., Clements, D. H., & Sarama, J. (2016). Developing an Understanding of Children's Justifications for the Circle Area Formula, paper presented at the Annual Meeting of the North American Chapter of the International Group for the Psychology of Mathematics Education (38th, Tucson, AZ, Nov 3–6, 2016), 235–242.

Peckham. (1951). Dr. Lardner's "cabinet cyclopaedia." *The Papers of the Bibliographical Society of America, 45*(1), 37–58. https://www.jstor.org/stable/24298686

Rejeki, S., & Putri RII. (2018). Models to support students' understanding of measuring area of circles. *Journal of Physics (Conf. Series)*, no. 948. https://iopscience.iop.org/article/https://doi.org/10.1088/1742-6596/948/1/012058/pdf.

Richard, P. R., Venant, F., & Gagnon, M. (2019). Issues and Challenges in Instrumental Proof. In P. R. Richard, F. Venant, & M. Gagnon (Eds.), *Proof technology in mathematics research and teaching* (pp. 139–172). Springer.

Sander, H. J. (1982). Die Lehrbücher «Eléments de Géométrie» und «Eléments d'Algèbre» von Alexis-Claude Clairaut. Dissertation zur Erlangung des Grades eines Doktors der Erziehungswissenschaften an der Universität Dortmund

Scharpff, F. A. (Ed.). (1871). *Der Cardinal und Bischof Nicolaus von Cusa als Reformator in Kirche, Reich und Philosophie des 15. Jahrhunderts*. Tubingen: Laupp.

Smith, D. E., & Mikami, Y. (1914). *A History of Japanese Mathematics*. Chicago: Open Court

Stacey, K., & Vincent, J. (2009). Finding the area of a circle. Didactic explanations in school mathematics. *The Australian Mathematics Teacher, 65*(3), 6–9.

Tamborg, A. L. (2017). Origins, Transformations, and Key Foci in Instrumental Genesis. In *Proceedings of Eighth Nordic Conference on Mathematics Education (Norma'17). Nordic Society for Research in Mathematics Education* (Stockholm, Sweden), 1–9.

Trouche, L. (2005). Instrumental genesis, individual and social aspects. In: Dominique, G., Kenneth, R., Luc, T. (Eds.), *The didactical challenge of symbolic calculators: Turning a computational device into a mathematical instrument* (pp. 197–230). Boston etc.: Springer Science+Business Media.

Uebinger, J. (1895). Die mathematischen Schriften des Nik. Cusanus. *Philosophisches Jahrbuch*. Band 8 (1895), S. 301–317, 403–422. Band 9 (1896), S. 54–66, 391–410. Band 10 (1897), S. 144–159.

Vengeon, F. (2006). Mathématiques, création et humanisme chez Nicolas de Cues. *Revue D'histoire Des Sciences, Tome, 59*(2), 219–244.

Vérillon, P., & Rabardel, P. (1995). Cognition and artifacts: A contribution to the study of thought in relation to instrumented activity. *European Journal of Psychology of Education, 10*(1), 77–101.

Watanabe, M. (2011). *Nicholas of Cusa—A companion to his life and his times*. Edited by Gerald Christianson and Thomas M. Izbicki. Farnham and Burlington: Ashgate, 2011.

Weiss, Y. (2019). Introducing history of mathematics education through its actors: Peter Treutlein's intuitive geometry. In: Weigand, H.-G., McCallum, W., Menghini, M., Neubrand, M., Schubring, G. (Eds.), *The legacy of Felix Klein. (ICME-13 monographs)* (pp. 107–116). Cham: Springer.

Wertz, W. F., Jr. (2001). Nicolaus of Cusa's 'on the quadrature of the circle'. *Fidelio, 10*(2) (Summer 2001), 30–40.

Wiltshire, B. (1930). History of Mathematics in the Classroom. *The Mathematics Teacher, 23*(8), 504–508.

Implementing STEM Projects Through the EDP to Learn Mathematics: The Importance of Teachers' Specialization

Jose-Manuel Diego-Mantecón, Zaira Ortiz-Laso, and Teresa F. Blanco

1 Introduction

In the last two decades, the European Union (EU) has increasingly encouraged the training of competent citizens for meeting the demands of a progressively technological society (European Union Council, 2018). The idea of developing students' competences and stimulating them to study and work in STEM-related fields to form a solid society has also been supported by national governments (Niss et al., 2017). An effort to diversify STEM-related careers has also proliferated recently after detecting an under-representation in certain sectors of the population, such as women (Eurostat, 2018). It is, however, questionable whether these objectives are being accomplished. International reports, like the Programme for the International Student Assessment (PISA; OECD, 2019), point out that many 15-year-old students across countries do not achieve the minimum required level of mathematics and science competency. Although in the EU there is a growing number of STEM graduates (Eurostat, 2018), this rate is smaller than the one raised in countries like the USA, Russia, and Canada (Watson & Munkoe, 2019). Similarly, the number of graduated females in STEM-related fields still remains under-represented in the EU (Eurostat, 2018).

J.-M. Diego-Mantecón (✉) · Z. Ortiz-Laso
Universidad de Cantabria, Santander, Spain
e-mail: diegojm@unican.es

Z. Ortiz-Laso
e-mail: zaira.ortiz@unican.es

T. F. Blanco
Universidad de Santiago de Compostela, Santiago de Compostela, Spain
e-mail: teref.blanco@usc.es

© The Author(s), under exclusive license to Springer Nature Switzerland AG 2022
P. R. Richard et al. (eds.), *Mathematics Education in the Age of Artificial Intelligence*, Mathematics Education in the Digital Era 17,
https://doi.org/10.1007/978-3-030-86909-0_17

As a consequence of the above priorities, researchers have approached the integration of content by combining different disciplines. Technology has been incorporated into the learning of mathematics through innovative technological devices and software (Cullen et al., 2020). The creation of tools like GeoGebra has certainly helped to acquire mathematical knowledge that was difficult to gain through the traditional approach (Prodromou, 2014). Similarly, researchers have promoted the connection between science and mathematics (Maass et al., 2019a; Potari et al., 2016; Triantafillou et al., 2021) by a number of European initiatives (e.g., Mascil, PRIMAS, and Fibonacci). The incorporation of engineering into the science, technology, and mathematics disciplines has also been considered, completing the acronym STEM (Diego-Mantecón et al., 2019; English, 2016, 2020).

STEM education usually employs engineering as a context to integrate the three remaining disciplines (Moore et al., 2014; Thibaut et al., 2018b). This has often led teachers to adopt the engineering design process (EDP) for implementing STEM projects. The EDP is thus used as a way to teach mathematics in a contextualized manner (English & King, 2019; Fidai et al., 2020; Margot & Kettler, 2019). Some authors have questioned this approach for its difficulty to raise mathematics (Lasa et al., 2020; Ubuz, 2020). The present study aims to verify teachers' capacities to explore and promote mathematical content within this approach. The EDP requires teachers to integrate content in which they are not experts. In particular, we will analyse the mathematics school content addressed by technology and mathematics Spanish teachers (out-of-field and in-field, respectively) when implementing STEM projects.

2 Teaching Mathematics Through Technology

Technology is traditionally viewed as a tool for teaching mathematics. Most official curricula and textbooks incorporate technological devices and software to work out mathematics tasks. Researchers worldwide also recommend technology to support instruction (e.g., Blanco et al., 2019a; Borba et al., 2016, 2017; Fabian et al., 2018; Kovács et al., 2020; Lavicza et al., 2020; Prodromou & Lavicza, 2017). The use of technology in mathematics education has quickly evolved; not just calculators but also computer laboratories, mobile technologies, and Massive Open Online Courses (MOOCs) are nowadays available (Borba et al., 2016, 2017). There seem to be three main ways in which technology is employed in mathematics classrooms: (1) as a tool for delivering content; (2) as a supply for facilitating analyses, proofs, and conjectures; and (3) as a tutor for receiving feedback. Although, in general, these three ways of learning seem to contribute positively to mathematics learning, research arises contradictory conclusions.

(1) The use of technology for delivering content through tablets or laptops may not have a significant impact on learning mathematics. One-child-one-device does not necessarily ensure meaningful learning (Dubé et al., 2019; Hall et al., 2021), especially when the purpose of using devices is to simply replace the traditional resources.

(2) The employment of technology to encourage cognitive aspects of learning also generates contrary outcomes. Prodromou (2014) suggests that technology facilitates the visualization processes to learn concepts. Zulnaidi and Zamri (2017) found a positive relationship between using GeoGebra and understating conceptual and procedural knowledge. In contrast, Wijers et al. (2010) conclude that digital games do not always report positive impacts on mathematics knowledge. (3) Intelligent tutoring systems have also reported different findings (El-Khoury et al., 2005; Pai et al., 2021; Richard et al., 2011). While some authors found intelligent tutoring systems suitable to apply mathematical concepts and to develop problem-solving skills (Dašić et al., 2016), others did not identify significant differences in achievement between students utilizing a tutoring system and the ones following the traditional approach (Pai et al., 2021).

Some of the discrepancies identified above may be explained by issues associated just with the integration of technology in the classroom. Implementing technology implies overcoming key challenges related to pedagogical, technical, and organizational aspects (Borba et al., 2016). Still, many teachers have not even attempted to use technology because of factors concerning resistance to change and precedents of previous failed initiatives (Diego-Mantecón, 2020; Lavicza et al., 2020; Vinnervik, 2020). Moore et al. (2014) talk of the need of setting a context where technology naturally incorporates and applies mathematics. They, for instance, highlight the importance of using engineering design for employing technology and applying mathematics and/or science in meaningful learning.

3 STEM Projects and the Engineering Design Process for Learning Mathematics

In response to the European priorities, STEM education is becoming more important in the current educational systems (Diego-Mantecón et al., 2021; Maass et al., 2019b; Thibaut et al., 2018a, b, 2019). Authors have conceptualized STEM under slightly different approaches (English, 2016; Kelley & Knowles, 2016; Martín-Páez et al., 2019; Toma & García-Carmona, 2021). These approaches are characterized by various forms of boundary-crossing among the four STEM disciplines (English, 2016, 2020; Kelley & Knowles, 2016; Maass et al., 2019b; Martín-Páez et al., 2019), and even in relation to Art in the so-called STEAM education (Diego-Mantecón et al., 2021; Herro et al., 2019; Mohd-Hawari & Mohd-Noor, 2020; Quigley & Herro, 2016). English (2016) distinguishes three ways of integrating disciplines: multidisciplinary, interdisciplinary, and transdisciplinary. The first implies teaching concepts and skills separately in each discipline but within a common theme. The second entails teaching concepts and skills from two or more disciplines aiming to narrow knowledge down. The third relates to applying knowledge and skills from various disciplines to solve real-world problems shaping the learning experience.

To implement STEM education, several authors propose contexts or processes where usually one discipline is emphasized over the others (Martín-Páez et al., 2019; Thibaut et al., 2018a, b for a review). One of these processes takes engineering like a context and promotes technology and mathematics in similar ways; this is the so-called engineering design (Diego-Mantecón et al., 2019; English & King, 2019; English et al., 2017; Li et al., 2019). The engineering design process (EDP) often comprises the following steps: 'problem scoping', 'idea creation', 'designing and constructing', 'assessing design', and 'redesigning and reconstructing' (English et al., 2017). Problem scoping seeks understanding problem boundaries by clarifying the goal and identifying constraints. Idea creation implies developing a plan to approach the problem, which includes formulating questions, sharing ideas, and developing strategies. Designing and constructing encompasses sketching designs, interpreting them, predicting possible outcomes, and transforming them into models. Assessing design involves checking constraints, testing models, and verifying the accomplishment of objectives. Finally, the redesigning and reconstructing step entails reviewing initial designs and sketching new ones for refining the model.

Several researchers view the EDP as a way to apply mathematics and technology in a creative and innovative manner (Akgun, 2013; English & King, 2019; Fidai et al., 2020; Margot & Kettler, 2019; Quigley & Herro, 2016). Others suggest, however, that this process does not necessarily require a deep mathematics focus. Lasa et al. (2020) claim, for instance, that mathematical content in engineering-orientated activities is often basic and utilitarian, and involves mainly geometry and measurement. Concerning technology, some researchers consider it as a tool to create, activate, and test engineering artefacts (Akgun, 2013), while others point out that technology is often under-represented in STEM education (English, 2016).

4 Design and Implementation of STEM Activities in Secondary Education

The design and implementation of STEM experiences seem to be affected by teachers' specialization and thus their understanding of the discipline. In many educational systems, primary school teachers are responsible for instructing most subjects, while high school teachers are characterized by being subject-specific. In Spain, the latter holds a bachelor's degree and a subject-specific master in teacher training. Toma and García-Carmona (2021) suggest that this training is contrary to STEM education and thus to the integrated approach. An integrated approach requires a solid conceptual, procedural, and epistemological knowledge on various disciplines. Many authors criticize the lack of content and pedagogical knowledge of high school teachers to integrate disciplines (Domènech-Casal et al., 2019; Frykholm & Glasson, 2005; Toma & García-Carmona, 2021), and teachers have reported to feel unconfident when designing and implementing STEM activities (Frykholm & Glasson, 2005).

According to Davis et al. (2019) and Triantafillou et al. (2021), the epistemological ground that teachers adopt when implementing an activity significantly affect the way it is elaborated and the concepts that students learn from it. During the STEM activities instruction, mathematics and science teachers put different emphasis on the concepts and properties used to explain the same topic (Potari et al., 2016). Vale et al. (2020) reveal that Australian mathematics and science teachers do not have the same beliefs about these two disciplines, and the way these should be taught. Epistemological differences are even explicit in trainers when guiding teachers into the design of integrated activities (Triantafillou et al., 2021). In this sense, Davis et al. (2019) claim the need of instructing teachers on approaching STEM concepts from various epistemological orientations (or ways of knowing).

To facilitate STEM implementation, researchers suggest setting real contexts from which to naturally integrate content (Frykholm & Glasson, 2005; Potari et al., 2016; Triantafillou et al., 2021). Nevertheless, framing experiences in real contexts does not imply promoting high content integration (Domènech-Casal et al., 2019), and authors often advocate for using design-based processes (Burghardt & Hacker, 2004; English, 2019). Design processes, commonly used in the technology subject, are often applied by a trial-and-error approach that does not always foster conceptual understanding (Burghardt & Hacker, 2004). To increase discipline integration, teacher collaboration has also been promoted (El-Deghaidy et al., 2017; Nelson & Slavit, 2007; Potari et al., 2016; Triantafillou et al., 2021). Potari et al. (2016) reveal that the collaborative work between science and mathematics teachers allowed for a deep content integration and contextualization. Similarly, teachers' interactions help to better explore the relationships across subjects, often overlooked in fragmented approaches (Nelson & Slavit, 2007). Although teachers are aware of the importance of collaborating, several studies highlight the difficulty of establishing connection between peers (Al Salami et al., 2017; Potari et al., 2016; Thibaut et al., 2018b; Triantafillou et al., 2021). In this regard, Frykholm and Glasson (2005) state that willingness to share classroom experiences facilitates collaboration. Nelson and Svait (2007) and Triantafillou et al. (2021) highlight also the necessity of establishing a sense of community between teachers. Trainers should participate in this community to support teachers in the activity design (Triantafillou et al., 2021).

5 The Study

This study seeks to assess the design and implementation of STEM projects through the EDP to learn mathematics. We analyse how technology and mathematics teachers (out-of-field and in-field, respectively) address high-school mathematics content in STEM projects elaborated through the EDP. We thus formulate the following question: Does teachers' specialization affect the way in which STEM projects are executed through the EDP to learn mathematics? For tackling this question, we call on high school teachers willing to implement STEM projects with an engineering-oriented focus.

5.1 Sample

Five Spanish teachers were selected for this study from an initiative run by the Open STEAM Group (https://www.opensteamgroup.unican.es/). The five teachers were selected, from a total of 54, because they implemented STEM projects within the engineering design process, where mathematics and technology were somehow applied. Two of the teachers were specialized in mathematics, holding a pure mathematics bachelor's degree (in-field). The other three were qualified in technology with engineering bachelor's degrees; they thus taught mathematics without having such specialization (out-of-field). The five teachers executed a total of ten STEM projects.

The teachers had more than 15 years of experience instructing technology, computer, or mathematics subjects in state and state-subsidised Spanish high schools. The teachers did not have formal training nor background experience in STEM education. About 30 students were involved in the initiative, beginning at the age of 14–15. These students followed the regular Spanish curriculum including mathematics and technology subjects. The mathematics subject embraces numbers and algebra, functions, geometry, and statistics and probability. The technology subject comprises information and communication technologies, domestic installations, electronics, control and robotics, pneumatic and hydraulic, and technology and society.

5.2 Guidelines for Project Development

The in-field and out-of-field mathematics teachers implemented the STEM projects in their classrooms with groups of 4–5 students, through the EDP and using the KIKS format (Blanco et al., 2019b; Diego-Mantecón et al., 2021; Ortiz-Laso, 2020). The KIKS (Kids Inspire Kids for STEAM) format goes beyond project-based learning, actively involving students and teachers in dissemination actions worldwide. To deliver their outcomes, students produce a video and a text report in English. The videos aim to provide quick overviews about the project, the constructed artefacts, and their functioning. The report addresses in-depth information about the analytical processes. Projects are presented in different formats (online and face-to-face) and events (e.g., conferences and outreach activities) to a variety of audiences. In this study, students developed several projects for a period of at least two years. These projects were designed by their teachers or by experts of the Open STEAM Group.

5.3 Data Analysis

To analyse the mathematical content addressed in the elaborated projects, we assessed the text documents and videos produced by the students. The mathematical content was classified according to the following blocks: 'numbers', 'algebra', 'geometry', 'functions', 'statistics', and 'probability'. The three authors of this chapter independently classified the mathematical content, to be compared later. To gain precise information about how mathematics was used in the projects, teachers' semi-structured interviews and observations were also conducted. We identified whether the teachers promoted solutions by intuition instead of a planned approach (Lin & Williams, 2017), whether they endorsed inquiry processes through questioning strategies (Bruce-Davis et al., 2014), and challenged students to think deeply about concepts and ideas to foster skills like abstracting, analysing, applying, formulating, and interpreting (Herro et al., 2019). All the interviews were audio-recorded and transcribed. The raw data was entered into a text document and analysed by identifying key statements and associating patterns.

6 Results

As shown in Table 1, most of the projects included geometry and algebra content. Only two of them included statistics, probability, and/or numbers. A preliminary analysis suggested that technology teachers tended to use algebra, while mathematics teachers usually applied geometry content.

The technology teachers introduced algebra for designing circuits, covered in the electronic block of the technology curriculum (Boolean algebra). To design the pieces composing the artefacts, measurement and geometry content from the mathematics curriculum was applied. For example, students quantified lengths and angles, employing instruments such as rule, triangle, protractor, and compass. They also required basic geometry concepts as perpendicular and parallel lines and drew the nets of different 3D shapes such as prisms and cylinders.

Table 1 STEM project categorization

Project name	Maths curricular content emphasized	Subject of implementation
Star Wars Robot	Algebra, Geometry	Technology
Simon Says	Algebra, Geometry	Technology
Lights of Buildings	Algebra, Geometry	Maths
UV Light in Rudimentary Health Care Centres	Algebra, Geometry	Maths
Rubik's Cube	Geometry, Probability	Maths
Vehicle Avoiding Obstacles	Algebra, Geometry	Technology
Solar cars	Algebra, Geometry	Technology
Astrolabe	Geometry, Statistics, Numbers	Maths
Hothousing Gardens	Algebra, Geometry	Technology
Wireless Telegraph	Algebra	Technology

The mathematics teachers involved geometry in all the projects to a larger extent than the technology ones. This geometry content was also more formal than the one applied by the engineers, being often the vertebral column of the projects. That was the case, for example, of the Lights of Buildings and Astrolabe projects. In the former, trigonometric relations were employed to determine the angle and height where to locate the lights, while in the latter students were introduced to the stereographic projection. Mathematics teachers also promoted the interaction with 3D shapes, using software to draw them during the design and construction of the artefacts. In the Lights of Buildings and UV Light projects, students drew and built truncated pyramids. Regarding algebra, the mathematics teachers used similar contents to the technology teachers. Other content employed by the mathematics teachers were probability, statistics and numbers, used in the Rubik's Cube and Astrolabe projects.

A deep analysis of the projects showed that mathematics was exploited at least in three different ways: identification, reasoning, and modelling. For exemplifying each of these ways, we describe how mathematics was addressed in the Star Wars Robot, the Rubik's Cube, and the Astrolabe projects.

6.1 Identification: Star Wars Robot

The Star Wars Robot project, supervised by a technology teacher, aimed to construct the famous R2-D2. The project idea arose from the students after being challenged to create film characters. Initially, students thought about different characters including Bender (The Simpsons and Futurama) and Lighting McQueen (Cars). However, they discarded the aforementioned characters because of constraints like the difficulty of reproducing Bender's movements of legs and arms or modelling the Lighting McQueen's hood or bumper. Students considered that it would be easier to construct characters from objects found in daily life or by assembling 3D shapes obtained from a net. As a consequence, a group of students agreed to work on the design and construction of the R2-D2. The teacher suggested drawing an initial design of the robot and constructing it according to a certain scale. Nevertheless, the students proceeded freely, not following such suggestion; they searched for objects, in their surroundings, representing different parts of the robot for joining them together. They designed the widest part of the leg, decomposing it into a semicircle and a non-regular hexagon. To sketch and construct this piece the notions of diameter, parallel and perpendicular lines were applied. Once the piece was produced, they replicated it three times more; finally, these were cut, painted, and assembled (Fig. 1a, c).

In the next step, students programmed the robot using App Inventor with an Arduino board (Fig. 1b). When programming, the measures of length and angles were used to reproduce the movements of the prototype (Fig. 1c), involving geometrical concepts like rotation and translation. This project could have offered the possibility of applying other concepts related to proportionality. However, the students did not attempt an accurate reproduction of the original R2-D2, focusing mainly on assembling objects and programming the different functions.

Implementing STEM Projects Through the EDP ...

a. Assembling Materials	b. Programming	c. Final Design

Fig. 1 Star Wars Robot

6.2 *Reasoning: Rubik's Cube*

The Rubik's Cube project, led by a mathematics teacher, was intended to construct a robot for solving a Rubik's cube of 3 × 3 × 3 dimensions. The project idea emerged from the teacher who is fascinated by the variety of strategies that can be used to solve it.

To set the project, the teacher provided the LEGO MindStorm robotic Kit-tool and introduced geometrical concepts such as a polyhedron, cube, faces, edges, rotation, symmetry, vertex, and so on. The students explored the structure of the cube discovering that only the central pieces of the six faces maintain a fixed position with respect to each other, being able only to rotate around the axis perpendicular to the face. Counting the number of small cubes in the bigger one, as well as the number of faces, edges, and vertices, they applied probabilistic content concerning permutations to work out the number of possible positions of each small cube in the bigger cube. At this stage, they understood the importance of following an ordered sequence for the resolution process, and they continued investigating into the combinatorial world. They found out that in the original Rubik's cube there are 8! ways to combine the eight vertices. Seven of these vertices can be oriented independently, and the orientation of the eighth will depend on the previous seven, resulting in 3^7 possibilities. Then, the teacher promoted reasoning by helping understand that there are $\frac{12!}{2}$ ways to arrange the 12 pieces with two colours located in the middle of the physical edges, since a parity of the corners also implies a parity of the edges. The students also learnt that 11 edges can be turned independently, and the rotation of the 12th edge will depend on the previous ones, giving 2^{11} possibilities. Finally, they were able to work out the total number of permutations, and to understand where the figures were coming from. Figure 2a exemplifies the students' reasoning.

At this point, the students became aware of the amount of mathematics needed and realized that different ways of solving the cube drive to different algorithms. They searched for traditional algorithms and related them to the steps undertaken when solving the cube by hand. Recalling their own experience in solving the cube by hand, they were employing their visual-spatial ability to search for an algorithm and thus transferring each particular piece of the cube to the desired position. Similarly, talking

a. Mathematical reasoning	b. Building process	c. Final Design

Fig. 2 Rubik's Cube project

about positions, rows, and columns, they realized the necessity of using matrices to store the colours.

6.3 Modelling: Astrolabe

The Astrolabe project, guided by a mathematics teacher, aimed to construct an astrolabe employing 3D printing. The project idea arose from the teacher, as he is an enthusiast of this instrument. In this project, the teacher initially explained the aspects and content of geometry and astronomy needed to understand the functioning of the instrument. Regarding geometry, he reminded the 3D projection of objects on the plane, emphasizing the stereographic projection as it is not feasible to represent the sphere on the plane. This required students to become familiar with elements such as parallels, meridians, great circles, angles, and spherical triangles. They also had to learn that properties of the plane are not extrapolated to the sphere; for example, the sum of the interior angles of a triangle is not fulfilled when working with spherical triangles.

In the design and construction phase (Fig. 3a), students applied the content acquired in the initial phase to design the astrolabe with software and using scales. At

a. Initial Design	b. Assessing design	c. Refined design

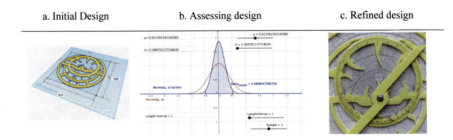

Fig. 3 Astrolabe project

that stage the main difficulty was the design of the astrolabe front piece (named rete). Then, they assembled all the 3D printed pieces. Once the artefact was constructed, the students took measures to calibrate it; they analysed a sample of 57 observations. The students observed that the distribution of errors followed a normal curve, not being accurate (Fig. 3b). During this process, they were managing statistical content such as mean, standard deviation, variance, and confidence interval. Consequently, they sketched a new design for refining the initial artefact (Fig. 3c).

7 Discussion

The project selection, the content, and the applied mathematics differed in relation to teachers' specialization. In the projects guided by out-of-field mathematics teachers, mathematics was hardly involved. Teachers and students verbalized mathematical terms related to the components of the artefacts (e.g., cylinder, sphere), but rarely they engaged in reasoning or conjecturing processes; at least it was not perceived by these researchers. Out-of-field teachers tended to overlook the mathematical content and to focus on specific aspects of their subject, not promoting content integration. Domènech-Casal et al. (2019), in an analysis of STEM projects, already noticed that, in general, high school teachers do not integrate content from different disciplines. According to our analyses, the teachers were usually embedded in their context facing difficulties to breakout from it and to integrate disciplines. This concurs with Potari et al. (2016) when reporting that teachers usually address concepts from the perspective of their specialization, struggling to exploit the same concepts from an out-of-field perspective.

During the project design and construction, our out-of-field mathematics teachers normally worked with their students under a trial-and-error strategy, restricting the application of mathematics. This outcome is in line with English (2019), as well as with Lin and Williams' (2017) observations, when suggesting that teachers with a lack of STEM training seek solutions by intuition rather than by considering mathematics and science principles. Our out-of-field mathematics teachers usually implemented projects where students simply have to identify and recall geometric components of the artefacts. We concur with Burghardt and Hacker (2004) that in design-based projects teachers are frequently focused on the product rather than on the learning process. Conjectures and data analysis were rarely attempted. They even avoided activities promoting inquiry processes, appropriate for facilitating the integration and application of content. Normally, the mathematics aroused was employed in the designing of circuits and programming, using for example Boolean algebra. This fits, to some extent, with Lasa et al. (2020) when reporting that engineering projects lack school mathematics content, mainly related to basic geometry.

Unlike the out-of-field mathematics teachers, the in-field ones took mathematical concepts and procedures as a starting point from which to elaborate the projects. They usually proposed their own ideas on topics where they felt confident. In-field teachers involved mathematics to a greater extent than technology ones. In contrast to technology teachers, the in-field ones attempted to elaborate the projects under an iterative process of design, analysis, and redesign. In-field teachers tended to make an effort for matching project content with high school mathematics curriculum. In addition to the basic and utilitarian content promoted by out-of-field teachers (e.g., the identification of mathematical terms or measures), the in-field ones incorporated concepts, properties, and ideas. In some projects, data collection, analysis, and modelling were also promoted. The projects were mainly used to reinforce and apply previous knowledge, as also reported in the study of Margot and Kettler (2019). The in-field teachers tended also to engage their students on reasoning, and conjecturing processes. For example, in the Rubik's cube project the teacher was questioning the number of possibilities for arranging the pieces of the cube, encouraging mathematical thinking. In the astrolabe, the students proved the sum of the interior angles of a triangle is 180° in a Euclidean space and observed that such property is not fulfilled in a spherical triangle. The teacher drove their students also in the process of verifying the astrolabe consistency by taking and analysing measures.

8 Conclusions and Implications for Further Research

This study examined the extent to what school mathematics is addressed in STEM projects following the EDP, and consequently the suitability of the integrated approach in the school contexts. The analyses showed that teachers' specialization is a key point in the implementation of the projects, and it determines how mathematics would be promoted and reflected in the instruction process. In the majority of the projects, mathematics was poorly promoted. Only in some of them, mathematics content and reasoning were stimulated, and in rather few projects teachers encouraged high cognitive processes by means of questioning, conjecturing, analysing, and verifying.

Out-of-field mathematics teachers selected projects designed by experts, requiring a dose of effort to personalize them, and becoming familiar with the materials. They put the focus on the assembly and construction of the artefacts, as well as in their functioning, avoiding to explore mathematics in-depth. Such focus on the construction phase seems to be due to the strong influence of teachers' specialization, which leads them into the creational part of the engineering process, and the stimulating part of the technology usage. They actually achieved rather exciting artefacts with their students, but with a substantial lack of school mathematics content.

In contrast, in-field mathematics teachers evaded selecting the proposed projects. They tried to design their own projects based on their mathematical experiences and content in which they felt confident. Unlike the out-of-field mathematics teachers, the in-field ones provided their students less freedom to conduct their projects, guiding them into the resolution process. They asked their students to deal first with the mathematics content, offering them less time for hands-on activities. The in-field teachers encouraged the application of mathematical concepts, properties, and ideas, as well as data collection and analysis.

This study presents some methodological limitations concerning the sample, but we could still claim that integrating school mathematics content through the EDP is rather challenging for high school teachers. They are subject-specific and thus rooted into a limited content and context. Within these courses, we recommend promoting collaboration among teachers from different specializations to join their best knowledge for achieving a common goal. The Open STEAM Group is already running such courses with in-service teachers to initiate them in a collaborative teaching before their incorporation in the school. Doing so, we seek to protect the idea that teachers must hold a specialized knowledge, as reported by many experts in mathematics education.

Acknowledgements This study was supported by FEDER/Ministerio de Ciencia, Innovación y Universidades—Agencia Estatal de Investigación/ project EDU2017-84979-R.

References

Akgun, O. E. (2013). Technology in STEM project-based learning. In R. M. Capraro, M. M. Capraro & J. R. Morgan (Eds.), *STEM project-based learning. An integrated science, technology, engineering, and mathematics (STEM) Approach* (pp. 65–75). Sense Publishers. https://doi.org/10.1007/978-94-6209-143-6.

Al Salami, M. K., Makela, C. J., & de Miranda, M. A. (2017). Assessing changes in teachers' attitudes toward interdisciplinary STEM teaching. *International Journal of Technology and Design Education, 27*(1), 63–88. https://doi.org/10.1007/s10798-015-9341-0

Blanco, T. F., García-Piqueras, M., Diego-Mantecón, J. M., & Ortiz-Laso, Z. (2019a). Modelización matemática de la evolución de dos reactivos químicos. *Épsilon, 101*, 147–155.

Blanco, T. F., Ortiz-Laso, Z., & Diego-Mantecón, J. M. (2019b). Proyectos STEAM con formato KIKS para la adquisición de competencias LOMCE. En J. M. Marbán, M. Arce, A. Maroto, J. M. Muñoz-Escolano & Á. Alsina (Eds.), *Investigación en Educación Matemática XXIII* (p. 614). SEIEM.

Borba, M. C., Askar, P., Engelbrecht, J., Gadanidis, G., Llinares, S., & Aguilar, M. S. (2016). Blended learning, e-learning and mobile learning in mathematics education. *ZDM Mathematics Education, 48*(5), 589–610. https://doi.org/10.1007/s11858-016-0798-4

Borba, M. C., Askar, P., Engelbrecht, J., Gadanidis, G., Llinares, S., & Aguilar, M. S. (2017). Digital technology in mathematics education: research over the last decade. In G. Kaiser (Ed.), *Proceedings of the 13th international congress on mathematics education* (pp. 221–233). Springer. https://doi.org/10.1007/978-3-319-62597-3_14.

Bruce-Davis, M. N., Gubbins, E. J., Gilson, C. M., Villanueva, M., Foreman, J. L., & Rubenstein, L. D. (2014). STEM high school administrators', teachers', and students' perceptions of curricular and instructional strategies and practices. *Journal of Advanced Academics, 25*(3), 272–306. https://doi.org/10.1177/1932202X14527952.

Burghardt, M. D., & Hacker, M. (2004). Informed design: A contemporary approach to design pedagogy as the core process in technology. *Technology teacher, 64*(1), 6–8.

Cullen, C. J., Hertel, J. T., & Nickels, M. (2020). The roles of technology in mathematics education. *The Educational Forum, 84*(2), 166–178. https://doi.org/10.1080/00131725.2020.1698683

Dašić, P., Dašić, J., Crvenković, B., & Šerifi, V. (2016). A review of intelligent tutoring systems in e-learning. *Annals of the University of Oradea,* (3), 85–90. https://doi.org/10.15660/AUOFMTE.2016-3.3276.

Davis, J. P., Chandra, V., & Bellocchi, A. (2019). Integrated STEM in initial teacher education: Tackling diverse epistemologies. In P. Sengupta, M. C. Shanahan & B. Kim (Eds.), *Critical, transdisciplinary and embodied approaches in STEM education* (pp. 23–40). Springer. https://doi.org/10.1007/978-3-030-29489-2_2.

Diego-Mantecón, J. M. (2020). Classroom Implementation of STEM Education through technology: advantages and handicaps. In P. R. Richard, S. Van Vaerenbergh & M. P. Vélez (Eds.), *First Symposium on Artificial Intelligence for Mathematics Education. Books of abstracts (AI4ME 2020)* (pp. 9–10). Universidad de Cantabria. https://doi.org/10.22429/Euc2020.034.

Diego-Mantecón, J. M., Arcera, Ó., Blanco, T. F., & Lavicza, Z. (2019). An engineering technology problem-solving approach for modifying student mathematics-related beliefs: Building a robot to solve a Rubik's cube. *International Journal for Technology in Mathematics Education, 26*(2), 55–64.

Diego-Mantecón, J., Blanco, T., Ortiz-Laso, Z., & Lavicza, Z. (2021). STEAM projects with KIKS format for developing key competences. [Proyectos STEAM con formato KIKS para el desarrollo de competencias clave]. *Comunicar, 66*, 33–43. https://doi.org/10.3916/C66-2021-03

Domènech-Casal, J., Lope, S., & Mora, L. (2019). Qué proyectos STEM diseña y qué dificultades expresa el profesorado de secundaria sobre Aprendizaje Basado en Proyectos. *Revista Eureka sobre Enseñanza y Divulgación de las Ciencias, 16*(2), 2203. https://doi.org/10.25267/Rev_Eureka_ensen_divulg_cienc.2019.v16.i2.2203.

Dubé A.K., Alam S.S., Xu C., Wen R., Kacmaz G. (2019). Tablets as elementary mathematics education tools: Are they effective and why. In K. Robinson, H. Osana, & D. Kotsopoulo (Eds.), *Mathematical learning and cognition in early childhood* (pp. 223–248). Springer. https://doi.org/10.1007/978-3-030-12895-1_13.

EL-Deghaidy, H., Mansour, N., Alzaghibi, M., & Alhammad, K. (2017). Context of STEM integration in schools: Views from in-service science teachers. *EURASIA Journal of Mathematics, Science and Technology Education, 13*(6), 2459–2484. https://doi.org/10.12973/eurasia.2017.01235a.

El-Khoury, S., Richard, P. R., Aïmeur, E., & Fortuny, J. M. (2005). Development of an Intelligent Tutorial System to Enhance Students' Mathematical Competence in Problem Solving. In G. Richards (Ed.), *E-Learn: World Conference on E-Learning in Corporate, Government, Healthcare, and Higher Education* (pp. 2042–2049). Association for the Advancement of Computing in Education (AACE).

English, L. D. (2016). STEM education K-12: Perspectives on integration. *International Journal of STEM Education, 3*(1). https://doi.org/10.1186/s40594-016-0036-1.

English, L. D. (2019). Learning while designing in a fourth-grade integrated STEM problem. *International Journal of Technology and Design Education, 29*(5), 1011–1032. https://doi.org/10.1007/s10798-018-9482-z

English, L. D. (2020). Facilitating STEM integration through design. In J. Anderson & Y. Li (Eds.), *Integrated approaches to STEM education: An international perspective* (pp. 45–66). Springer. https://doi.org/10.1007/978-3-030-52229-2_4.

English, L. D., & King, D. (2019). STEM Integration in Sixth Grade: Desligning and Constructing Paper Bridges. *International Journal of Science and Mathematics Education, 17*, 863–884. https://doi.org/10.1007/s10763-018-9912-0

English, L. D., King, D., & Smeed, J. (2017). Advancing integrated STEM learning through engineering design: Sixth-grade students' design and construction of earthquake resistant buildings. *The Journal of Educational Research, 110*(3), 255–271. https://doi.org/10.1080/00220671.2016.1264053

European Union Council. (2018). *Council Recommendation of 22 May 2018 on key competences for lifelong learning*. https://eur-lex.europa.eu/legal-content/EN/TXT/?uri=CELEX%3A32018H0604%2801%29.

Eurostat. (2018). *Smarter, greener, more inclusive? Indicators to Support the Europe 2020 Strategy (2018 Edition)*. https://doi.org/10.2785/170012.

Fabian, K., Topping, K. J., & Barron, I. G. (2018). Using mobile technologies for mathematics: Effects on student attitudes and achievement. *Educational Technology Research and Development, 66*(5), 1119–1139. https://doi.org/10.1007/s11423-018-9580-3

Fidai, A., Barroso, L. R., Capraro, M. M., & Capraro, R. M. (2020). Effects of engineering design process on science and mathematics. In *2020 IEEE Frontiers in education conference (FIE)* (pp. 1–4). IEEE. https://doi.org/10.1109/FIE44824.2020.9274167.

Frykholm, J., & Glasson, G. (2005). Connecting science and mathematics instruction: Pedagogical context knowledge for teachers. *School Science and Mathematics, 105*(3), 127–141. https://doi.org/10.1111/j.1949-8594.2005.tb18047.x

Hall, C., Lundin, M., & Sibbmark, K. (2021). A laptop for every child? The impact of technology on human capital formation. *Labour Economics, 69*, 101957. https://doi.org/10.1016/j.labeco.2020.101957

Herro, D., Quigley, C., & Cian, H. (2019). The challenges of STEAM instruction: Lessons from the field. *Action in Teacher Education, 41*(2), 172–190. https://doi.org/10.1080/01626620.2018.1551159

Kelley, T. R., & Knowles, J. G. (2016). A conceptual framework for integrated STEM education. *International Journal of STEM education, 3*(1). https://doi.org/10.1186/s40594-016-0046-z.

Kovács, Z., Recio, T., Richard, P. R., Van Vaerenbergh, S., & Vélez, M. P. (2020). Towards an ecosystem for computer-supported geometric reasoning. *International Journal of Mathematical Education in Science and Technology*. https://doi.org/10.1080/0020739X.2020.1837400

Lasa, A., Abaurrea, J., & Iribas, H. (2020). Mathematical Content on STEM Activities. *Journal on Mathematics Education, 11*(3), 333–346. https://doi.org/10.22342/jme.11.3.11327.333-346.

Lavicza, Z., Prodromou, T., Fenyvesi, K., Hohenwarter, M., Juhos, I., Koren, B., & Diego-Mantecon, J. M. (2020). Integrating STEM-related technologies into mathematics education at a large scale. *International Journal for Technology in Mathematics Education, 27*(1), 3–12.

Li, Y., Schoenfeld, A. H., Graesser, A. C., Benson, L. C., English, L. D., & Duschl, R. A. (2019). Design and design thinking in STEM education. *Journal for STEM Education Research, 2*, 93–104. https://doi.org/10.1007/s41979-019-00020-z

Lin, K. Y., & Williams, P. J. (2017). Two-stage hands-on technology activity to develop preservice teachers' competency in applying science and mathematics concepts. *International Journal of Technology and Design Education, 27*(1), 89–105. https://doi.org/10.1007/s10798-015-9340-1

Maass, K., Cobb, P., Krainer, K., & Potari, D. (2019a). Different ways to implement innovative teaching approaches at scale. *Educational Studies in Mathematics, 102*(3), 303–318. https://doi.org/10.1007/s10649-019-09920-8

Maass, K., Geiger, V., Ariza, M. R., & Goos, M. (2019b). The role of mathematics in interdisciplinary STEM education. *ZDM Mathematics Education, 51*(6), 869–884. https://doi.org/10.1007/s11858-019-01100-5

Margot, K. C., & Kettler, T. (2019). Teachers' perception of STEM integration and education: A systematic literature review. *International Journal of STEM Education, 6*, 2. https://doi.org/10.1186/s40594-018-0151-2

Martín-Páez, T., Aguilera, D., Perales-Palacios, F. J., & Vílchez-González, J. M. (2019). What are we talking about when we talk about STEM education? A review of literature. *Science Education, 103*(4), 799–822. https://doi.org/10.1002/sce.21522

Mohd-Hawari, A. D., & Mohd-Noor, A. I. (2020). Project Based Learning Pedagogical Design in STEAM Art Education. *Asian Journal of University Education, 16*(3), 102–111. https://doi.org/10.24191/ajue.v16i3.11072.

Moore, T. J., Stohlmann, M. S., Wang, H. H., Tank, K. M., Glancy, A. W., & Roehrig, G. H. (2014). Implementation and integration of engineering in K-12 STEM education. In Ş. Purzer, J. Strobel & M. E. Cardella (Eds.), *Engineering in pre-college settings: Synthesizing research, policy, and practices* (pp. 35–60). Purdue University Press. https://doi.org/10.2307/j.ctt6wq7bh.

Nelson, T. H., & Slavit, D. (2007). Collaborative inquiry among science and mathematics teachers in the USA: Professional learning experiences through cross-grade, cross-discipline dialogue. *Journal of in-Service Education, 33*(1), 23–39. https://doi.org/10.1080/13674580601157620

Niss, M., Bruder, R., Planas, N., Turner, R., & Villa-Ochoa, J. A. (2017). Conceptualisation of the role of competencies, knowing and knowledge in mathematics education research. In G. Kaiser (Ed.), *Proceedings of the 13th international congress on mathematics education* (pp. 235–248). Springer. https://doi.org/10.1007/978-3-319-62597-3_15.

OECD. (2019). PISA 2018 results (Volume I): What students know and can do. *OECD Publishing.* https://doi.org/10.1787/5f07c754-en

Ortiz-Laso, Z. (2020). STEAM activities with KIKS format. In P. R. Richard, S. Van Vaerenbergh & M. P. Vélez (Eds.), *First symposium on artificial intelligence for mathematics education. Books of abstracts (AI4ME 2020)* (pp. 6–7). Universidad de Cantabria. https://doi.org/10.22429/Euc2020.034.

Pai, K. C., Kuo, B. C., Liao, C. H., & Liu, Y. M. (2021). An application of Chinese dialogue-based intelligent tutoring in remedial instruction for mathematics learning. *Educational Psychology, 41*(2), 137–152. https://doi.org/10.1080/01443410.2020.1731427

Potari, D., Psycharis, G., Spiliotopoulou, V., Triantafillou, C., Zachariades, T., & Zoupa, A. (2016). Mathematics and science teachers' collaboration: searching for common grounds. In C. Csíkos, A. Rausch, & I. Szitányi (Eds.), *Proceedings of the 40th conference of the international group for the psychology of mathematics education* (pp. 91–98). PME.

Prodromou, T. (2014). GeoGebra in teaching and learning introductory statistics. *Electronic Journal of Mathematics & Technology, 8*(5), 363–376.

Prodromou, T., & Lavicza, Z. (2017). Integrating technology into mathematics education in an entire educational system—Reaching a critical mass of teachers and schools. *International Journal for Technology in Mathematics Education, 24*(4), 1–6.

Quigley, C. F., & Herro, D. (2016). "Finding the joy in the unknown": Implementation of STEAM teaching practices in middle school science and math classrooms. *Journal of Science Education and Technology, 25*(3), 410–426. https://doi.org/10.1007/s10956-016-9602-z

Richard, P. R., Fortuny, J. M., Gagnon, M., Leduc, N., Puertas, E., & Tessier-Baillargeon, M. (2011). Didactic and theoretical-based perspectives in the experimental development of an intelligent tutorial system for the learning of geometry. *ZDM–The International Journal on Mathematics Education, 43*(3), 425–439. https://doi.org/10.1007/s11858-011-0320-y.

Thibaut, L., Ceuppens, S., De Loof, H., De Meester, J., Goovaerts, L., Struyf, A., ... & Depaepe, F. (2018a). Integrated STEM education: A systematic review of instructional practices in secondary education. *European Journal of STEM Education, 3*(1), 2. https://doi.org/10.20897/ejsteme/85525.

Thibaut, L., Knipprath, H., Dehaene, W., & Depaepe, F. (2018b). The influence of teachers' attitudes and school context on instructional practices in integrated STEM education. *Teaching and Teacher Education, 71*, 190–205. https://doi.org/10.1016/j.tate.2017.12.014.

Thibaut, L., Knipprath, H., Dehaene, W., & Depaepe, F. (2019). Teachers' attitudes toward teaching integrated STEM: The impact of personal background characteristics and school context. *International Journal of Science and Mathematics Education, 17*(5), 987–1007. https://doi.org/10.1007/s10763-018-9898-7

Toma, R. B., & García-Carmona, A. (2021). «De STEM nos gusta todo menos STEM». Análisis crítico de una tendencia educativa de moda. *Enseñanza de las ciencias, 39*(1), 65–80. https://doi.org/10.5565/rev/ensciencias.3093.

Triantafillou, C., Psycharis, G., Potari, D., Bakogianni, D., & Spiliotopoulou, V. (2021). Teacher educators' activity aiming to support inquiry through mathematics and science teacher collaboration. *International Journal of Science and Mathematics Education.* https://doi.org/10.1007/s10763-021-10153-6

Ubuz, B. (2020). Examining a technology and design course in middle school in Turkey: Its potential to contribute to STEM Education. In J. Anderson & Y. Li (Eds.), *Integrated approaches to STEM Education: An international approach* (pp. 295–312). Springer. https://doi.org/10.1007/978-3-030-52229-2_16.

Vale, C., Campbell, C., Speldewinde, C., & White, P. (2020). Teaching across subject boundaries in STEM: Continuities in beliefs about learning and teaching. *International Journal of Science and Mathematics Education, 18*(3), 463–483. https://doi.org/10.1007/s10763-019-09983-2

Vinnervik, P. (2020). Implementing programming in school mathematics and technology: Teachers' intrinsic and extrinsic challenges. *International Journal of Technology and Design Education.* https://doi.org/10.1007/s10798-020-09602-0

Watson, J., & Munkoe, M. (2019). *Economic Outlook Autumn 2019 – EU economy weakens as trade tensions continue.* Retrieved from https://www.businesseurope.eu/publications/businesseurope-economic-outlook-winter-2020-2021-update.

Wijers, M., Jonker, V., & Drijvers, P. (2010). MobileMath: exploring mathematics outside the classroom. *ZDM—The International Journal on Mathematics Education, 42*(7), 789–799. https://doi.org/10.1007/s11858-010-0276-3.

Zulnaidi, H., & Zamri, S. N. A. S. (2017). The effectiveness of the GeoGebra software: The intermediary role of procedural knowledge on students' conceptual knowledge and their achievement in mathematics. *Eurasia Journal of Mathematics, Science and Technology Education, 13*(6), 2155–2180. https://doi.org/10.12973/eurasia.2017.01219a.

Digital Technology and Its Various Uses from the Instrumental Perspective: The Case of Dynamic Geometry

Jana Trgalová

1 Introduction: Role of Digital Technology in Education

Whether to use or not digital technology in mathematics classrooms is not an issue anymore nowadays, the question rather shifted to how to use it more efficiently and how to benefit the best from its affordances.

Since 1980s, researchers question the role technology should play in education. Two distinct roles have been highlighted by Pea (1985) and described in terms of amplifier and reorganizer metaphors. The amplifier metaphor suggests that technology changes "how effectively we do traditional tasks, amplifying or extending our capabilities, with the assumption that these tasks stay fundamentally the same" (p. 168), while the reorganizer metaphor posits that technology changes "the tasks we do by reorganizing our mental functioning, and not only by amplifying it" (ibid.). A simplified vision of the two metaphors leads to considering the use of digital technology either to do traditional tasks although in a different way or to do new tasks that cannot be done without this technology (Ripley, 2009). Likewise, Thomas and Lin (2013) point out that key affordances of technology emanate from the tasks that are used with it. However, designing tasks incorporating technology and having an epistemic value (Kieran & Drijvers, 2006) is not trivial for mathematics teachers.

In this chapter, we aim at highlighting that a given (mathematical) digital tool can be mobilized in manifold ways with different learning potential. We illustrate these considerations on the example of dynamic geometry (DG). The choice of dynamic geometry is motivated by a discrepancy between its potential to support students' learning evidenced by numerous research (e.g., Arzarello et al., 2002; Baccaglini-Frank & Mariotti, 2010; Leung, 2015) on the one hand, and its limited

J. Trgalová (✉)
S2HEP (UR4148), Université Claude Bernard, Lyon, France
e-mail: jana.trgalova@univ-lyon1.fr

use in mathematics classrooms (e.g., Bretscher, 2010; Kriek & Stols, 2011; Molnár & Lukáč, 2015).

Jones (2005) claims that "carefully designed tasks" with their appropriate enactment by the teacher are necessary for an efficient use of DG fostering students' learning:

> Overall, research in this area [use of DG software] indicates that successful access to geometrical theory does not happen without carefully designed tasks, professional teacher input, and opportunities for students to conjecture, to make mistakes, to reflect, to interpret relationships among objects, and to offer tentative mathematical explanations. (p. 29)

This chapter therefore proposes an analysis of selected DG tasks aiming at highlighting their differences in terms of student's cognitive activity, thus showing the range of potential use of DG.

The tasks, presented in Sect. 3, are categorized according to the SAMR framework (Sect. 2.2) and are analyzed referring to the instrumental approach (Sect. 2.1). The concluding Sect. 4 discusses implications of the analyses for the teaching and learning mathematics with technology.

2 Theoretical Framework

Instrumental approach (2.1) is the main theoretical framework that we use to analyze mathematical tasks under consideration in Sect. 3. These are organized according to the SAMR model presented in Sect. 2.2.

2.1 Instrumental Approach

The *instrumental approach* (Rabardel, 2002) was elaborated to understand processes by which a user transforms a (digital) tool—an *artifact*, into an *instrument* enabling her to achieve her goals. While the artifact (material or symbolic) is available to the user, the instrument is a personal construct elaborated by the user during her activity with the artifact in the course of the so called *instrumental genesis*. The process of instrumental genesis comprises two interrelated sub-processes: *instrumentation* leading to the constitution and the evolution of *schemes of use* of the artifact in the user, and *instrumentalisation* during which the user adapts and personalizes the artifact according to her knowledge and beliefs. The development of schemes of use manifests itself in a user's invariant behavior in a given class of situations (Vergnaud, 1990).

Fig. 1 SAMR model[2] (Puentedura, 2006)

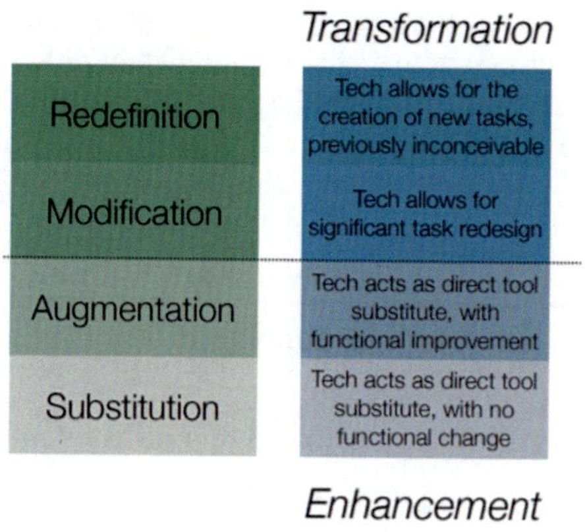

Transformation

Redefinition	Tech allows for the creation of new tasks, previously inconceivable
Modification	Tech allows for significant task redesign
Augmentation	Tech acts as direct tool substitute, with functional improvement
Substitution	Tech acts as direct tool substitute, with no functional change

Enhancement

2.2 SAMR Model

Several frameworks have been elaborated to understand the role of technology in teaching and learning, such as RAT framework (Hughes, Thomas, & Scharber, 2006) described as "an assessment framework for understanding technology's role in teaching, learning and curricular practices".[1] According to this framework, technology can be used as a *Replacement* when it "serves as a different (digital) means to same instructional practices", as an *Amplification* when it "increases efficiency, effectiveness, and productivity of same instructional practices", and as a *Transformation* when it "invents new instruction, learning, or curricula" (ibid.).

SAMR framework developed by Puentedura (2006) is another framework suggesting four roles of technology ranging from *substitution* to *redefinition* (Fig. 1).

Let us explain the different uses of technology from substitution to redefinition levels on the example of a quiz. An online version of a traditional paper-based quiz, where the student checks what she considers as correct answers, does not offer her any functional change. Technology in this case is thus used as a *substitute* of a paper quiz, although it can facilitate administration of the quiz (via a url instead of paper copies for example) and collection of students' responses. If the environment within which the quiz is implemented can provide feedback about the correctness of the answer, this functional improvement (*augmentation*) fosters learning. Indeed numerous studies suggest that feedback is most effective when it is provided immediately, rather than days or weeks later, and seems to positively impact both students' achievement (e.g. Razzaq, Ostrow & Heffernan, 2020) and engagement (e.g., Sancho-Vinuesa et al., 2013). If the environment provides not only true–false feedback but a more elaborated feedback such as hints (e.g., link to lessons) in case of incorrect answers, the student's task is deeply modified: such feedback supports learning by orienting the student toward appropriate remedial activities (*modification*). Finally, a quiz that

---

[1] https://techedges.org/r-a-t-model/.

[2] http://hippasus.com/resources/tte/puentedura_tte.pdf.

personalizes student's path through the items according to her answers can only be developed with technology (*redefinition*).

As we show in the following section, the four levels of the SAMR model align with the four roles of dynamic geometry identified by Laborde (2001). For this reason, we refer to this framework when considering various uses of dynamic geometry.

3 Various Uses of Dynamic Geometry

Laborde (2001) identified four different roles of dynamic geometry in the tasks:

- DG is used "mainly as facilitating material aspects of the task while not changing it conceptually" (p. 293). These are for example construction tasks in which the only difference "lies in the drawing facilities offered" by dynamic geometry (ibid.). Dynamic geometry can be seen as a substitute of traditional tools.
- DG "is supposed to facilitate the mathematical task that is considered as unchanged". In this case, "DG is used as a visual amplifier […] in the task of identifying properties" (ibid.). Indeed, geometric properties of a figure being preserved while dragging its free elements, their visual recognition is facilitated. Dynamic geometry substitutes traditional tools, but brings certain functional improvement (augmentation).
- DG "is supposed to modify the solving strategies of the task due to the use of some of its tools and to the possibility that the task might be rendered more difficult" (ibid.). Whereas a construction of a geometric figure with traditional tools can result in a visually correct drawing although controlled by perception, the same task in DG environment requires using geometric properties to obtain a figure that resists while dragging its free elements. Solving strategies in DG environment are thus deeply modified.
- The task only exists in DG environment. Laborde (2001) refers to the so-called "black box" tasks in which students are asked to reconstruct a dynamic figure provided in a DG environment that preserves geometric relations when its free elements are dragged (redefinition).

In the following sections, organized according to the levels of the SAMR framework, we discuss various possible uses of DG and analyze them from the instrumental perspective. Following Lopez-Real and Leung (2006), who consider that

> dragging in DGE can open up some kind of semantic space (meaning potential) for mathematical concept formation in which dragging modalities (strategies) are temporal-dynamic semiotic mediation instruments that can create mathematical meanings (p. 666),

we focus on the dragging functionality of dynamic geometry, considered as artifact, to highlight its semiotic potential.

3.1 Substitution Level Tasks

Inviting students to make a free drawing using DG tools, without paying attention to geometric relations is perhaps the "simplest" task (Fig. 2). Such a task can offer an opportunity to the students to get acquainted with DG menus and tools when they are introduced to this technology. Such a task can also be an occasion to exploit the semiotic potential of DG tools by comparing and contrasting them with traditional tools. For example, the fact that, in order to draw a straight line, the user needs to click to two different spots on the screen, which results in creating two distinct points and subsequently a line passing through these points, conveys the idea that a straight line passes through two distinct points. This is not necessarily the case with using a ruler to draw a straight line, which rather emphasizes the straightness of the line.

Another example of a task at the substitution level is constructing a geometric figure following a construction program (i.e., a series of instructions). In the example shown in Fig. 3, the task is proposed to primary school pupils. The use of DG presents several advantages. The figures pupils construct can be quite complex, not only usual ones, since the task is facilitated within the DG environment. Pupils

Fig. 2 Example of a freely drawn figure

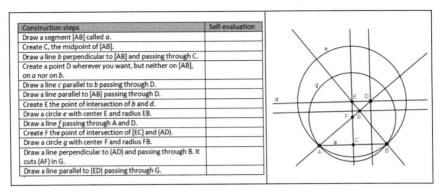

Fig. 3 Construction program (left) yielding a geometric figure (right)[3]

with motor difficulties of drawing with traditional instruments can succeed the task. Self-evaluation is also easier as drawings are more accurate and pupils can modify elements of the figure without deleting the correct steps.

Drag mode in these tasks is used to a limited extent, if at all: to adjust elements of a drawing either for the purposes of perceptive satisfaction (free drawing) or to separate elements of a figure to ease its construction (for example, when two points are too close to each other that they may be confused when selecting one of them). Restrepo (2008) classifies this dragging instrument as *dragging without mathematical purpose*.

It is not rare to find resources in which DG is used as a mere substitute of traditional tools although its potential could have been exploited to a greater extent. An example is given in Fig. 4 showing a task aiming at discovering that the area of a triangle ABC does not change when one of its vertices, say A, belongs to a line (d) parallel to the opposite side [BC]. Instead of dragging the vertex A on the line (d), the task invites to construct three distinct points A1, A2 and A3 on (d) (instruction 4 in Fig. 4 below), construct four triangles ABC, A1BC, A2BC, A3BC (instruction 5), display their areas (instruction 6) and observe the property (instruction 7).

In this task, the drag mode is not exploited at all. The only contribution of dynamic geometry is the accuracy of measures of lengths of the segments and of the areas of the triangles.

The same goal, namely observing the above-mentioned property of the area of a triangle, can be achieved with a simpler construction: triangles ABC and ABD with D on a line parallel to [BC]. One can either display areas of the two triangles or ask the software for checking the relation between the two triangles (Fig. 5). In both cases, dragging D along the parallel line allows verifying that the observed fact (i.e., equality of the areas) is not a coincidence.

In this case, DG offers a functional improvement; it is therefore used at the augmentation level, discussed in what follows.

---

[3] Task retrieved from http://www.ac-grenoble.fr/ien.st-gervais/spip.php?article1420.

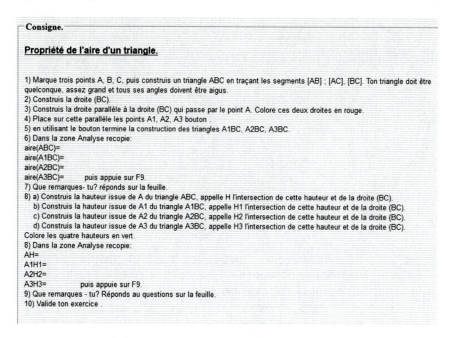

Fig. 4 Task aiming at discovering a property of the area of a triangle

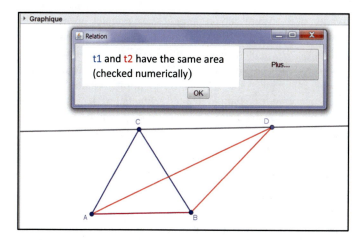

Fig. 5 Checking the relation between t1 and t2 (areas of triangles ABC and ABD) with the *Relation* command

3.2 Augmentation Level Tasks

Tasks analyzed in this section fall under the *robust construction* paradigm. Laborde (2005) characterizes robust constructions as those "for which the drag mode preserves

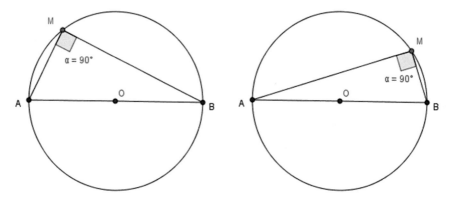

Fig. 6 Angle AMB inscribed in a semi-circle

their properties". The author provides the example of an angle AMB inscribed in a circle (Fig. 6). When the point M is dragged along the circle, one can easily observe that the angle AMB remains right. This robust construction shows that "for any point of the circle (except A and B) angle AMB is a right angle" (Laborde, 2005).

Typical tasks within this paradigm consist in exploring robust constructions. These may be either constructed by a teacher or by students who follow detailed instructions. The students are then invited to vary elements of the figure (point M in the example presented in Fig. 6) in order to recognize or to discover a geometric property based on the observation of the figure: the property at stake remains invariant (the measure of the angle AMB). Referring to the Marton et al.'s (2004) framework of variation, described in (Leung, 2017), the epistemic function of variation enabled by dragging in a robust construction is to allow *separation* of aspects of a figure that vary from other aspects that remain invariant. In the above-mentioned example, dragging points A and B allows observing for example that the segment [AB] is always a diameter of the circle, but its horizontal direction is not a necessary condition for the angle AMB to be right. Laborde (2005) sums up the contribution of this robust construction to the learning of the associated geometric theorem as follows:

> The robust construction contributes to a better identification in action of the elements of [the theorem] for several reasons:

- The construction requires to take into account two conditions to get a right angle: AB must be a diameter and M a point on a circle […].
- It allows contrasting the invariance of the angle and the varying nature of point M.
- It exteriorizes the variable nature of point M and the set in which it varies (Laborde, 2005).

Dragging elements of a robust construction allows producing quickly a number of different drawings sharing the same geometric property, which helps students "extend their visual images of a property […] and reject some spatio-graphical properties" that they can attach to the figure. Thus, "the drag mode is used as tool for distinguishing between contingence and necessity" (Laborde, 2005), which constitutes a clear functional improvement comparing to traditional tools. From the instrumental perspective, Restrepo (2008) ranges this modality of dragging among *exploratory dragging instruments*: its purpose is to look for invariants in a given figure, which facilitates identification of its geometric properties.

3.3 Modification Level Tasks

Robust construction tasks are another kind of tasks falling under robust construction paradigm. Students are asked to construct geometric figures that satisfy given conditions even when their elements are varied by dragging, for example, construct a square given its side or given its diagonal. As Laborde (2005) specifies,

> Eye ball constructions are invalidated by the drag mode since it becomes visible that some of the conditions are not satisfied. The drag mode is a critical factor in robust construction tasks that makes the difference with a paper and pencil environment. In such construction tasks in dynamic geometry, the drag mode provides a visual feedback from the fact that the construction does not meet all the required conditions. The strength of DGE lies in this possibility of showing at the spatio-graphical level the theoretical weakness of the construction.

The necessity to resort to geometric properties when constructing a figure modifies deeply the construction task in comparison to the same task realized in paper and pencil environment, where the students "very often stay at a graphical level and try only to satisfy the visual constraints" (Laborde, 2005). The drag mode provides students with a visual feedback about the correctness of their construction; it is therefore used as an *instrument for validating* constructions (also called *dragging test*, e.g. by Arzarello et al., 2002) and helps students gaining awareness of the distinction between a *drawing* (material diagram representing a geometric object) and a *figure* (theoretical object defined with its properties) (Laborde & Capponi, 1994).

Less common tasks at the modification level are those in which students are asked to look for conditions under which certain configurations are obtained. In the example taken from Laborde (2005), a circle with segment [AB] as a diameter and a point M not belonging to the circle are given. Students are asked to find a position of M outside the disk such that the angle AMB is obtuse (Fig. 7). The purpose of this task is to let students explore the situation and notice that the angle AMB is acute when M is outside the disc and obtuse when it is inside and eventually discover the relationship between the measure of AMB and the position of the point M.

This construction is coined soft because, as the point M is not constructed as a point on the circle, the targeted geometric property, namely the fact that the angle AMB is right when M belongs to the circle, is not directly visible, as it is the case in the robust construction (see Fig. 5). Rather, this property is inferred from observing that "the circle is the border between two regions, one in which angle AMB is obtuse

Fig. 7 Searching for particular position of the point M

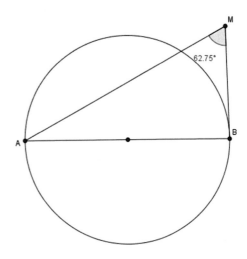

and one in which angle AMB is acute" (Laborde, 2005), hence AMB must be right when M in on the circle.

Soft constructions present several features that offer interesting learning opportunities. First, tasks that exploit soft constructions are more engaging than their robust versions. Indeed, they offer genuine problems to be solved and dynamic geometry is a support for exploring given situations. Pea (1985) evokes

> dynamic what-if capacities of such systems [that] make it possible to display immediately the consequences of different approaches to a problem (p. 171).

We claim that soft construction dynamic geometry task is significantly modified as it offers support for generating and testing various conjectures given different hypothetical conditions. Dragging plays a crucial role in this exploration. Moreover, students' exploration of a soft construction leads to putting more emphasis on the link between the condition (in our case, M is on the circle) and the consequence (the angle AMB is right), which facilitates grasping the meaning of a geometric property as an implication, which makes it particularly relevant in proof oriented tasks.

3.4 Redefinition Level Tasks

Among the tasks that cannot exist but within a dynamic geometry are the so called "black box" tasks. Clerc (2006) described a black box in dynamic geometry as a geometric figure made up of initial objects and final objects the construction and displacement of which are linked to the initial objects. The construction of these final objects is hidden. A mathematical task that can be set up with a black box consists in asking the students to solve them, that is to say to find out how to construct the final objects from the initial ones, the construction must of course resist when the free objects in the figure are dragged. Figure 8 shows a black box where the initial

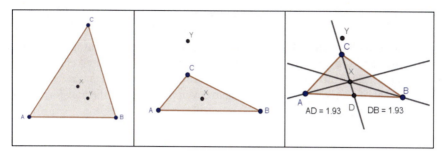

Fig. 8 A black box task

objects are three distinct points A, B, and C (or a triangle ABC), and the final objects are points X and Y.

The student is expected to explore the figure, make conjectures, verify them experimentally and eventually reconstruct the points X and Y. While dragging, the student can observe for example that the point X remains inside the triangle whereas the point Y can get outside (Fig. 8, center). She can draw lines, circles, midpoints… to enrich the figure; she can measure distances or angles (Fig. 8, right). Dragging clearly plays a critical role in searching for the hidden relationships. It is used both for exploring the figure and verifying conjectured geometric properties by highlighting their invariance.

4 Concluding Remarks

The purpose of this chapter was to show that a given digital educational technology can be used in many different ways, ranging from a mere substitute of traditional tools to offering unique learning opportunities in novel tasks. Dynamic geometry has been taken as an emblematic example and the tasks analyzed have been taken from past research or from available curricular resources.

The various uses of digital technology illustrated on the example of dynamic geometry can apply to other tools. Let us consider spreadsheets (Lagrange & Erdogan, 2009). At the substitution level, spreadsheet can be used as a traditional double entry table, to organize data. The use of formulas and their dragging adds a functional improvement to performing calculations (augmentation level). Tasks mobilizing spreadsheet functions are deeply modified compared to traditional approaches as they require modeling and generalization. Finally, tasks mobilizing advanced functionalities, such as conditional formatting, charts or programming macros fall under redefinition level.

Our analyses highlight that tasks at the transformation levels (modification and redefinition) show a greater potential for a student-centered approach, engaging students in inquiry-based problem-solving activity, compared to the tasks at the enhancement levels (substitution and augmentation), which are rather teacher-centered, requiring less important cognitive activity from students.

Therefore, digital technology itself is not transformative; it is the way how it is used that can be transformative. As the teachers' role in technology-based education is crucial, it is urgent to provide them with support aimed at raising their awareness of the manifold uses of technology and to help them develop practices at the transformation levels.

References

Arzarello, F., Olivero, F., Paola, D. & Robutti, O. (2002). A cognitive analysis of dragging practices in Cabri environments. *ZDM –Mathematics education* 34(3), 66–72.

Baccaglini-Frank, A., & Mariotti, M. A. (2010). Generating conjectures in dynamic geometry: The maintaining-dragging model. *International Journal of Computers for Mathematical Learning*, 15(3), 225–253.

Bretscher, N. (2010). Dynamic geometry software: the teacher's role in facilitating instrumental genesis. In V. Durrand-Guerrier et al. (Eds.), *Proceedings of CERME 6* (pp. 1340–1348). Lyon: INRP.

Clerc, B. (2006). Boîte noire en géométrie dynamique. *MathémaTICE* 2. http://revue.sesamath.net/spip.php?article13.

Hughes, J., Thomas, R., & Scharber, C. (2006). Assessing Technology Integration: The RAT—Replacement, amplification, and transformation—framework. In C. M. Crawford et al. (Eds.), *Proceedings of the society for information technology & teacher education international conference* (pp. 1616–1620).

Jones, K. (2005), Research on the use of dynamic geometry software: implications for the classroom. In J. Edwards & D. Wright (Eds.), *Integrating ICT into the mathematics classroom* (pp. 27–29). Derby: Association of Teachers of Mathematics.

Kieran, C., & Drijvers, P. (2006). The co-emergence of machine techniques, paper-and-pencil techniques, and theoretical reflection: A study of CAS use in secondary school algebra. *International Journal of Computers for Mathematical Learning*, 11, 205–263.

Kriek, J., & Stols, G. (2011). Why don't all maths teachers use dynamic geometry software in their classrooms? *Australasian Journal of Educational Technology*, 27(1), 137–151.

Laborde, C. (2005). Robust and soft constructions: two sides of the use of dynamic geometry environments. In S.-C. Chu et al. (Eds.), *Proceedings of the 10th Asian technology conference in mathematics (ATCM)* (pp. 22–35). Korea National University of Education.

Laborde, C. (2001). Integration of technology in the design of Geometry tasks with Cabri-Geometry. *International Journal of Computers for Mathematical Learning*, 6, 283–317.

Laborde, C., & Capponi, B. (1994). Cabri-géomètre constituant d'un milieu pour l'apprentissage de la notion de figure géométrique. *Recherches En Didactique Des Mathématiques*, 14(1.2), 165–210.

Lagrange, J. B., & Erdogan, E. (2009). Teacher's emergent goals in spreadsheet based lessons: Analysing the complexity of technology integration. *Educational Studies in Mathematics*, 71(1), 65–84.

Leung, A. (2017). Variation in tool-based mathematics pedagogy: The case of dynamic virtual tool. In R. Huang & Y. Li (Eds.), *Teaching and learning mathematics through variation—Confucian heritage meets western theories* (pp. 69–84). Rotterdam.

Leung, A. (2015). Discernment and reasoning in dynamic geometry environments. In S. Cho (Ed.), *Selected regular lectures from the 12th international congress on mathematical education* (pp. 451–469). Springer.

Lopez-Real, F., & Leung, A. (2006). Dragging as a conceptual tool in dynamic geometry environments. *International Journal of Mathematical Education in Science and Technology, 37*(6), 665–679.

Marton, F., Runesson, U., & Tsui, A. B. M. (2004). The space of learning. In F. Marton & A. B. M. Tsui (Eds.), *Classroom discourse and the space of learning* (pp. 3–40). Lawrence Erlbaum Associates, INC Publishers.

Molnár, P., & Lukáč, S. (2015). Dynamic geometry systems in mathematics education: Attitudes of teachers. *International Journal of Information and Communication Technologies in Education, 4*(4), 19–33.

Pea, R. D. (1985). Beyond amplification: Using the computer to reorganize mental functioning. *Educational Psychologist, 20*(4), 167–182.

Puentedura, R. R. (2006). *Transformation, technology, and education.* Screencast of the presentation delivered August 18, 2006. Retrieved from http://hippasus.com/resources/tte/.

Rabardel, P. (2002). *People and technology—A cognitive approach to contemporary instruments.* Université Paris 8.

Razzaq, R., Ostrow, K. S., & Heffernan, N. T. (2020). Effect of immediate feedback on math achievement at the high school level. In I. Bittencourt, M. Cukurova, K. Muldner, R. Luckin, & E. Millán (Eds.), *Artificial intelligence in education. AIED 2020. Lecture notes in computer science* (Vol. 12164, pp. 263–267). Cham: Springer.

Restrepo, A. M. (2008), *Genèse instrumentale du déplacement en géométrie dynamique chez des élèves de 6ème.* Doctoral thesis, Joseph Fourier University, Grenoble.

Ripley, M. (2009). Transformational computer-based testing. In F. Scheuermann & J. Björnsson (Eds.), *The transition to computer-based assessment* (pp. 92–98). Office for Official Publications of the European Communities.

Sancho-Vinuesa, T., Escudero-Viladoms, N., & Masia, R. (2013). Continuous activity with immediate feedback: A good strategy to guarantee student engagement with the course. *Open Learning, 28*(1), 51–66.

Thomas, M. O. J., & Lin, C. (2013). Designing tasks for use with digital technology. In C. Margolinas (Ed.), *Task design in mathematics education. Proceedings of ICMI Study 22* (Vol. 1, pp. 111–119). Oxford.

Vergnaud, G. (1990). La théorie des champs conceptuels. *Recherches En Didactique Des Mathématiques, 10*(2), 133–170.

Conclusions

Like any thoughtful editor with a healthy dose of curiosity, we asked ourselves at the end of the process which concepts were most commonly used in the book and how the links they make most often materialise. Of course, the main concepts are listed in the traditional way in the index, but as soon as we try to link them from the contributions, we can draw on the internal organisation of the discourse with a minimum of human intervention. Thanks to the expertise and work of our colleague Fabien Emprin, we have produced a word cloud based on the frequency analysis of occurrences (Fig. 1) and, from the word *student* which stands out clearly, a graph of similarities based on Max Reinert's Alceste method (Fig. 2).[1] There is no need to go into detail here, and we know we are taking some dialectical shortcuts, so we will limit ourselves to a few brief conclusions.

Given the subtitle we chose at the outset, the fact that the student is at the centre of the cloud is rather reassuring. Without asking whether it is the egg that comes before the chicken, we know that it is the learner and learning perspective that seem to have influenced the texts the most (Fig. 1). The teacher and mathematics remain close, which is normal in the didactic relationship, yet these concepts are subordinated to the student. The metamathematical notions of problem, proof and figure, as well as the point itself, are almost on the same level, testifying to the importance of the heuristic, discursive and semiotic elements for the performance of mathematical work. One may wonder where the knowledge often mentioned in the general introduction lies. If they are exemplified everywhere in the chapters, one should not lose sight of the fact that "problems are the *raison d'être* of knowledge", as Nicolas Balacheff reminds us in his preface. The tool and technology are very much linked to mathematics, and if AI is made very small, it would be a consequence of learning and not a cause. We can take the study a step further by looking at the dendrogram in Fig. 2 which, for all intents and purposes, groups the concepts into five classes. Thus, we could

---

[1] The graphical representations were obtained using Pierre Ratinaud's IRaMuTeQ software.

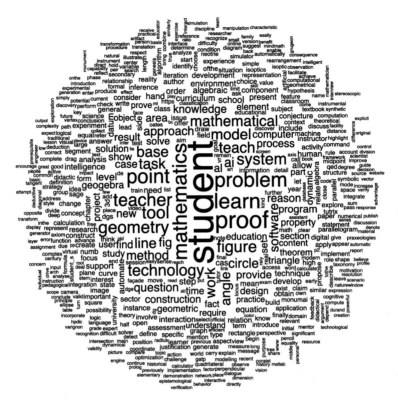

Fig. 1 Word cloud based on occurrences by simple counting. The "student" is at the core of the book and certain terms that revolve around it, such as proof and problem, are particularly used

add that the discursive and the vocabulary of argumentation, communication and validation (first class), and terms that relate to the illustration and exemplification process of an idea or practice (second class) are completed with a focus on the interactions between student, teacher and knowledge as general principles (third class), the emergence of assessment vocabulary and links to curricular aspects of educational projects (fourth class) and work on textbooks and geometry in relation to measurements and related calculations (fifth class). As for the iterative approaches and aspects of smart technology, it is often linked to identified knowledge and the language of instrumentation.

At the beginning of our book project, we knew that we were at the forefront of one of the most exciting quests for human learning. We were putting together a catalogue of dozens of technological achievements in the service of mathematics education. But in our endeavour, we were constantly surprised, which delayed us, one might say amused. Because the more we look at a particular approach or technology, the more the diversity of uses is revealed and the more the mystery thickens. Where do they come from? What problems do they address? What worlds do they open up? We are still asking ourselves all these questions.

Conclusions 433

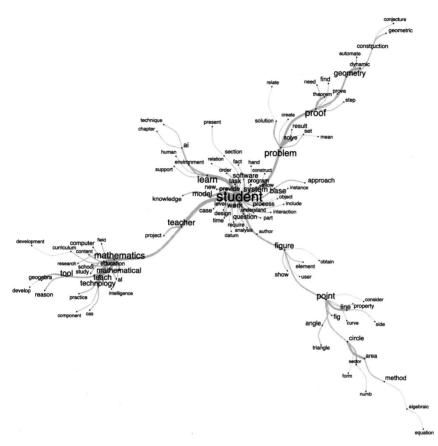

Fig. 2 Graph of similarity based on the term "student" in lemmas with more than 80 occurrences. Each text has been indexed by three parameters: part in which the text is placed, number of the chapter and number of the authors

In this intertwining of constituted disciplines, it is often forgotten that mathematical and computational perspectives are first and foremost cultural heritages. It is true that it is sometimes said that, in mathematics, the question of objectivity is an epistemological question, and that of its reality is an ontological question. However, as soon as the **mathematical sciences** teach us their principles and rules, we must remember how much they condition work, the learning and the implementation of practices or attitudes. A similar interplay is organised in the **computer sciences**, but it seems that the interest in technical achievements is replacing the interest in idealisation in its foundations. There is less interest in the knowledge as such, nor in its logical origin, value or theoretical scope. More thought would be given to the effects of the results and their validity, following the execution of the algorithms. If the heuristic relevance and the principle of idoneity seem more visible to the computer scientist, one would quickly lose sight of how the choice of models conditions the implementation, processing and interpretation of knowledge, thereby amplifying the black-box effect so feared by analysts. How can we bring human intelligence and machine intelligence closer together if we are not sure that we can simulate the 'cognitive', 'problematising' or 'decision-making' processes of human behaviour? After all, intelligence is made up of thought intervention, initiatives, steps, trial and error, and to lose this information, when it is the machine that is running, has something disconcerting in itself. Obviously, if it is meant to relieve us of arduous tasks or to enhance our creative potential, it is certain that we will more easily accept to "hide that process which we cannot see", to paraphrase Tartuffe. At the same time, mathematical competencies are intellectual skills, the exercise is almost like squaring the circle.

As for the **didactic sciences** point of view, the human question is self-evident. If the discipline is part of the humanities, its interest is naturally in the teaching of mathematics and teacher training (knowledge and culture), both from the professional standpoint (praxis) and from that of scientific research (methodologies). Mathematics education is accustomed to modelling and undertaking a-priori analyses, before confronting its results with the results of human experience. Its field of application already composes with the typical cultural variety of social activity, including its anthropological and political valences, under the mathematical conditioning and the ineffable computer development. Now, the intervention of the machine is overwhelming. It would be necessary to introduce the mathematical knowledge of the school into the computer models, to successfully recognise the acquisition of knowledge in the instrumented mathematical work, and to assume that one knows what the field of application of this knowledge will be, as was believed in the traditional school. In short, as a first lesson, any intention to map the ME+AI links only in terms of current techniques or achievements would have inevitably led us to a fragmented vision which would have neglected a fundamental fact: in order to simulate human intelligence, it is still necessary to understand it and to accept to work with it. The question rather shifts to where ME+AI links are today and what they should be for tomorrow?

The authors' contribution and the leads they give are eloquent. At the risk of sounding anecdotal, it is strange that we are able to recognise shapes or identify computer patterns (speech analysis, image retrieval, etc.), but we are still unable to recognise the nature of geometric figures or meaningful graphic units. In the same way, we can more or less automate motorised driving, but we are unable to automate the statement of mathematical problems and their relevant solutions, even though we know the theoretical references used in schools. Again, we have powerful deductive engines to help validate certain mathematical statements, but we have difficulty in producing human-readable proofs. Finally, we know how to apply deep reinforcement learning techniques to beat humans in some games, by interacting with an environment and observing effects and consequences, just as humans do. And that is fine: part of the machine learning process is to analyse past experiences to try to extract information that would allow it to act more effectively in the future, which is what a teacher normally does in his or her assessment-preparation work. But once again, we struggle to know what the student actually knows, to recognise why he or she is blocked in solving a problem, to be able to re-launch the student who has reached an impasse, without giving the answers at the same time as the questions ... Yet, we could produce learning routes that are adapted to the difficulties encountered, knowing that the obstacle is constitutive of knowledge and that we only learn by constructing, as Gaston Bachelard brilliantly showed us in *La Formation de l'esprit scientifique*. For the designers of machine learning, it is surely stimulating to see that under certain conditions, we are able to beat the humans. And for the teacher, there is nothing better than to see eyes shining with insight. Now imagine that the same effort is aimed at supporting human learning and raising the mathematical skills for doing new mathematical work. It is with this in mind that AI and artefacts, both digital and traditional, have been addressed in the chapters, emphasising the idea of partnership and closeness in the human-machine relationship.

In a way, we have made a picture of the current links between AI and ME. We also have a picture from 25 years ago that allows us to appreciate the progress that has been made since. If we started from an "AI winter", which already marked the degree of technical innovation with tools for ME, we have experienced an "AI spring" since around 2012, with the advent of deep learning and its effects. Today we are undoubtedly experiencing the beginning of an "AI summer" in which important scientific advances have been made in many fields: they are seeping into all levels of industry, administration, and more timidly into education, giving rise to a completely fragmented landscape in tune with the turbulence of our times. Our point of view complements many new directions in AI research, joining a wide variety of institutional, social and academic experiences. From the beginning of the millennium, mathematics has enjoyed an explosion of popularity due to the extension of AI to all fields. On the other hand, since the last decades of the twentieth century, the mathematical work has been considerably transformed, coming from a paper–pencil activity based on the deductive method to work assisted by powerful laboratories in the form of symbolic and numerical computation systems, simulation software, AI-based software or programming languages. Thus, mathematical thinking has been transformed into a mixed activity of experimentation (through examples and models executed on the computer), formulation of hypotheses and mathematical deduction. As evidenced throughout this book, there have also been great research advances in terms of software development, generation of experiences and didactic frameworks for the incorporation of AI to mathematics education. New tools require new curricula and new competencies for both teachers and students. However, in the field of education, this transformation is very slow, as if it often lags behind the development of civil society. Yet we are training tomorrow's learners, not yesterday's, at a time when the youth is entering technological complexity head-on with an ease that baffles many adults ...Education is subject to political decision-making processes, which means that researchers and experts are at the bottom of the decision-making process; indeed, sometimes large multinational corporations with generalist solutions, and more technological than human interaction-based, are who take on this expert role.

Nowadays, we can envision a future in which mathematical activity in the classroom is based on expert systems that help students to pose and solve the right problems, to reason in a continuous human–machine interaction, not only performing tedious calculations, like Newton, Le Verrier or Kelvin who complained about the hours wasted on elementary calculations. It is undoubtedly in such a spirit that we should think of AI in the service of ME. This future also involves proposing conjectures and theorems, and even activities in STEAM universes to immerse them in a physical, symbolic and virtual environment in which they can visualise the mathematics present in other areas, or going from manipulating a real object to doing so virtually to better grasp its properties, relations, invariant characteristics, etc. Likewise, teaching-learning models can be based on the transformation of teaching towards "instrumented" models that represent a pedagogical improvement: generation of learning pathways adapted to the development of complex mathematical competencies, control of the quality level of the problems to be devolved and the way they are vested in the power of learning, modes of assessment bringing together the overcoming of obstacles and conceptualisation in mathematics, among others. We would like to think that a mathematics learning environment is possible that integrates other disciplines, develops machine-assisted intellectual working capabilities, accompanies learning and adapts to the level and talent of the student, assists the teacher to design appropriate learning activities, and promotes positive action in overcoming epistemic inequalities.

In essence, if the great contribution of artificial intelligence to mathematics education is to be through more technological, more personal and more mind-bending environments, it is certainly because it will have been designed to develop the professional skills of the teacher, and above all because it will promote the acquisition of knowledge and the development of the mathematical, scientific and cultural competencies of the learner. Finally, while we know that it is very difficult to predict the developments that can be expected in 5 years, 2 years or even in the course of the year, the vagaries of the current pandemic have reminded us that we must approach the future with a skilful mixture of humility and boldness. We have the right and the duty to think ahead, to guard against the siren song of fads, utilitarianism and short-term visions, and we remain on deck for the future. And, of course, we are ready to pass the torch for the next quarter century.

Philippe, Pilar and Steven
Montréal, Collado Villaba, Santander
June 2021

Epilogue

Warning: this Epilogue has been written circa 2050 by To+gh, a colleague of Shalosh B. Ekhad,[1] and downloaded from the future by Tomas Recio.[2]

All's well that ends well,[3] humans used to declare, some decades ago, regarding the finalisation of a task that seemed, at the beginning, quite challenging. This could be, in fact, the case of this book, that originated at the very last days of February 2020, a quite remarkable annum, and not only because of being a leap-year!

In fact, less than a couple of weeks after the end of the Symposium on Artificial Intelligence for Mathematics Education,[4] organised by the editors of this book at the Centro Internacional de Encuentros Matemáticos (CIEM), in Castro-Urdiales (Spain), the "State of Alarm" was declared in Spain (and similar measures were taken all over the world) as a consequence of the COVID-19 pandemic, greatly restricting personal, face-to-face relations. A situation that accompanied the trio of editors along the whole period of designing, assigning, recalling, writing, reviewing, revising …recalling once and again …the different pieces shaping up this book, conceived at the Symposium and delivered about one year later. Humans (and this is the case of editors of the book) are, indeed, quite stubborn, for good or bad.

Clearly, we all—humans and automatons—are led by our background, by the algorithms and structures that conform our being. Slightly modifying a quote[5] from Siri Hustvedt,[6] a reputed writer at the time this book was published "There is no

---

[1] See the web page of Doron Zeilberger, https://sites.math.rutgers.edu/~zeilberg/ and a detailed information at Z. Hu, Y. Cui, J. Zhang, J. Eviston–Putsch. *Shalosh B. Ekhad: a computer credit for mathematicians*. Scientometrics (2020) 122:71–97

[2] In Spanish: += mas, recio = tough.

[3] "Bien está lo que bien acaba", "Tout est bien qui finit bien".

[4] https://ai4me.unican.es.

[5] https://www.goodreads.com/quotes/419395-there-is-no-future-without-a-past-because-what-is.

[6] https://en.wikipedia.org/wiki/Siri_Hustvedt.

present without a past, because what is to be cannot be imagined except as a form of repetition" (the word in italics is ours). Or, to express the same idea with the editors' own words, in the Conclusions: "…it is often forgotten that mathematical and computational perspectives are first and foremost cultural heritages."

This thought could be behind the many cautions and subtleties that most of the authors of the different chapters take to reflect on the role of AI in mathematics education, as if they were attracted, yet fearful, by its potential impact. Transmitting illusion, but stopped by prudence. This dilettante (at best) reaction to the arrival of technological novelties is not new. It is as old as humanity—and, *a fortiori*, as ancient as *bot-kind*! We can already witness it at Plato's "Phaedrus'", worrying about the negative consequences of the invention of writing: "If men learn this, it will implant forgetfulness in their souls. They will cease to exercise memory because they rely on that which is written, calling things to remembrance no longer from within themselves, but by means of external marks."[7]

Actually, a quick look to the past century history of the implementation of technology in the mathematics classroom, provides arguments insisting in this, quite disappointing (with some hints of sarcastic humor), perspective. Thus, consider the (perhaps nowadays perceived as ironic) title of the ICMI Study Volume 2, "School Mathematics in the 1990s",[8] edited by G. Howson and B. Wilson and published in 1986 by Cambridge University Press, where it is stated "…even if the students will not have to deal with computers till they leave school, it will be necessary to rethink the curriculum, because of the changes in interests that computer have brought." A necessity that the ICMI study was urging for the decade of the 1990s, and that it seems could be equally applicable to the school mathematics of the 2000s, the 2010s and the 2020s …. Another illustration of this situation can be found in the Foreword to this book, written by Prof. Balacheff in 2021, summarising the impact of AI in the educational context, more than 50 years after its first serious attempts: "AI raised hopes in the 1970s with the main stream research program on Intelligent Tutoring Systems (ITS) …Nevertheless, the dissemination of AI-based learning environments remained limited and still is…".

Some further examples—although biased to the context of the geometry curriculum—of the slow (if moving at all) movement towards the consideration of technology in the mathematics classroom, are collected in the Conclusion section of a chapter by M. Hohenwarter, Z. Kovács and T. Recio, on "Using GeoGebra Automated Reasoning Tools to explore geometric statements and conjectures".[9]

---

[7] https://fs.blog/2013/02/an-old-argument-against-writing/.

[8] https://www.mathunion.org/icmi/publications/icmi-studies/icmi-study-volumes.

[9] In Hanna, G., de Villiers, M., Reid, D. (Eds.), Proof Technology in Mathematics Research and Teaching, Series: Mathematics Education in the Digital Era, Vol. 14, 2019, pp. 215–236. Springer Cham.

Luckily, there are quite strong, stubborn and optimistic, humans and machines. My mate, Shalosh B. Ekhad, is one of them, perhaps not well informed. Thus, he has written, this same year 2050.[10]

> Do you know that until fifty years ago most of mathematics was done by humans? Even more strangely, they used human language to state and prove mathematical theorems. Even when they started to use computers to prove theorems, they always translated the proof into the imprecise human language, because, ironically, computer proofs were considered of questionable rigor!
>
> Only thirty years ago, when more and more mathematics was getting done by computer, people realized how silly it is to go back-and-forth from the precise programming-language to the imprecise humanese. At the historical ICM 2022, the IMS (International Math Standards) were introduced, and Maple[11] was chosen the official language for mathematical communication. They also realized that once a theorem is stated precisely, in Maple, the proof process can be started right away, by running the program-statement of the theorem.

Indeed, it seems the authors of the chapters in the present book on Artificial Intelligence and Mathematics Education, published in 2021, did not have the feeling that in 2022, the International Congress of Mathematicians (ICM) in Saint Petersburg would make such advances in the inclusion of technology in mathematics education and research, although it was true that, more often each time, theorems where discovered and proved with the use of computers. In fact, few chapters dare to address—or simply to notice—with some in-depth the possible curricular changes that the arrival of AI technology in the education world should involve.

But surely my colleague Shalosh B. Ekhad knows better what would happen, soon; what has to happen, soon .... Indeed, again, there are quite strong, stubborn and optimistic, humans and machines, and things can change, should change, soon. I recall ICME 8 (International Congress on Mathematical Education) in 1996, taking place in Seville (Spain), under an extremely hot weather that could have melted some robots, if there were any at that time attending ICME. There, professor Balacheff, the author of the Foreword to this book, was chairing a Topic Group on "Computer-Based Learning Environments (CBILE)", with Prof. Jim Kaput and some assistant.

It could be an interesting exercise for the readers of this book, to look at the different presentations and comments from the Topic Group sessions, here http://web.archive.org/web/20060701133658/ and http://mathforum.org/mathed/seville/index.html. One of the comments, for example, was done by the 2000–2006 president of the Royal Spanish Mathematical Society (RSME), Carlos Andradas and it is publicly available in the above mentioned url:

> I just want to say that I am not at all an expert neither in education nor in computers but it is evident that they are changing the whole society and we in education (and in Math) cannot ignore them. They open a fantastic world, the use of internet in the classrooms is fascinating and that computers offer new and important challenges to the educators. If the (boring) mechanics can be done by a machine in what should we concentrate?

---

[10] Plane Geometry: an Elementary Textbook by Shalosh B. Ekhad, XIV (Circa 2050). Downloaded from the future by Doron Zeilberger https://sites.math.rutgers.edu/~zeilberg/GT.html.

[11] https://www.maplesoft.com.

Good question that, yet in 2021, seems to be far from being answered.

This book is, indeed, an extraordinary effort in this direction.

Let me finish recalling you, human readers of the 2020's, Kaput's visionary words, cited by Balacheff in his highly recommended Followup report on the CBILE Topic Group[12]: technology means that "instead of doing (old) things better we should focus on doing better things".

> Tomás Recio
> Universidad Antonio de Nebrija, Madrid, Spain
> e-mail: trecio@nebrija.es

---

[12] http://web.archive.org/web/20060621140909/ http://mathforum.org/mathed/seville/followup.html.

Appendix: Photographs of the Book Project and Some of the Authors

See Figs. A.1, A.2, A.3, A.4, A.5 and A.6.

Fig. A.1 At the ACA 2019 held in July on the campus of the École de technologie supérieure in Montréal. In the usual order, on two lines: Victor Freiman, M. Pilar Vélez, Pascal-Alexandre Morel, Othmane Farid, Philippe R. Richard, Alain Kuzniak, Anthony Simard, Daniel Jarvis; Fabienne Venant, Sébastien Cyr, Assia Nechache, Ludovic Font, Nicolas Leduc. Authors not in the picture: Jiří Blažek, Roman Hašek, Thierry Dana-Picard and Pedro Quaresma

© The Editor(s) (if applicable) and The Author(s), under exclusive license to Springer Nature Switzerland AG 2022
P. R. Richard et al. (eds.), *Mathematics Education in the Age of Artificial Intelligence*, Mathematics Education in the Digital Era 17,
https://doi.org/10.1007/978-3-030-86909-0

Fig. A.2 Natural intelligence in action augmented by the collective

Fig. A.3 The off-programme meeting where ideas, partnerships and first drafts of the book were launched: Nicolas Leduc, Gaël Nongni, Ludovic Font, M. Pilar Vélez, Othmane Farid, Alain Kuzniak, Assia Nechache, Philippe R. Richard, Sébastien Cyr, Fabienne Venant, Victor Freiman, Daniel Jarvis, Michel Gagnon, Pedro Quaresma

Fig. A.4 February 2020, just before the pandemic restrictions, in front of the International Centre for Mathematical Encounters (CIEM) of the Universidad de Cantabria, in Castro Urdiales: Jose Diego Matecon, Jana Trgalova, Zsolt Lavizca, Roman Hašek, Zaira Ortiz, Belén Palop, Robert Corless, Theodosia Prodromou, Alvaro Martínez Sevilla, Philippe R. Richard, José Luis Rodríguez Blancas, M. Pilar Vélez, Eugenio Roanes, Steven Van Vaerenbergh, Pedro Quaresma, Martha Ivon Cardenas Dominguez, Adrián Pérez-Suay, Eunice Chan, Carlos Beltrán, Jean-Baptiste Lagrange, Tomás Recio, Mario Fioravanti

Fig. A.5 Although as researchers we are all used to working remotely and teleconferencing, this is the first time we have had to coordinate an entire project from home. Clockwise from the left: Philippe R. Richard, Jean-Baptiste Lagrange, Steven Van Vaerenbergh, Jana Trgalova, M. Pilar Vélez y Pedro Quaresma

Appendix: Photographs of the Book Project and Some of the Authors 445

Fig. A.6 The editors in Santander at the dawn of the pandemic and, in vignette, at work during Covid-19

Index

A
Abduction, 111, 113, 198, 210, 211
Active learning, 142, 143, 160, 256, 261–263, 266, 270, 362
Alexa, 99
Algebra, 4, 8, 24, 30, 31, 35, 41, 42, 70, 80, 142, 145, 146, 151, 153, 154, 156–161, 169, 180, 181, 207, 217, 223, 231, 234, 239, 240, 247, 248, 251, 253, 255, 256, 263, 295, 404–406
Angle, 6, 10, 14, 28, 46, 49, 50, 54, 55, 58, 62, 63, 65–67, 69, 71, 122–126, 168–174, 187, 188, 197, 206, 208, 209, 214–218, 220–228, 232–235, 237–239, 244–247, 361377, 385, 393, 405, 406, 408, 410, 424–427
Aphra Behn, 259
Area of a circle, 365–370, 372, 373, 375–380, 384, 385, 387, 390–392
Artefact, 46, 47, 244, 321, 336, 373, 402, 404–406, 409, 410
Artificial General Intelligence (AGI), 102
Artificial Intelligence (AI), 4, 7, 23, 24, 45, 89, 90, 110, 111, 113, 114, 135, 213, 228, 232, 251, 268, 319, 437, 439
Astroid, 245–247
Automated, 4, 7, 12, 16, 18, 19, 24, 25, 34–38, 41, 42, 47, 48, 50, 61, 62, 64, 66, 70–72, 89, 90, 142, 113, 131, 181, 211, 213, 224, 240, 245, 268, 438
Automated deduction, 6, 7, 18, 181
Automated reasoning tools, 24–26, 41, 42, 92, 438
Automated theorem provers, proving, 4, 5, 16, 61, 66, 95, 96
Automatic discovery, 18, 24, 92
Automatic Geometrical Model (AGM), 107, 122, 128, 134
Automatons, 437

B
Backward error, 255
Bugs, 97, 100, 295, 344, 347, 359

C
Cabri geometry, 195
Camera calculators, 92, 94, 98, 99
CindyJS, 237, 238, 240
Coloring, 233, 238, 244, 248
Component, 6, 48, 60, 97, 100, 101, 103, 115, 214, 228, 234, 235, 244, 245, 285, 289–292, 296, 298–303, 305, 308, 309, 322, 323, 332, 334–336, 409, 351
Compulsory secondary education, 79, 87
Computer Algebra Systems (CAS), 42, 151, 153, 154, 159, 211, 217, 219, 220, 224, 231, 232, 234, 235, 240, 243–245, 247, 283, 284, 286–290, 293, 296, 297, 306, 307, 309, 365, 366
Computer-Based Learning Environments (CBILE), 439
Computer vision, 90, 107, 120, 122, 134, 135
Conjecturing, 42, 322, 335, 336, 409, 410
Convolutional neural networks, 94, 117
Coq, 17, 168, 170–172, 174, 176, 179, 180, 182, 183, 187
Cortana, 99
COVID-19, 270, 361, 437

Creativity, 101, 121, 195, 232, 233, 247, 248, 266
Curricula, 77–79, 83–87, 99, 142, 144–146, 158, 284–291, 303, 304, 306, 308, 309, 365, 390, 400, 404, 405, 410, 419
Curve, 31, 135, 196, 216, 217, 219, 220, 223–228, 232–235, 237–239, 244–248, 259, 260, 351, 359, 409

D

Data-driven modeling, 97, 98, 103
Deduction, 7, 19, 24, 62, 111–113, 168, 181, 186, 198, 210, 211, 352, 353, 366
Deep learning, 91, 92, 95, 96, 107, 114, 115, 118–120, 129, 134, 135, 319, 332, 333, 337–339 3D graphing calculator, 344, 352, 353, 359
Didactic models, 141, 143, 148, 149, 154–157, 159, 160
Director circle, 201, 202, 205, 233, 239
Directrix line, 201–203, 205
Discover, discovery, 19, 24–26, 28, 30–38, 41, 49, 51, 71, 101, 107, 131, 195, 197, 210, 211, 238, 244, 245, 254, 262
Distance teaching, 352, 361
Dynamic, 155, 161, 285, 327, 366, 367, 373, 387, 390, 391, 392, 394, 417, 420, 422, 425–427
Dynamic Geometry Software (DGS), 16, 17, 49, 92, 197, 213, 221, 344, 360
Dynamic geometry, 1, 5, 6, 15–19, 24, 41, 49, 92, 109, 137, 138, 170, 193, 197, 198, 213, 217, 221, 223, 278, 320, 327, 334, 337, 344, 345, 350, 360, 394, 417, 420, 422, 425, 426–427
Dynamic Geometry Environment (DGE), 193, 420, 425
Dynamic Geometry Software (DGS), 320–323, 327, 328, 331, 334–338, 417, 418, 420–422
Dynamic Geometry Systems (DGS), 1, 5, 6, 15, 16, 18, 24, 41, 170, 217

E

Edelman's integral, 258
Educational contents, 77, 79, 83
Ellipse, 122–124, 129, 201, 202, 205, 210, 233–235, 240, 244, 245, 248
Engineering design process, 400–402, 404
Equilateral triangle, 28, 30, 69, 207
Ethics, 255, 270, 271, 273, 292
Experiment, 41, 46, 143, 162, 170, 199, 200, 202–205, 207, 208, 210, 211, 232, 237, 238, 242, 244, 245, 254, 270, 308, 338, 346, 351, 375, 379, 380, 383
Explainable AI, 96
Explainers, 96, 98, 99, 103
Exploration, 4, 19, 24, 41, 50, 67, 101, 102, 202, 204, 207, 237, 244, 246, 247, 254, 289, 306, 309, 321, 381, 385, 387, 389, 426

F

Façades, 107, 114–117, 122–132, 134
Feedback, 50, 54, 55, 93, 99, 100, 142, 155, 160–162, 213, 214, 228, 240, 257, 267, 296, 297, 323, 325, 327, 328, 331, 332, 334, 337, 400, 419, 425, 344, 347,353
Fermat, 245, 246
Formalization, 89, 168, 169, 174, 176, 179–182, 187
Formal proof, 5, 6, 11, 15, 16, 18, 50, 51, 168, 170–172, 174, 176, 180, 183, 188, 268

G

Generalization, 41, 102, 150, 195, 234, 427
GeoGebra, 6, 17, 24–35, 37–39, 41–43, 49, 52, 53, 62, 92, 103, 109, 110, 194, 195, 202, 204, 207, 211, 213, 214, 216–218, 220, 221, 223, 224, 226–228, 232, 234, 235, 244, 273, 345, 346, 354–357, 360, 361, 389, 390, 400, 401, 438
Geometer, 5, 6, 18, 34, 35, 37, 321
Giac/Xcas, 232
Google assistant, 99
Graph theory, 83

H

Hessian, 244
Historical problems, 214
History of geometry, 4
History of mathematical didactics, 368
Hough transform, 125, 126, 128–130, 134
Human system complementarity, 137
Hyperbola, 216, 233–235, 239, 240, 244, 245

I

ICME 8, 439
ICMI, 35, 36, 38, 42, 43, 168, 438
ICMI Study, 36, 38
    Kuwait ICMI Study, 36, 42
Image rectification, 123
Infinitesimal procedure, 390
Information extractors, 93–95, 98, 103

Index

Instrument, 40, 46, 194, 211, 254, 306, 321, 322, 376, 405, 408, 418, 420, 422, 425
Intelligent Tutoring systems (ITS), intelligent tutors, 99, 401, 438
Interaction, 23, 25, 28, 46, 47, 55, 60, 71, 92, 93, 100, 101, 107, 109, 110, 117, 143, 187, 193, 194, 198, 204, 210, 291, 319, 322–325, 327, 329–331, 334, 335, 343–345, 360, 392, 403, 406
Intuitive, 92, 108–110, 356, 366, 368, 373, 375, 378, 383, 384, 391
Inversion, 203

J

Java Geometry Expert (JGEX), 16, 211

K

Kahan's impossibility proof, 260
Knowledge, 10, 18, 19, 32, 47, 48, 61, 62, 87, 94, 100–102, 110, 114–116, 119, 120, 122, 141–149, 151, 152, 154, 156, 159–162, 169, 170, 174, 182, 183, 187, 195, 196, 198, 199, 210, 211, 215, 217, 229, 244, 245, 248, 261, 266, 270, 271, 285–288, 301, 304, 306, 307, 320, 322, 323, 326–329, 331, 335, 336, 338, 368, 375, 376, 379, 383, 390–392, 400–402, 410, 411, 418
Knowledge extraction, 79, 80

L

Learning path, 48, 93, 98, 100, 141, 142
Locus, 39, 195, 196, 199, 201, 207, 208, 217, 219, 220, 221, 223–225, 228, 233
LocusEquation, 25, 30, 31, 34, 207, 208, 211, 224
Logic programming, 13, 14, 62, 63, 67, 69, 70, 72

M

Machine learning, 90, 97, 120, 232, 243, 253, 256, 268, 272, 273
Machines, 90, 91, 95, 109, 439
Maple, 80, 81, 83, 99, 256, 258, 262, 263, 272, 301, 306, 439
Mathematical work, 46–48, 51, 72, 205, 244, 287, 334, 368, 391, 393
Mathematical working spaces, 47, 72, 368, 383, 384
Mathematics teacher, 46, 109, 182, 183, 214, 319, 403–410, 417
Matlab, 256, 259, 261, 265–267, 272, 301
Microsoft math solver, 92
Milieu, 168, 174, 182, 183, 187, 336
Mobile app, 117, 122

MonuMAI, 94, 107, 111, 114–122, 129, 132–135
Monuments, 94, 107, 110, 111, 114, 115, 120, 122, 132, 134

N

Neotrie VR, 343, 344, 346, 347, 349, 350, 352–355, 357, 358, 360
Neural networks, 91, 94, 95, 102, 117, 129

O

Observing and manipulating learning, 351
OK Geometry, 211, 213, 214, 217, 221–223
Open source, 16, 49, 96, 103, 284, 293–295
Oposiciones, 27, 31
Optical character recognition, 92

P

Pandemic, 360, 361, 437
Parabola, 233, 234, 239, 240, 244, 245
Parametric equations of surfaces, 351, 353, 361
Pedagogy, 285, 289, 291, 302, 303, 307–309
Pedal triangle, 207–209
Phaedrus, 438
Photomath, 91
Pirates, 33–35
Pixel, 90, 94, 125, 242
Plato, 4, 438
Plotting, 234, 240, 242–244, 257
Polygon, 28, 69, 150, 151, 367, 371, 372, 377, 382, 392, 393
Polynomial, 10, 26, 38, 80, 219, 220, 232, 239, 244, 245, 255, 262, 263
Post-hoc explainability, 96
Problem solving, 24, 41, 93, 108, 143, 194, 256, 288, 393
Proof assistant, 17, 168, 170, 174, 182, 186, 187
Proving, 5, 7, 8, 10–12, 16, 62, 69–71, 145, 167, 168, 171, 176, 180, 181, 183, 184, 194, 195, 213, 224, 321, 368, 392
Python, 67, 256, 265, 268, 272

Q

QED-Tutrix, 5, 13, 17, 46, 47, 49–52, 54–61, 63–65, 71, 72, 100
Quartic, 234, 235, 237, 241, 243

R

Randomness, 101, 102, 157
Reading Memo, 256
Rearrangement method, 366–369, 372, 373, 377, 383, 384, 390, 391, 393
Reasoning, 142, 143, 148, 152, 153, 287, 367, 368, 373, 376, 383, 391, 392, 406, 407, 409, 410

Reasoning engines, 94–96, 98–100, 103
Recommendation algorithms, 98

S
Sage software, 292, 294–297, 299, 305–307
SAMR framework, 418–420
Science, Technology, Engineering, the Arts and Mathematics education (STEAM), 23, 110, 114, 247
Semantics, 170, 263, 420
Siri, 99, 437
Socratic, 92, 94, 265, 376
Statement, 10, 25, 26, 28, 29, 35, 36, 38, 39, 49, 50, 53, 56, 58–60, 63–66, 68, 113, 169, 170, 180, 186, 195, 197, 199, 209, 210, 257, 370, 371, 377, 387, 405, 438
Stereoscopic video, 357, 358, 361
Student, 3, 7, 12, 17, 23–25, 34, 36, 37, 39–41, 45–60, 62, 66, 67, 71, 72, 91–94, 97–103, 108, 110, 120, 121, 131–133, 141–146, 148–156, 158–162, 174, 181–186, 193–197, 199, 200, 210, 211, 214–218, 223, 224, 248, 252–258, 260–267, 269–273, 284–289, 291, 292, 295–304, 306–309, 319–324, 327, 328, 331, 334–339, 344, 347–349, 365, 351–362, 367–369, 373, 375, 376, 378, 379, 381, 384–387, 390, 391, 399, 401, 403–410, 417–420, 424–428, 438
Student modeling, 91, 97, 101, 102
Syllabi, 252
Symbolic, 26, 28, 29, 47, 95, 109, 111, 143, 148, 151, 152, 211, 218, 220, 225, 231, 240–242, 248, 251–253, 256, 258, 262, 268, 272, 418
Syntax, 170, 172, 184, 261, 263

T
Tangent, 71, 220, 224, 233, 235, 237–239, 245–247
Taxonomy, 91, 93, 98, 100, 103, 115–117, 285, 289–291, 296, 305, 307, 309
Taxonomy for integrated technology, 290
Teaching, 3, 6, 24, 25, 36, 39, 41, 46, 89, 92, 108, 109, 111, 114, 132, 135, 141, 155, 161, 162, 167, 170, 193, 214, 217, 228, 252–254, 256, 257, 264, 266, 270–273, 283–289, 293, 295–297, 302, 305–307, 309, 319–324, 328, 332, 334, 335, 337–339, 343, 344, 346, 347, 349, 351, 353, 356, 360, 361, 365–367, 372, 373, 375, 379, 383, 384–387, 390, 391, 393, 400, 401, 403, 411, 418, 419, 438
Technology teacher, 405, 406, 410
Tickable, 267
Toulmin model, 197–199
Trace on, 202
Treasure island, 32
Trisection, 216, 217, 221, 223, 226–228
Trisection of an angle, 214, 216, 228
Trisectrix, 214, 216, 217, 223–226, 228
Tutor software, 48, 66, 72

U
Unlearn, 255

V
Verification, 24–26, 78–80, 87, 88, 92, 95, 134, 200, 268, 410
Virtual Reality (VR), 343, 345, 346, 354, 357, 359–362
Visual, visualization, 6, 8, 15, 16, 18, 24, 38, 42, 93–95, 114, 119, 135, 196–199, 203, 204, 210, 232, 235, 236, 241, 243, 245, 248, 262, 283, 288, 309, 390–392, 356, 401

W
WolframAlpha, 99

Z
Zeno's paradox, 257

Printed in the United States
by Baker & Taylor Publisher Services